NEW TRENDS IN QUANTUM SYSTEMS IN CHEMISTRY AND PHYSICS

Progress in Theoretical Chemistry and Physics

VOLUME 7

The titles published in this series are listed at the end of this volume.

New Trends in Quantum Systems in Chemistry and Physics

Volume 2

Advanced Problems and Complex Systems

Paris, France, 1999

Edited by

Jean Maruani

CNRS, Paris, France

Christian Minot

UPMC, Paris, France

Roy McWeeny

Università di Pisa, Italy

Yves G. Smeyers

CSIC, Madrid, Spain

and

Stephen Wilson

Rutherford Appleton Laboratory, Oxfordshire, United Kingdom

SPRINGER-SCIENCE+BUSINESS MEDIA, B.V.

Library of Congress Cataloging-in-Publication Data

ISBN 978-1-4020-0412-4 ISBN 978-0-306-46950-3 (eBook)
DOI 10.1007/978-0-306-46950-3

Originally published by Kluwer Academic Publishers in 2001
Softcover reprint of the hardcover 1st edition 2001

Progress in Theoretical Chemistry and Physics

A series reporting advances in theoretical molecular and material sciences, including theoretical, mathematical and computational chemistry, physical chemistry and chemical physics

Aim and Scope

Science progresses by a symbiotic interaction between theory and experiment: theory is used to interpret experimental results and may suggest new experiments; experiment helps to test theoretical predictions and may lead to improved theories. Theoretical Chemistry (including Physical Chemistry and Chemical Physics) provides the conceptual and technical background and apparatus for the rationalisation of phenomena in the chemical sciences. It is, therefore, a wide ranging subject, reflecting the diversity of molecular and related species and processes arising in chemical systems. The book series *Progress in Theoretical Chemistry and Physics* aims to report advances in methods and applications in this extended domain. It will comprise monographs as well as collections of papers on particular themes, which may arise from proceedings of symposia or invited papers on specific topics as well as initiatives from authors or translations.

The basic theories of physics – classical mechanics and electromagnetism, relativity theory, quantum mechanics, statistical mechanics, quantum electrodynamics – support the theoretical apparatus which is used in molecular sciences. Quantum mechanics plays a particular role in theoretical chemistry, providing the basis for the valence theories which allow to interpret the structure of molecules and for the spectroscopic models employed in the determination of structural information from spectral patterns. Indeed, Quantum Chemistry often appears synonymous with Theoretical Chemistry: it will, therefore, constitute a major part of this book series. However, the scope of the series will also include other areas of theoretical chemistry, such as mathematical chemistry (which involves the use of algebra and topology in the analysis of molecular structures and reactions); molecular mechanics, molecular dynamics and chemical thermodynamics, which play an important role in rationalizing the geometric and electronic structures of molecular assemblies and polymers, clusters and crystals; surface, interface, solvent and solid-state effects; excited-state dynamics, reactive collisions, and chemical reactions.

Recent decades have seen the emergence of a novel approach to scientific research, based on the exploitation of fast electronic digital computers. Computation provides a method of investigation which transcends the traditional division between theory and experiment. Computer-assisted simulation and design may afford a solution to complex problems which would otherwise be intractable to theoretical analysis, and may also provide a viable alternative to difficult or costly laboratory experiments. Though stemming from Theoretical Chemistry, Computational Chemistry is a field of research in its own right, which can help to test theoretical predictions and may also suggest improved theories.

The field of theoretical molecular sciences ranges from fundamental physical questions relevant to the molecular concept, through the statics and dynamics of isolated molecules, aggregates and materials, molecular properties and interactions, and the role of molecules in the biological sciences. Therefore, it involves the physical basis for geometric and electronic structure, states of aggregation, physical and chemical transformations, thermodynamic and kinetic properties, as well as unusual properties such as extreme flexibility or strong relativistic or quantum-field effects, extreme conditions such as intense radiation fields or interaction with the continuum, and the specificity of biochemical reactions.

Theoretical chemistry has an applied branch – a part of molecular engineering, which involves the investigation of structure–property relationships aiming at the design, synthesis and application of molecules and materials endowed with specific functions, now in demand in such areas as molecular electronics, drug design or genetic engineering. Relevant properties include conductivity (normal, semi- and supra-), magnetism (ferro- or ferri-), optoelectronic effects (involving nonlinear response), photochromism and photoreactivity, radiation and thermal resistance, molecular recognition and information processing, and biological and pharmaceutical activities, as well as properties favouring self-assembling mechanisms and combination properties needed in multifunctional systems.

Progress in Theoretical Chemistry and Physics is made at different rates in these various research fields. The aim of this book series is to provide timely and in-depth coverage of selected topics and broad-ranging yet detailed analysis of contemporary theories and their applications. The series will be of primary interest to those whose research is directly concerned with the development and application of theoretical approaches in the chemical sciences. It will provide up-to-date reports on theoretical methods for the chemist, thermodynamician or spectroscopist, the atomic, molecular or cluster physicist, and the biochemist or molecular biologist who wish to employ techniques developed in theoretical, mathematical or computational chemistry in their research programmes. It is also intended to provide the graduate student with a readily accessible documentation on various branches of theoretical chemistry, physical chemistry and chemical physics.

Contents

Part VIII. Condensed Matter

Preface

These two volumes collect thirty-eight selected papers from the scientific contributions presented at the Fourth European Workshop on *Quantum Systems in Chemistry and Physics* (QSCP-IV), held in Marly-le-Roi (France) in April 22-27, 1999. A total of one hundred and fifteen scientists attended the workshop, 99 from Europe and 16 from the rest of the world. They discussed the state of the art, new trends, and future evolution of the methods and applications.

The workshop was held in the old town of Marly-le-Roi, which lies to the West of Paris between the historic centres of Saint-Germain-en-Laye and Versailles. Participants were housed at the National Youth Institute, where over sixty lectures were given by leading members of the scientific community; in addition, over sixty posters were presented in two very animated sessions. We are grateful to the oral speakers and to the poster presenters for making the workshop such an stimulating experience. The social programme was also memorable - and not just for the closing banquet, which was held at the French Senate House. We are sure that participants will long remember their visit to the 'Musée des Antiquités Nationales': created by Napoleon III at the birthplace of Louis XIV, this museum boasts one of the world finest collections of archeological artifacts.

The Marly-le-Roi workshop followed the format established at the three previous meetings, organized by Prof. Roy McWeeny at San Miniato Monastery, Pisa (Italy) in April, 1996 (the proceedings of which were published in the Kluwer TMOE series); Dr Steve Wilson at Jesus College, Oxford (United Kingdom) in April, 1997 (which resulted in two volumes in Adv. Quant. Chem.); and Prof. Alfonso Hernandez-Laguna at Los Alixares Hotel, Granada (Spain) in April, 1998 (for which proceedings appeared in the present series). These meetings, sponsored by the European Union in the frame of the Cooperation in Science and Technology (COST) chemistry actions, create a forum for discussion, exchange of ideas and collaboration on innovative theory and applications.

Quantum Systems in Chemistry and Physics encompasses a broad spectrum of research where scientists of different backgrounds and interests jointly place special emphasis on quantum theory applied to molecules, molecular interactions and materials. The meeting was divided into several sessions, each addressing a different aspect of the field: 1 - Density matrices and density functionals; 2 - Electron correlation treatments; 3 - Relativistic formulations and effects; 4 - Valence theory (chemical bond and bond breaking); 5 - Nuclear motion (vibronic effects and flexible molecules); 6 - Response theory (properties and spectra); 7 - Reactive collisions and chemical reactions, computational chemistry and physics; and 8 - Condensed matter (clusters and crystals, surfaces and interfaces).

Density matrices and density functionals have important roles in both the interpretation and the calculation of atomic and molecular structures and properties. The fundamental importance of electronic correlation in many-body systems makes this topic a central area of research in quantum chemistry and molecular physics. Relativistic effects are being increasingly recognized as an essential ingredient of studies on many-body systems, not only from a formal viewpoint but also for practical applications to molecules and materials involving heavy atoms. Valence theory deserves special attention since it improves the electronic description of molecular systems and reactions from the point of view used by most laboratory chemists. Nuclear motion constitutes a broad research field of great importance accounting for the internal molecular dynamics and spectroscopic properties.

Also very broad and of great importance in physics and chemistry is the topic of response theory, where electric and magnetic fields interact with matter. The study of chemical reactions and collisions is the cornerstone of chemistry, where traditional concepts like potential-energy surfaces or transition complexes appear to become insufficient, and the new field of computational chemistry finds its main applications. Condensed matter is a field in which progressive studies are performed, from few-atom clusters to crystals, surfaces and materials.

We are pleased to acknowledge the support given to the Marly-le-Roi workshop by the European Commission, the Centre National de la Recherche Scientifique (CNRS) and Université Pierre et Marie Curie (UPMC). We would like to thank Prof. Alfred Maquet, Director of Laboratoire de Chimie Physique in Paris, Prof. Alain Sevin, Director of Laboratoire de Chimie Théorique in Paris, and Dr Gérard Rivière, Secretary of COST-Chemistry in Brussels, for financial and logistic help and advice. Prof. Gaston Berthier, Honorary Director of Research, and Prof. Raymond Daudel, President of the European Academy, gave the opening and closing speeches. The supportive help of Ms Françoise Debock, Manager of INJEP in Marly-le-Roi, is also gratefully acknowledged. Finally, it is a pleasure to thank the work and dedication of all other members of the local organizing team, especially Alexandre Kuleff, Alexis Markovits, Cyril Martinsky and, last but not least, Ms Yvette Masseguin, technical manager of the workshop.

Jean Maruani and Christian Minot
Paris, 2000

Part VI

Response Theory: Properties and Spectra

ON GAUGE INVARIANCE AND MOLECULAR ELECTRODYNAMICS

R. G. WOOLLEY
Department of Chemistry and Physics,
Nottingham Trent University,
Clifton Lane, Nottingham NG11 8NS, U.K.

Abstract. A general account of the interaction of electric charges with electromagnetic radiation is presented, the charges being treated in the non-relativistic approximation. The two main approaches in the literature, based on either the potentials of the electromagnetic field ('minimal coupling') or the polarization fields of the charges (e.g., the 'multipolar Hamiltonian') are presented in a unified framework, that characterizes them as simply different viewpoints of the same general structure. This structure is very suitable for a discussion of the gauge invariance of the theory. The final quantum mechanical Hamiltonian presented here has recently been shown to provide the basis for a proof of the gauge invariance of its associated S-matrix to all orders of perturbation theory, provided the S-matrix is taken on the energy-shell. Some outstanding problems are described briefly.

1. Introduction

The quantum theory of atoms and molecules interacting with electromagnetic radiation has a very long history; it was the subject of the first successful attempts at field quantization in the late 1920's. It has been presented in a variety of ways of which only two have become well-known. Firstly, there is the 'minimal coupling' form using a vector potential for the radiation, generally subject to the Coulomb gauge condition. There is also the 'multipolar Hamiltonian' in which electric and magnetic multipole moments are coupled directly to the (transverse) electric and magnetic fields. This latter form can be generalized by replacing the multipole series by closed form 'polarization fields', and when this is done one sees that the polarization fields are subject to the same kind of arbitrariness as that implied by the gauge transformations of the vector potential. Some definite choice of vector potential, or of the polarization fields, has always seemed necessary in order to have a practical scheme for calculations of absorption, emission and scattering phenomena; it is thus crucial to decide what calculations can be made that are independent of the choices made

J. Maruani et al. (eds.), New Trends in Quantum Systems in Chemistry and Physics, Volume 2, 3–21.

for the gauge (or the polarization fields), and for this one needs a suitably general theory.

In covariant QED gauge invariance is a straightforward matter; the field potential A_μ occurs coupled linearly to the 4-current j^μ which satisfies the equation of continuity $\nabla_\mu j^\mu = 0$, and this means that the gauge transformations of the potential are easily disposed of. It is therefore perhaps worth noting explicitly at the outset why there is something to discuss for the non-relativistic theory. The point is simply that in molecular electrodynamics we do *not* calculate with the covariant formalism and take the non-relativistic limit at the end. Instead we start again with the atomic/molecular Hamiltonian (a collection of a fixed number of charges) and the Hamiltonian for free radiation which are coupled by a gauge-dependent interaction. The complete Hamiltonian is easily shown to be gauge invariant; furthermore the equation of continuity must still be satisfied but it does not seem possible to use it straightforwardly as in covariant QED to show that perturbation theory is gauge invariant, and as we shall see there are examples in the literature of gauge-dependent perturbation calculations (e.g., of spectral lineshapes).

This paper presents an account of the dynamics of electric charges coupled to electromagnetic fields. The main approximation is to use non-relativistic forms for the charge and current density. A quantum theory requires either a Lagrangian or a Hamiltonian formulation of the dynamics; in atomic and molecular physics the latter is almost universal so the main thrust of the paper is the development of a general Hamiltonian. It is this Hamiltonian that provides the basis for a recent demonstration that the S-matrix on the energy shell is gauge-invariant to all orders of perturbation theory.

2. Charges, currents and polarization fields

In electrodynamics the coupling between matter and the electromagnetic field is mediated by the charge density, $\rho(\mathbf{x},t)$, and the current density, $\mathbf{j}(\mathbf{x},t)$, which together satisfy an equation of continuity,

$$\frac{d\rho(x,t)}{dt} + \Delta . j(x,t) = 0 , \tag{1}$$

that expresses local conservation of charge. In the electrodynamics of bulk matter it is usual to express the charge and current density in terms of two *polarization* fields. In the classical formulation of Lorentz [1] this description has a definite physical motivation, being accompanied by an explicit division of the charges into 'bound' and 'free' charge densities. But we shall not make any assumptions here about localization; thus we write

$$\rho(x,t) = - \Delta . P(x,t) \tag{2a}$$

$$j(x,t) = \frac{dP(x,t)}{dt} + \Delta \wedge M(x,t) \tag{2b}$$

with ρ(**x**,t) being the *total* charge density. Equation (2) is to be taken to apply to a single charge, and then extended by linearity to a collection of N charges. These are classical equations which eventually will have to be reinterpreted as operator relations for a quantum theory; the polarization fields will become operators, because ρ(**x**,t) and **j**(**x**,t) do. We retain the classical terminology however; thus **P**(**x**,t) and **M**(**x**,t) are called the electric polarization and magnetic polarization (`magnetization') fields respectively. Such a polarization field description has commonly been regarded as being particularly `physical' or `natural' (e.g. through its multipolar representation) even though the pair {**P**, **M**} are not defined uniquely by (2) [2]. It should be noted that *in a quantum theory* the distinction between `bound' and `free' charges is one that is carried by the solutions of the Schrödinger equation for the charges (bound-state versus continuum wavefunctions) and is not directly associated with the operators in the Schrödinger representation we anticipate using in atomic and molecular physics.

To begin with we concentrate on equation (2a), which is purely static. We define the Green's function **g**(**x**,**x**′) for the divergence operator:

$$\Delta_x . g(x,x') = - \delta^3(x - x'), \tag{3}$$

in terms of which a formal solution of (2a) is

$$P(x,t) = \int d^{3'}x \, g(x,x')\rho(x',t). \tag{4}$$

Here ρ(**x**,t) for point charges can be expressed as a sum of Dirac delta functions in the usual way. The *longitudinal* component of **g** is well-defined; it can be written as [3]:

$$\mathbf{g}(\mathbf{x},\mathbf{x}')^{\parallel} = \nabla_{\mathbf{x}} \frac{1}{4\pi|\mathbf{x}-\mathbf{x}'|}. \tag{5}$$

Although the differential equation (3) is consistent with solutions involving (**x** - **x**′), these are only a subset of the solution set, so treating **x** and **x**′ as separate variables, its Fourier transform on **x** yields the longitudinal solution

$$\mathbf{g}(\mathbf{p},\mathbf{x}')^{\parallel} = -\frac{i\mathbf{p}}{p^2} e^{i\mathbf{p}.\mathbf{x}'}. \tag{6}$$

The longitudinal component of $\mathbf{g}(\mathbf{p},\mathbf{x}')$is directed along $\mathbf{p}$ so we introduce an orthogonal triad of unit vectors $\{\varepsilon_i, i = 1,2,3\}$ with $\varepsilon_3 = \mathbf{p}/p$; then the transverse component lies in the plane containing $\{\varepsilon_1,\varepsilon_2\}$,

$$g(p,x')^{\perp} = \varepsilon_1 f(p,x') + \varepsilon_2 h(p,x'), \tag{7}$$

in terms of two scalar functions f and h. Since $\mathbf{g}(\mathbf{p},\mathbf{x}')^{\perp}$ satisfies

$$p\, \varepsilon_3 \cdot g(p,x')^{\perp} = 0 \tag{8}$$

$f(\mathbf{p},\mathbf{x}')$ and $h(\mathbf{p},\mathbf{x}')$ can behave as p^{ν}, $\nu \geq -1$ near $p = 0$; otherwise their singularities are controlled only by the requirement that the Fourier transforms required for (2) exist. The transverse component of the Green's function $\mathbf{g}(\mathbf{x},\mathbf{x}')$ is thus essentially arbitrary. In view of (4) these characteristics are inherited by the longitudinal and transverse components of the electric polarization field, $\mathbf{P}(\mathbf{x},t)$ (and its transform $\mathbf{P}(\mathbf{p},t)$). We note specifically that the `long-wavelength' limit, $\mathbf{p} \to 0$, of $\mathbf{P}(\mathbf{p},t)$ need not exist. A general account of the use of polarization fields in electrodynamics should reflect this situation which may strongly constrain any proposed physical interpretation.

Given a particular choice of electric polarization field, $\mathbf{P}(\mathbf{x},t)$, we can use it to solve for $\mathbf{M}(\mathbf{x},t)$ in (2b). If $\mathbf{j}(\mathbf{x},t)$ and $\mathbf{P}(\mathbf{x},t)$ are separated into longitudinal and transverse parts, a simple calculation shows that the longitudinal contributions cancel identically because of the equation of continuity, (1). We then find,

$$\begin{aligned} M(x,t) = \int d^3x' \frac{1}{4\pi\_x - x'\_} &(\Delta' \wedge j(x',t) \\ - \int d^3x''\, \Delta' \wedge g(x',x'') &\frac{d\rho(x'',t)}{dt}). \end{aligned} \tag{9}$$

Equations (4) and (9) are general forms for the polarization fields which display their arbitrary content through their dependence on the Green's function $\mathbf{g}(\mathbf{x},\mathbf{x}')$.

3. Electrodynamics in Lagrangian form

A Lagrangian for a collection of charged particles in an electromagnetic field can be written down directly using the polarization fields

$$L = L_p + \int d^3x\, P(x,t).E(x,t) + \int d^3x\, M(x,t).B(x,t), \tag{10}$$

where L_p is the kinetic energy contribution of the particles. It is easily verified with the aid of some specific choice of the polarization fields that (10) with the Euler-Lagrange equations reproduces the Lorentz Force Law [4] provided that equation (11) holds for the fields,

$$\Delta \wedge E(x,t) = -\frac{\partial B(x,t)}{\partial t}, \quad \Delta . B(x,t) = 0. \tag{11}$$

Suppose we adopt (4) as the general form for the electric polarization field; then the part of the Lagrangian potential depending on $\mathbf{E}(\mathbf{x},t)$ in full is

$$V_{E,\rho} = \int d^3x \int d^3x' \, E(x,t).g(x,x')\,\rho(x',t). \tag{12}$$

Turning the argument round we could regard (12) as the fundamental form for this part of the Lagrangian potential and take (4) as a convenient abbreviation for one of the integrations that leaves the electric field explicit in the Lagrangian. We could as well 'simplify' (12) by defining a scalar functional of the electric field,

$$\phi(x',t) = \int d^3x \, E(x,t).g(x,x'), \tag{13}$$

and present (12) in a form with the charge density explicit,

$$V_\rho = \int d^3x' \, \phi(x',t)\,\rho(x',t). \tag{14}$$

If we put $\mathbf{g}(\mathbf{x},\mathbf{x}')^{|}$, equation (5), into (13) and call the result $V(\mathbf{x}',t)$, we have at once,

$$\mathbf{E}(\mathbf{x},t)^{|} = -\nabla V(\mathbf{x},t) \tag{15}$$

and $V(\mathbf{x},t)$ is recognized as the conventional *scalar potential* for the electromagnetic field. However once there is a non-zero transverse component in $\mathbf{g}(\mathbf{x},\mathbf{x}')$, ϕ is not in general integrable.

In similar fashion (9) with (10) yields

$$V_{B,j,\rho} = \int d^3x \int d^3x' \, \frac{B(x,t).(\Delta' \wedge j(x',t))}{4\pi\,|\,x-x'\,|} - \int d^3x \int d^3x' \int d^3x'' \, \frac{B(x,t).(\Delta' \wedge g(x',x''))}{4\pi\,|\,x-x'\,|}\frac{d\rho(x'',t)}{dt} \tag{16}$$

and in complete analogy with (12) we can introduce a vector valued functional of the magnetic field, and express (16) in `simplified' form with the charge and current densities explicit. It is convenient to make an integration by parts in (16) and set

$$A(x,t) = \int d^3x' \frac{1}{4\pi\,|\,x - x'\,|}\left(\Delta' \wedge B(x',t)\right) \tag{17}$$

so that

$$\begin{aligned} V_{A,j,\rho} &= \int d^3x\, A(x,t).\, j(x,t) \\ &- \int d^3x \int d^3x'\, A(x,t).\, g(x,x')\, \frac{d\rho(x',t)}{dt}. \end{aligned} \tag{18}$$

The vector defined by (17) is a familiar object in electrodynamics; it is indeed the *Coulomb gauge vector potential* [5], satisfying

$$\nabla.\mathbf{A}(\mathbf{x},t) = 0, \quad \nabla \wedge \mathbf{A}(\mathbf{x},t) = \mathbf{B}(\mathbf{x},t). \tag{19}$$

Furthermore with the aid of (12) we verify that

$$-\frac{\partial A(x,t)}{\partial t} = E(x,t)^{\perp}. \tag{20}$$

Thus, if we separate the electric field into its longitudinal and transverse parts, we easily obtain the full Lagrangian potential as

$$\begin{aligned} V = &- \int d^3x\, V(x,t)\, \rho(x,t) + \int d^3x\, A(x,t).\, j(x,t) \\ &- \frac{d}{dt}\int d^3x\, A(x,t).\, P(x,t) \end{aligned} \tag{21}$$

where V($\mathbf{x}$,t) satisfies (15), $\mathbf{A}$($\mathbf{x}$,t) is the Coulomb gauge vector potential, (17), and (4) has been reintroduced into the last term.

Suppose we give the vector potential a longitudinal component by writing,

$$\mathbf{a}(\mathbf{x},t) = \mathbf{A}(\mathbf{x},t) + \nabla\chi(\mathbf{x},t), \tag{22}$$

where $\mathbf{A}$($\mathbf{x}$,t) is the purely transverse Coulomb gauge vector potential, and

$$\chi(x;t) = \int d^3x' a(x',t).g(x',x)^{\parallel} \tag{23}$$

(so $\mathbf{a}^{|} = \nabla\chi$), and simultaneously transform V(**x**,t) by setting

$$\phi(x,t) = V(x,t) - \frac{\partial\chi(x,t)}{\partial t}. \tag{24}$$

Equations (22), (24) are usually referred to as the equations defining the *gauge transformations* of the scalar and vector potentials for the fields. If we set

$$E(x,t) = -\Delta\phi(x,t) - \frac{\partial a(x,t)}{\partial t}, \quad B(x,t) = \Delta \wedge a(x,t) \tag{25}$$

then **E**(**x**,t) and **B**(**x**,t) so defined are unchanged by (22), (24) and automatically satisfy the Maxwell equations (11). In terms of the *arbitrary* pair of potentials {**a**, ϕ} the Lagrangian potential has exactly the same form as (21) for the special potentials {**A**, V}, i.e.

$$\begin{aligned} V = &- \int d^3x\, \phi(x,t)\, \rho(x,t) + \int d^3x\, a(x,t).\, j(x,t) \\ &- \frac{d}{dt}\int d^3x\, a(x,t).\, P(x,t) \end{aligned} \tag{26}$$

is valid in any gauge, and is invariant under arbitrary gauge transformations of the potentials. It has long been known that the integral in the last term of (26) is the generator of the Power-Zienau-Woolley transformation, that relates the Coulomb gauge and multipole Hamiltonians [6-9] with **a** = **A**, the Coulomb gauge vector potential, and **P**(**x**,t) is given as a multipole series which may be summed into a particular line integral form. The true significance of this integral, however, is that the time derivative in (26) is required if the Lagrangian is to display its full invariance under arbitrary gauge transformations of the potentials. It is a standard result in classical mechanics that under very mild assumptions, the addition of a total time derivative to a Lagrangian does not affect the classical equations of motion; equivalently, if the time derivative is written as dF/dt, F is the generator of a canonical (unitary) transformation in the classical (quantum) Hamiltonian formalism. For this reason the customary Lagrangian for charged particles in a given field is based only on the first two terms in (26) [10].

A particular set of potentials can be specified by imposing a linear functional constraint on the vector potential. Such a constraint is usually referred to as a *gauge condition*; a general gauge condition is provided by the equation

$$\int d^3x'\, a(x',t).\, g(x,x') = 0, \tag{27}$$

which causes the total time derivative in (26) to vanish. The interpretation of this equation is that, for every choice of the Green's function **g(x,x′)**, there is a vector potential **a(x′**,t) such that (27) is true. To see how this works we separate the vector potential into its longitudinal ($\mathbf{a}^{\parallel}$) and transverse (**A**, equation (17)) components, so that (27) gives

$$\int d^3x'\, a(x',t).g(x',x)^{\parallel} = -\int d^3x'\, A(x',t).g(x',x)^{\perp}. \tag{28}$$

The left-hand side of (28) is just the gauge function in (23), and we can therefore say that the vector potential specified by,

$$a(x',t) = A(x',t) - \Delta_{x'} \int d^3x''\, A(x'',t).g(x'',x') \tag{29}$$

satisfies (27) identically. In this equation **A(x**,t) is the Coulomb gauge vector potential which is regarded as fixed, so that the gauge freedom in the potential **a(x′**,t) results from the freedom to choose the transverse component of the Green's function **g(x,x′**). The choice $\mathbf{g}(\mathbf{x},\mathbf{x}')^{\perp} = 0$ yields the Coulomb gauge vector potential, (17),

$$\mathbf{a}(\mathbf{x},t) \equiv \mathbf{A}(\mathbf{x},t). \tag{30}$$

Equation (29) is particularly useful for the quantum theory because we know how to represent the Coulomb gauge vector potential operator in terms of photon annihilation and creation operators; since the Green's function **g(x,x′**) remains a c-number, (29) gives an operator representation of the vector potential in an arbitrary gauge.

4. Electrodynamics in Hamiltonian form

The Lagrangian of the previous section is easily extended to give a total Lagrangian for a closed system of charges and fields by including a contribution describing free electromagnetic radiation. This extra term is [10]

$$L_{field} = -\frac{1}{2}\varepsilon_0 \int d^3x\, \{E(x,t).E(x,t) - c^2 B(x,t).B(x,t)\} \tag{31}$$

so that the total Lagrangian has the structure

$$L = L_p + L_{int} + L_{field} \tag{32}$$

with L_{int} either as in (10), or (26).

The general Hamiltonian formalism was developed principally by Dirac [11-13];

its application to molecular Lagrangians of the form of (32) has been described previously [4, 7, 14, 15]. The scalar potential is a redundant variable, because its time derivative does not appear in the Lagrangian, and is eliminated from the Hamiltonian, which reads

$$\begin{aligned} H &= \Sigma_\alpha \frac{1}{2m_\alpha}\left(p_\alpha - e_\alpha a(r_\alpha)\right)^2 \\ &+ \frac{1}{2}\varepsilon_0^{-1}\int d^3x \,\{\pi(x).\pi(x) + \varepsilon_0^2 c^2 B(x).B(x)\} \end{aligned} \tag{33}$$

where $\{\mathbf{p}_\alpha\}$, $\pi(\mathbf{x})$ are the particle and field conjugate momenta respectively. They have canonical Poisson-Brackets with their coordinates

$$\begin{aligned} \{r_\alpha^r, p_\beta^s\} &= \delta_{\alpha\beta}\delta_{rs} \\ \{a(x,t)^r, \pi(x',t)^s\} &= \delta_{rs}\delta^3(x - x'). \end{aligned} \tag{34}$$

At this stage the gauge of the vector potential is left free with the consequence that the evolution generated by the Hamiltonian equations of motion

$$\frac{dC}{dt} = \{C,H\} \tag{35}$$

for any dynamical variable C, is subject to a non-integrable equation of constraint,

$$\Omega(x) = \rho(x) + \Delta.\pi(x) \approx 0 \tag{36}$$

where $\rho(\mathbf{x})$ is the charge density, and $\approx$ is Dirac's `weak equality′ sign [11-13].

In atomic and molecular physics we require the Coulomb energies to be made explicit. However when the Hamiltonian (33) is cast in the form

$$\mathrm{H} = \mathrm{H}_{\mathrm{atom}} + \mathrm{H}_{\mathrm{int}} + \mathrm{H}_{\mathrm{rad}} \tag{37}$$

the terms in (37) are individually gauge dependent. The principle of gauge invariance claims that physical observables must be independent of the choice of gauge, so it is important to consider how gauge invariance of calculated quantities can be established. First however we must deal with the constraint (36).

To do this we bring in another equation of constraint which has the property that its Poisson bracket with $\Omega(\mathbf{x})$ does not vanish. In Dirac's terminology, this yields a canonical scheme with two *second-class* constraints. The extra constraint can be taken to be the general gauge condition (27), so the two second-class constraints are

$$\Omega(x)_1 = \int d^3x'\, a(x').g(x',x) \tag{38a}$$

$$\Omega(x)_2 = \rho(x) + \Delta.\pi(x) \tag{38b}$$

Their mutual Poisson brackets calculated using (34) yield a non-singular matrix with elements

$$\begin{aligned} K(x,x'')_{12} &\equiv \{\Omega(x)_1, \Omega(x'')_2\} = -\delta^3(x-x'') \\ K(x,x'')_{21} &= \delta^3(x-x'') \\ K(x,x'')_{11} &= K(x,x'')_{22} = 0. \end{aligned} \tag{39}$$

Its inverse is given by

$$\begin{aligned} K(x,x'')^{-1}{}_{12} &= -K(x,x'')^{-1}{}_{21} = \delta^3(x-x'') \\ K(x,x'')^{-1}{}_{11} &= K(x,x'')^{-1}{}_{22} = 0. \end{aligned} \tag{40}$$

The *Dirac bracket* of two dynamical variables j and k which we write as [j,k]* is defined using this inverse matrix as

$$[j,k]^* = \{j,k\} - \int d^3x \int d^3x' \{j, \Omega(x)_r\}\, K(x,x')^{-1}_{rs}\, \{\Omega(x')_s, k\} \tag{41}$$

the brackets on the rhs being calculated using (34). Using (39) - (41) shows that the Dirac bracket of any dynamical variable with a constraint ($\Omega(\mathbf{x})_i$, i = 1, 2) vanishes identically. The non-zero Dirac brackets of the variables in the Hamiltonian (33) are easily found to be:

$$[r^r_\alpha, p^s_\beta]^* = \delta_{\alpha\beta}\delta_{rs} \tag{42a}$$

$$[a(x)^r, \pi(x')]^* = \delta_{rs}\delta^3(x-x') - \nabla^r_x g(x',x)^s \tag{42b}$$

$$[r^r_\alpha, p^s_\beta]^* = \delta_{\alpha\beta}\delta_{rs} \tag{42c}$$

The Dirac brackets are like the original Poisson brackets, being antisymmetric and satisfying the Jacobi identity, and are the basis for quantization using the usual correspondence [11-13]

$$[\hat{a}, \hat{c}] = i\hbar[a,c]^*. \tag{43}$$

We can now set

$$\Omega(x)_l = \int d^3x'\, a(x').g(x,x') = 0 \tag{44}$$

as the condition that fixes the gauge of the vector potential in the Hamiltonian (33), and in (42), according to the choice of the Green's function **g(x,x′)**. The gauge condition (44) implies that the vector potential in *any* gauge can be represented in terms of the purely transverse Coulomb gauge vector potential, **A(x**,t), and the Green's function **g(x″,x′)** as in (29),

$$\begin{aligned} a(x,t)_r &= A(x,t)_r - \Delta_{x,r}\int d^3x'\, A(x',t).g(x',x), \\ &= \int d^3x' \,[a(x,t)_r, \pi(x',t)_s]^*\, A(x',t)_s^{\perp} \end{aligned}$$

in terms of the Dirac bracket (42b). The operator properties of **a(x**,t) can therefore be implemented through the operator representation of **A(x**,t), while its arbitrariness is carried by the c-number **g(x,x′)**. Accordingly, we write the quantum theory in terms of **A(x**,t) and its conjugate and we then have to establish what kind of calculations can be carried out that yield answers independent of the arbitrary Green's function **g(x,x′)**. We see from (42) that every gauge has its own set of Dirac brackets; correspondingly after quantization, every gauge has its own set of commutation relations and its own Hilbert space. Since canonical (unitary) transformations should leave the Dirac brackets (commutators) invariant, the passage from one gauge to another cannot be effected by such transformations. Hence we shall not be able to rely on the nice properties of canonical (unitary) transformations in seeking to check gauge invariance.

The properties of the dynamical variables are fixed by the brackets (42), and it is easily verified that separating $\pi(\mathbf{x})$ as

$$\pi(\mathbf{x}) = \pi(\mathbf{x})^{\perp} + \mathbf{P}(\mathbf{x}) \tag{45}$$

is consistent with (42) if $\pi(\mathbf{x})^{\perp}$ is interpreted as the conjugate of the Coulomb gauge vector potential **A(x)** (associated purely with electromagnetic radiation), and **P(x)** is the electric polarization field given by (4). We complete the identification of $\pi(\mathbf{x})$ by computing its equation of motion; using (33), (35) and (42) we obtain, independently of the gauge,

$$\frac{\partial \pi(x,t)}{\partial t} = -\varepsilon_0 c^2\, \Delta \wedge B(x,t) + j(x,t) \tag{46}$$

so that the conjugate momentum $\pi(\mathbf{x})$ can be recognized as the *total* electric field, to within a constant, $\pi(\mathbf{x}) = -\varepsilon_0 \mathbf{E}(\mathbf{x},t)$, i.e. (46) is one of Maxwell's equations.

With this identification of the transverse field variables, the second term in (33) can be written as

$$H_2 = \frac{1}{2}\varepsilon_0 \int d^3x \{E(x,t)^{\perp}.E(x,t)^{\perp} + c^2 B(x,t).B(x,t))\}$$
$$- \int d^3x\, E(x,t)^{\perp}.P(x,t) + \frac{1}{2}\varepsilon_0^{-1}\int d^3x\, P(x,t).P(x,t) \tag{47}$$

where $\mathbf{E}(\mathbf{x},t)^{\perp}$ is determined by (45). The first term in (47) is the Hamiltonian for radiation, which after quantization can be interpreted in terms of photons in the usual way [6]:

$$H_{rad} = \frac{1}{2}\varepsilon_0 \int d^3x \{E(x,t)^{\perp}.E(x,t)^{\perp} + c^2 B(x,t).B(x,t)\}$$
$$= \Sigma_{k,\lambda}\hbar\omega_\lambda(\hat{n}_{k,\lambda} + \frac{1}{2}). \tag{48}$$

The term in (47) linear in $\mathbf{E}(\mathbf{x},t)^{\perp}$ can be put in the form:

$$- \int d^3x\, E(x,t)^{\perp}.P(x,t) = - \Sigma_\alpha e_\alpha \tag{49}$$

using (4), and so couples charges to electromagnetic radiation via the (arbitrary) *transverse* part of the Green's function $\mathbf{g}(\mathbf{x},\mathbf{r}_\alpha)$. Finally, if we define

$$G(x',x'') = \int d^3x\, g(x,x').g(x,x'') \tag{50}$$

the last term in (47) is

$$\frac{1}{2}\varepsilon_0^{-1}\int d^3x\, P(x,t).P(x,t) = \frac{1}{2}\varepsilon_0^{-1}\Sigma_{\alpha,\beta}e_\alpha e_\beta G(r_\alpha,r_\beta) \tag{51}$$

again using (4). With the aid of (5) the longitudinal part of the Green's function yields

$$\mathbf{G}(\mathbf{x}',\mathbf{x}'')^{\parallel} = \int d^3x\mathbf{g}(\mathbf{x},\mathbf{x}')^{\parallel}.\mathbf{g}(\mathbf{x},\mathbf{x}'')^{\parallel} = \frac{1}{4\pi|\mathbf{x}'-\mathbf{x}''|} \tag{52}$$

so that separating the Green's function into its longitudinal and transverse parts makes the Coulombic interaction energy of the charges explicit:

$$\frac{1}{2}\varepsilon_0^{-1}\int d^3x\, P(x).P(x) = \frac{1}{2}\Sigma_{\alpha,\beta}\frac{e_\alpha e_\beta}{4\pi\,\varepsilon_0|r_\alpha - r_\beta|} + \frac{1}{2}\varepsilon_0^{-1}\Sigma_{\alpha,\beta}e_\alpha e_\beta G(r_\alpha, r_\beta)^\perp \tag{53}$$

with

$$G(x', x'')^\perp = \int d^3x\, g(x,x')^\perp . g(x,x'')^\perp . \tag{54}$$

Of course, the last term in (53) can have *any* value, $0 \leq G(\mathbf{x}',\mathbf{x}'')^\perp \leq \infty$, and for this reason cannot have any physical significance. Taking the atomic Hamiltonian as

$$H_{atom} = \Sigma_\alpha \frac{p_\alpha^2}{2m_\alpha} + \frac{1}{2}\Sigma_{\alpha,\beta}\frac{e_\alpha e_\beta}{4\pi\,\varepsilon_0 |\, r_\alpha - r_\beta \,|} \tag{55}$$

we may collect all this together and write

$$H_{int} = H - H_{atom} - H_{rad}$$

$$= -\Sigma_\alpha \frac{e_\alpha}{2m_\alpha}\big(p_\alpha . a(r_\alpha) + a(r_\alpha).p_\alpha - e_\alpha a(r_\alpha).a(r_\alpha)\big) - \Sigma_\alpha e_\alpha \int d^3x\, E(x,t)^\perp . g(x, r_\alpha) + \frac{1}{2}\varepsilon_0^{-1}\Sigma_{\alpha,\beta}e_\alpha e_\beta G(r_\alpha, r_\beta)^\perp \tag{56}$$

with $\mathbf{a}(\mathbf{r}_\alpha)$ given by (29). Equations (48), (55), (56) supplemented by the commutation relations:

$$[r_\alpha^r, p_\beta^s] = i\hbar\,\delta_{\alpha\beta}\delta_{rs} \tag{57a}$$

$$[A(x,t)^r, E(x',t)^{\perp s}] = -i\hbar\varepsilon_0^{-1}\delta(x-x')_{rs} \tag{57b}$$

constitute the basis for a quantum theory of atoms and molecules interacting with electromagnetic radiation, with the gauge of the vector potential arbitrary.

For practical calculations it is often convenient to use the occupation number representation. In terms of boson annihilation ($a_{\mathbf{k},\lambda}$) and creation ($a_{\mathbf{k},\lambda}^*$) operators satisfying

$$[a_{k,\lambda}, a_{k',\lambda'}^*] = \delta_{\lambda,\lambda'}\,\delta(k - k') \tag{58}$$

the radiation field is described as a collection of photons of wavevector **k** (and frequency ω = kc) and polarization λ (= 1,2). The conjugate field variables in (57b) can then be given a Schrödinger operator representation,

$$A(x) = \sqrt{\frac{\hbar}{2V\varepsilon_0\omega}}\,\Sigma_{k,\lambda}\,\varepsilon(k)_\lambda\{a_{k,\lambda}\,e^{ik.x} + a^*_{k,\lambda}\,e^{-ik.x}\} \tag{59}$$

$$E(x)^\perp = i\sqrt{\frac{\hbar}{2V\varepsilon_0}}\,\Sigma_{k,\lambda}\,\varepsilon(k)_\lambda\{a_{k,\lambda}\,e^{ik.x} + a^*_{k,\lambda}\,e^{-ik.x}\} \tag{60}$$

V is the quantization volume, and **k**/k, ε_1, ε_2 form an orthogonal triad.

5. Gauge invariant calculation

In the previous section we presented a general Hamiltonian for the interaction of electromagnetic radiation with atoms and molecules that can be put in the form:

$$H = H_0 + H_{\text{int}} \tag{61}$$

While H_0 has the familiar form of the sum of the non-relativistic atomic/molecular Hamiltonian, (55), based purely on Coulombic interactions, and the Hamiltonian for free radiation (48), H_{int} has the unfamiliar feature of involving the essentially arbitrary Green's function **g(x,x′)** because no gauge for the vector potential is specified. In particular the form (55) does *not* require the Coulomb gauge condition. Of course, overall H is gauge-invariant, and observables must be as well, so we need to consider gauge-invariant calculation.

A formal description of an experiment can be given briefly as follows. At time t_0 a state Ψ_i of the full Hamiltonian H is prepared; this state then evolves in time according to the time-dependent Schrödinger equation

$$i\hbar\frac{\partial\Psi}{\partial t} = H\Psi \tag{62}$$

and the experiment consists of measuring what this state looks like some later time t, say Ψ_f. The probability amplitude for the transition i $\rightarrow$ f is

$$a_{if} = \langle\Psi_f|\Psi_i\rangle \tag{63}$$

according to the usual rules of quantum mechanics. As a preliminary to discussing the solution of (62) we first transform to the *interaction representation* in which operators carry the time dependence induced by the free Hamiltonian H_0, and the state satisfies a modified evolution equation driven solely by H_{int} (in the new representation), explicitly:

$$i\hbar\frac{\partial\Phi}{\partial t} = V(t)\Phi \tag{64}$$

where

$$V(t) = e^{iH_0t/\hbar} H_{\text{int}}\, e^{-iH_0t/\hbar}. \tag{65}$$

A formal solution of (64) incorporating the initial condition $\Phi(t_0) = \Psi(t_0)$ is

$$\Phi(t) = U(t,t_0)\,\Phi(t_0) \tag{66}$$

where the unitary time-development operator is a solution of the differential equation (67), which may be solved by iterations to give the Dyson series (68):

$$U(t,t_0) = 1 - \frac{i}{\hbar}\int_{t_0}^{t} d\tau V(\tau)\, U(\tau,t_0), \tag{67}$$

$$U(t,t_0) = 1 + \sum_{n=1}^{\infty}\frac{(-i/\hbar)^n}{n!}\int_{t_0}^{t}dt_1\int_{t_0}^{t}dt_2\ldots\ldots\int_{t_0}^{t}dt_n\; T\{V(t_1)V(t_2)\ldots..V(t_n)\} \tag{68}$$

The chronological operator T is required because of the general non-commutativity of an operator taken at two different times [13, 16, 17].

The issue of a gauge invariant calculation can now be formulated succinctly: the interaction representation perturbation V(t) obtained from (65) is a functional of the arbitrary Green's function $\mathbf{g}(\mathbf{x},\mathbf{x}')$ because H_{int} (56) is, and thus $U(t, t_0)$ (the time development operator) is too, Eqn (69). Consider now two arbitrary gauges specified by $\mathbf{g}(\mathbf{x},\mathbf{x}')^1$ and $\mathbf{g}(\mathbf{x},\mathbf{x}')^2$; then generally:

$$U(t,t_0) \equiv U(t,t_0,[g]), \tag{69}$$

$$U(t,t_0,[g^1]) \neq U(t,t_0,[g^2]) \;\; \mathit{if} \;\; g^1 \neq g^2 \tag{70}$$

since t, t_0 are arbitrary. Physical observables are obtained from matrix elements of U, so a transition $\Phi_i \rightarrow \Phi_f$ is a physical process if and only if

$$\langle \Phi_f | U(t,t_0,[g^1]) | \Phi_i \rangle \equiv \langle \Phi_f | U(t,t_0,[g^2]) | \Phi_i \rangle \qquad (71)$$

for all **g**. In practice, one can only obtain approximate forms for $U(t,t_0)$ and care is required to ensure that approximations used preserve the gauge invariance of the matrix elements and the unitarity of U.

One particular special case is of course of most interest, namely $t_0 = -\infty$, $t = +\infty$, in which case $U(-\infty,+\infty)$ is known as the S-matrix. The calculation of the S-matrix using perturbation theory provides a natural framework for quantum theory of light scattering and optical birefringence [18,19]; the main feature of the calculation is that the transition probability for any specified optical process can be related to an element of the S-matrix evaluated with energy conservation i.e. *on-the-energy-shell*. If complete sets of eigenstates are introduced between the operators $\{V(t_i)\}$ in (68), there result time-independent matrix elements multiplied by simple exponentials containing photon-molecule energy differences. The time integrations are then trivial; in n^{th} order perturbation theory they yield (n-1) energy denominators, and one integration gives a factor of $\delta(E_i - E_f)$, so that

$$\langle \Phi_f | S[g] | \Phi_i \rangle \propto \delta(E_i - E_f) \qquad (72)$$

for all functions $\mathbf{g}(\mathbf{x},\mathbf{x}')$. For every order n of perturbation theory one can then prove by induction a result of the form

$$\langle \Phi_f | S^n[g^1] | \Phi_i \rangle - \langle \Phi_f | S^n[g^2] | \Phi_i \rangle \propto \delta(E_i - E_f) \langle \Phi_f | [H_0, \Omega[g^1, g^2]] | \Phi_i \rangle \qquad (73)$$

which vanishes between eigenstates of H_0 provided the operator Ω is regular at $E_i = E_f$. Hence the on-energy-shell transition probability is gauge-invariant, provided that in each order of perturbation theory the full interaction (56) is retained in the intermediate steps of the calculation [21].

Many instances of this general result are known from low-order perturbation theory calculations of course; the very extensive literature of light scattering is essentially based on the choice of one of two gauges;: the Coulomb gauge theory, corresponding to $\mathbf{g}(\mathbf{x},\mathbf{x}')^{\perp} = 0$, and the multipole Hamiltonian, which is derived from a special choice of integration path C in the line-integral form [16]:

$$g(x,x') = \int_C^{x'} dz\, \delta(z-x) \qquad (74)$$

The identity of the Kramers-Heisenberg formula obtained in the two gauges is a celebrated example [22]. It should be noted, however, that the popular `multipolar Hamiltonian'

$$H_{int} \approx -\mathrm{d}.\mathrm{E}^{\perp} - \mathrm{m}.\mathrm{B} - \mathrm{Q}.\nabla\mathrm{E}^{\perp} - \ldots \tag{75}$$

does not provide a (miraculous !) relief from the gauge invariance question that is explicit when H_{int} is written in terms of a vector potential. To see this, consider the Fourier transform (on **x**) of (74), which reads

$$g(k, x') = \int_C^{x'} dz e^{ik.z} \tag{76}$$

The electric dipole approximation - or `long-wavelength limit' is obtained by taking $\mathbf{k} \to 0$. For a *finite* path, C, Lebesgue's dominated convergence theorem justifies the interchange of the order of integration and the limit $\mathbf{k} \to 0$, so that

$$\lim_{k \to 0} g(k, x') \sim x' \quad \textit{finite paths } C \tag{77}$$

where $\mathbf{x}'$ is measured from the starting point of the finite path C. However the requirement for a finite path is simply a selection of a class of gauges; there are infinitely many acceptable choices of *infinite* paths for which the limit does not exist. The choice of the path C has *nothing* to do with whether `the charges are bound' - a notion involved in the classical interpretation of the electric polarization field and a classical picture of an atom / molecule. All that matters, and this is essential, is that the perturbation operator H_{int} should be defined on the whole of the Hilbert space associated with H_0 describing bound *and* continuum states for the charges and photons. The gauge-dependence of (75) thus undermines any `physical' interpretation in terms of `electric-field'photons and `electric multipoles', and corresponding magnetic terms [20].

Another important class of calculations is concerned with the theory of the natural line-shape, resonance fluorescence and related aspects of emission spectroscopy. Their characteristic feature is the specification of $t_0 = 0$ in (66), (67) and the requirement only for solutions with $t > 0$. Following the initial development of the quantum theory of the natural lineshape [23], a much more general theory of radiation damping was developed by Heitler and coworkers [24], and later reconsidered by Power and Zienau [6]. They gave a detailed discussion of the natural shape of emission lines and of resonant absorption using both the Coulomb gauge Hamiltonian ((56) with $\mathbf{g}^{\perp} = 0$), and the multipolar Hamiltonian (75). In a conventional low-order perturbation treatment of the damping theory these two interaction Hamiltonians give different predictions for the energy variation of the lineshape; for example, it is the latter interaction, (75), that leads to a result in accord with the experimental evidence of Lamb on resonant stimulated emission from the metastable $2S_{1/2}$ level of the hydrogen atom (the classic Lamb shift measurement experiments), as emphasized by Lamb [25]. While it is surely desirable that there should be a calculation in agreement with the experimentally determined lineshape, the theoretical position is unsatisfactory, since both calculations involve particular choices of gauge. The crucial theoretical question is how should these phenomena be dealt with in a *gauge invariant* fashion, i.e., so that the choice of the arbitrary Green's function $\mathbf{g}(\mathbf{x},\mathbf{x}')$ is irrelevant.

From the present perspective, an obvious weakness of these calculations is the restriction to the one-photon subspace of Fock space. Because every gauge (choice of $\mathbf{g}(\mathbf{x},\mathbf{x}')$) leads to its own commutation relations (42), each has its own Fock space so projection on a one-photon subspace is not gauge-invariant. We have previously shown [21] that the one-photon part of H_{int} (i.e linear in the charge e) in an arbitrary gauge is related to the Coulomb gauge interaction ($\mathbf{g}^{\perp} = 0$) by

$$V_1[g] = V_1[0] + \frac{i}{\hbar}[H_0, A[g]] \tag{78}$$

where

$$A[g] = \int d^3x \int d^3x' A(x).g(x,x')\rho(x') \equiv \int d^3x P(x).A(x) \tag{79}$$

from (4). The second term in (78) contributes matrix elements V_{fi} proportional to $(E_f - E_i)$ which vanish for real transitions by energy conservation, but otherwise are arbitrary; $|f\rangle$ differs from the state $|i\rangle$ by one photon. The remainder of H_{int}, which is quadratic in e and also contains $\mathbf{g}(\mathbf{x},\mathbf{x}')$, couples $|i\rangle$ to states with the same number of photons or differing by two photons. It seems plausible that cancellation of $\mathbf{g}(\mathbf{x},\mathbf{x}')$ requires consideration of all contributions from H_{int}, but how this comes about remains to be shown, and it may be that entirely different methods of calculation are required.

The proof of the gauge invariance of the S-matrix requires the use of *complete sets of states* in the matrix multiplications in the perturbation expansion in order that a delicate cancellation of gauge dependent terms takes place to leave the expression (73); a characteristic feature of Heitler's theory of spectral lineshapes [24] is that the initial unperturbed state at $t_0 = 0$ is to be omitted from all the matrix multiplications, suggesting that this cancellation may not occur in this formalism. Moreover, as one moves away from resonance the energy conservation condition does not apply, so that for example off-diagonal matrix elements of the momentum and position operators cannot be interchanged freely using the velocity-dipole formula, which implies different frequency dependencies for the lineshape, as noted by Power and Zienau [6]. The perturbation expansion for the S-matrix [21] cannot be expected to converge for resonant scattering, because of small denominators; however, isolation of the pole term for a more accurate treatment means that the non-resonant part of the S-matrix has one or more states excluded from the matrix multiplications and it is not obvious that the required cancellation of gauge-dependent terms will still take place. It is a curious and surprising situation that at the end of the twentieth century, non-relativistic quantum electrodynamics is not a completed subject.

References

1. Lorentz, H.A.: *The Theory of the Electron*, 2nd Ed., Teubner, Leipzig, 1916.
2. Hirst, L.L.: Rev. Mod. Phys., **69** (1997) 607-627.
3. Schwinger, J.: *Particles, Sources and Fields*, p. 260, Addison-Wesley, 1970.
4. Woolley, R.G.: Ann. Inst. Henri Poincaré, **23A** (1975) 365-378.
5. Belinfante, F.J.: Phys. Rev. **128** (1962) 2832-2837.
6. Power, E.A., and Zienau, S.: Phil. Trans. Roy. Soc. (London), **A251** (1959) 427-454.
7. Woolley, R.G.: Proc. Roy. Soc. (London), **A321** (1971) 557-572.
8. Loudon, R.: *Quantum Theory of Light*, 2nd Edition, Oxford University Press, 1983.
9. Cohen-Tannoudji, C., Dupont-Roc, J., and Grynberg, G.: *Photons and Atoms: Introduction to Quantum Electrodynamics*, Wiley, New York, 1989.
10. Power, E.A.: *Introductory Quantum Electrodynamics*, Longmans, Harlow, 1964.
11. Dirac, P.A.M.: Ann. Inst. Henri Poincaré, **13** (1952) 1-42.
12. Dirac, P.A.M.: *Lectures on Quantum Mechanics*, Academic Press, 1964.
13. Weinberg, S.: *The Quantum Theory of Fields*, Vol. 1, Foundations, Cambridge University Press, 1995.
14. Woolley, R.G.: Adv. Chem. Phys., **33** (1975) 153-233.
15. Atkins, P.W., and Woolley, R.G.: Proc. Roy. Soc. (London), **A319** (1970) 549-563.
16. Roman, P: *Advanced Quantum Theory*, Addison-Wesley Publishing Co. Inc., Reading, Mass. 1965.
17. Berestetskii, V.B., Lifshitz, E.M., and Pitaevskii, L.P.: *Relativistic Quantum Theory*, Vol. 4 of *Course of Theoretical Physics*, Part 1, §73, Pergamon Press, Oxford, 1971.
18. Atkins, P.W. and Barron, L.D.: Proc. Roy. Soc. (London), **A304** (1968) 303-317.
19. Atkins, P.W. and Barron, L.D.: Proc. Roy. Soc. (London), **A306** (1968) 119-134.
20. Woolley, R.G.: Int. J. Quant. Chem. (1999), in the press.
21. Woolley, R.G.: Mol. Phys., **94** (1998) 409-416.
22. Healy, W.P.: Phys. Rev., A**16** (1977) 1568.
23. Weisskopf, V., and Wigner, E.P.: Z. Physik, **63** (1930) 54; **65** (1930) 18.
24. Heitler, W.: *The Quantum Theory of Radiation*, 3rd Edition, Oxford University Press (1954).
25. Lamb, W.E.: Phys. Rev., **85** (1952) 259.

QUANTUM MECHANICS OF ELECTRO-NUCLEAR SYSTEMS TOWARDS A THEORY OF CHEMICAL REACTIONS

O. TAPIA
Department of Physical Chemistry, Uppsala University, Box 532, S-75121 Uppsala, Sweden

Abstract. A new electronuclear separability approach is used to split global translational motion from the molecular hamiltonian (H_m) and to define state dependent molecular frames; an operator is obtained coupling the global translational momentum to the operators for linear momenta of nuclei and electrons. The method prompts for a different and hopefully sounder description of chemical processes. The electronic wave functions do not depend upon the instantaneous nuclear positions; they determine stationary geometric arrangements of sources of external Coulomb potential. For processes conserving charge and particle number, such as those intervening in a chemical reaction, reactants and products are eigenstates of H_m with momenta quantized by taking the system to be in a volume V with periodic boundary conditions. The physical and chemical processes are represented by changes in populations among the eigenstates of H_m. Such changes must be produced by a coupling to an external field (e.g. electromagnetic). One of the very new results is that chemical processes may even be affected by sound fields. Quantum mechanical conservation principles enter in a natural manner to describe chemical processes. They help define selection rules. In particular, parity plays a central role. If both reactant and product channels have the same parity, the theory requires the existence of a transition structure with different parity to mediate the interconversion. This rule is important since most of the chemical reactions in the ground electronic states of reactants and products belong to this class. Chemical processes can be described in the same general terms as Franck-Condon spectroscopy.

1. Introduction

To describe a chemical reaction from a physical standpoint at the nonrelativistic level, one must first construct the Hilbert space associated with all quantum states related to the system defined by its molecular hamiltonian, H_m. For the isolated system the time-dependent Schrödinger equation

J. Maruani et al. (eds.), New Trends in Quantum Systems in Chemistry and Physics, Volume 2, 23–47.

$$i\hbar\partial\Psi(\Gamma,t)/\partial t = H_m(\Gamma)\Psi(\Gamma,t) \tag{1}$$

is separated, thereby leading to the standard time-independent Schrödinger equation

$$H_m(\Gamma)\,\Phi(\Gamma) = \varepsilon\,\Phi(\Gamma) \tag{2}$$

where $\Gamma = (\rho, R)$ are the configuration space coordinates in the laboratory frame. For a system inside a given volume V and decomposable into n-electrons and m-nuclei, the Hilbert space is comprised of all (direct products) states related to clusterings conserving the number of particles. Reactant and products, for a given chemical reaction, are just particular clusters; each one can be decomposed into the same number of particles; each being in a particular electronic quantum state. A chemical reaction is a process changing the electronic structure of the system. This can be seen as a change between stationary states of H_m. From this point of view, an adiabatic interconversion between different electronic states is not allowed. The problem is with the current theoretical approach based upon the Born-Oppenheimer approximation (BO) [1, 2] where a chemical reaction is accompanied by a *smooth* change in the overall electronic state (see page 210 in ref.[3]); the process of interconversion in the ground electronic state is then described as adiabatic wanderings on the reactant BO energy hypersurface; such variations smear the relationship between eigenstates and chemical species. This work attemps to overcome the adiabatic picture by emphasizing the quantum mechanical nature of the full process and get a more clear relationship between chemical species and eigenstates.

The quantum theory of molecular structure developed here and the standard BO approach rely on the separability between electronic and nuclear configurational degrees of freedom. However, the way this is achieved differs radically between the approaches. In the treatment described here, the nuclei are seen to be trapped by an attractor generated by the stationary electronic wave function (nuclei "follow the electronic states"); the electronic wave function does not depend upon the instantaneous positions of the nuclei as early proposed by this author [4]; a change of electronic state, characterizing a chemical reaction with reactants and products in their ground electronic states, is described as a Franck-Condon like process.

2. Theory

$H_m(\Gamma)$ generates the time evolution of any state according to eq.(1). The problem is the construction of a base of molecular states making chemical processes apparent. The base diagonalizing H_m should cover all possible stationary states for the system inside the given volume V. For example, at sufficiently high energies one has the set of free electrons and nuclei. Thereafter at lower energies one would find ions, molecules, supermolecules, free particles, etc. One would find that sets formed by these species gather together in all possible manners, where each set fulfils the particle conservation condition.

One may argue that for molecular systems in their ground electronic state, an eigenstate has no obvious relationship to a chemical species. The problem of isomerism is often invoked to sustain this criticism. In the BO approach, such a criticism is sustainable [5-7] and, since current molecular theory is a part of the BO approximation scheme [8], it is the whole molecular theory which would require a new foundation. This amounts to defining the molecular structure problem from a different perspective, as it is discussed in the following sections.

2.1. ELECTRO-NUCLEAR SEPARABILITY MODEL

The standard quantum chemical model for the molecular hamiltonian H_m contains, besides purely electronic terms, the Coulomb repulsion among the nuclei V_{NN} and the kinetic energy operator K_N. The electronic terms are the electron kinetic energy operator K_e and the electron-electron Coulomb repulsion interaction V_{ee} and interactions of electrons with the nuclei, these latter acting as sources of external (to the electrons) potential designated as V_{eN}. The electronic hamiltonian H_e includes V_{NN} and is defined as

$$H_e(\Gamma) = K_e + V_{ee}(\rho) + V_{eN}(\rho, R) + V_{NN}(R) \tag{3}$$

At the level of eq.(3) one can think of the electrons as being submitted to a Coulomb field *independent* of the dynamical characteristics of its source, i.e. the nuclei may have different masses and spin, what counts is the charge distribution of the sources. There is a problem if one uses the hamiltonian $H_e(\Gamma)$ to obtain electronic wave functions by solving the Schrödinger equation for arbitrary R-configurations; the electronic wave functions must have symmetries that are not apparent by looking at the structure of hamiltonian eq.(3), except for obvious atomic and diatomic cases. One solution to this difficulty would be to take wave functions fulfilling strict symmetry conditions (e.g. parity, rotational invariance, etc), eliminating the general R-dependence. This approach was adopted in ref. [4]. Here we elaborate on and sharpen the theoretical issues found there.

In principle the introduction of symmetry constraints, i.e. invariances permits to cast the separability problem into a manner differing from the standard approach. This can be achieved by defining an auxiliary electronic problem. The dynamical variables R are replaced by parameters α *locating general external sources of Coulomb potential* $\alpha=(\alpha_1,...,\alpha_m)$, equivalent to that set up by the charged nuclei.

Sources of external potential can be produced in a number of ways in which there is no need for special massive nuclei. As they identify external sources they are classical variables, namely, position coordinates for the sources. There is no quantum dynamics related to them yet. Symmetry constraints can be naturally defined. We formally write $H_e(\rho;\alpha)$ to distinguish this situation from the standard approach. Since the primacy is given to the electronic wave function, and no Schrödinger equation is available at this point, its existence has to be taken as working hypotheses.

2.2. ELECTRONIC WAVE FUNCTIONS

The complete group of an electronic system is defined by the set of all unitary operators which commute with the hamiltonian describing it. The hamiltonian $H_e(\rho;\alpha)$ must have symmetry properties, thereby introducing constrains in the α-coordinate space. Operators considered here are those of angular momentum, those which transform the spatial coordinates of the system (including Coulomb sources), and those permuting equivalent sites within the system [9]; α may play the role of a symmetry label.

Let us assume then the existence of a stationary electronic state $\Xi(\rho)$ transforming according to one of the irreducible representations of the symmetry group to which it belongs. We contend that this wave function determines a stationary geometric configuration α^o of the external Coulomb sources which is invariant under the group operations. With this wave function a potential energy $E_e(\alpha; \alpha^o)$ is obtained as the expectation value of $H_e(\rho; \alpha)$ with respect to the stationary electronic wave function:

$$E_e(\alpha; \alpha^o) = \langle \Xi(\rho, \alpha^o) | H_e(\rho;\alpha) | \Xi(\rho, \alpha^o) \rangle \tag{4}$$

The brackets indicate integration over the electronic configuration and spin coordinates. The double dependence is given by the hamiltonian and the particular electronic wave function, which is written as $\Xi(\rho;\alpha^o)$ to remind us that it is the electronic state which enforces a stationary Coulomb field configuration. We now look for reasonable properties that such wave functions should have:

i) Continuity: the configuration of the external sources generating the Coulomb potential can be varied around a given point submitted to constraints if required by symmetry. These coordinates must be taken as parameters in the sense discussed by Goscinski and Mujica [10]. The energy functional $E_e(\alpha; \alpha^o)$ is assumed to be continuous and stationary in the neighborhood of α^o. To simplify notation, $E_e(\alpha^o; \alpha^o) = E_e(\alpha^o)$ represents a stationary point on this energy hypersurface.

ii) Boundary conditions: these wave functions should fulfill the boundary conditions implied by the given physical problem.

iii) Symmetry: as noticed above, the electronic state transforming under a given irreducible representation cannot depend on arbitrary geometries of the external sources which may have any other symmetry or no symmetry at all.

iv) Uniqueness: while an electronic state defines a unique external Coulomb potential via the stationary geometry, in turn the particular potential might not be sufficient to uniquely determine a stationary electronic state; the same α^o may be associated with other electronic states transforming according to different irreducible representations of the same symmetry group (implying different constraints) and, thereby, representing a subset of all excited states (for polyatomic molecules some of them perhaps having different stationary geometries). *It is then $\Xi(\rho)$ which must determine such configuration and the constraints for which the total energy is stationary.* This hierarchy departs from the standard frozen nuclei approach where the electrons, as pointed out by Witkowski [11], may show retardation effects in following the nuclear motions. In the present approach there are stationary electronic states, not a set of particles (electrons). The nuclear motion will be controlled by relaxation processes in a given electronic state potential energy function.

Now, using the variational principle with eq.(4) it can be shown that the electronic wave function $\Xi(\rho)$ fulfills the Schrödinger equation

$$H_e(\rho;\alpha^o)\,\Xi(\rho) = E_e(\alpha^o)\,\Xi(\rho) \tag{5}$$

Eq. (5) is exact and it can be solved analytically for some one-electron problems [12]; for many electron systems the existence of $\Xi(\rho)$ is postulated. For the n-th stationary state with energy $E_{en}(\alpha^o{}_n)$, the electronic state is represented with the symbol $\Xi_n(\rho;\alpha^o{}_n)$. Observe, $\alpha^o{}_n$ implies some symmetry constraints imposed by the electronic wave function, so that geometry variations for that state are represented by α_n; this is to be taken as a symmetry label implying some definite constraints for the sources. $E_e(\alpha_n;\,\alpha^o)$ will be the symmetry adapted potential energy surface. It is this functional that can be used for finding numerical solutions to the electronic problem, using standard quantum chemical methods.

For the ground electronic states, there is a one-to-one relation beween the wave function and the stationary Coulomb potential. This derives from eq.(5) following a line of analysis Gilbert used to study the density functional theory approach [13].

A complete set $\{\Xi_n(\rho;\,\alpha^o{}_n)\}$ of electronic stationary states can be used to work out solutions for the nuclear dynamics that follow from the Schrödinger eq.(2), as shown below. Before doing this, one needs to establish a connection between the electronic Schrödinger equation in the frame $\alpha^o{}_n$ to one in the laboratory fixed frame.

2.3. REFERENCE FRAMES

The electronic stationary wave function defines its own reference frame, that can be constructed using the coordinates $\alpha^o{}_n$ and, without loss of generality, the electron coordinates can be taken with respect to one such fixed frame. The coordinates $\alpha^o{}_n$ are the nearest one can get to a model of a rigid molecule. It is sufficient to put the corresponding masses at the vertices where the charges are. This is a point mass model; it is not a nuclear dynamics object yet. In this manner, the idea of molecular structure appears in a sense used in chemistry. This time the drawbacks in defining the concept of molecular shape, noticed by Woolley and Sutcliffe [6], do not apply. The property is here related to electronic stationary states thereby taking the problem from the opposite side. One has here a direct relationship between electronic structure and a chemical species. The powerful topological techniques developed by Mezey can be used to analyze molecular shape [14].

Properties related to angular momentum follow. The spin coordinates are implicit in the whole argument. Insofar as symmetry properties are concerned, it is the electronic total spin angular momentum S^2 and its projection along the arbitrary z-axis of the rigid frame M_S which factor out the state function. The same is valid for the electronic orbital angular momentum $L_e{}^2$, M_{Le}. Introducing a rotated rigid frame, $\alpha^o{}_n{}''(\alpha,\beta,\gamma)$, the electronic hamiltonian contains only rotationally invariant distances and scalar products which are independent of Euler angles α, β, γ, and, in the body-fixed coordinates, the form remains invariant. The rotational angular momentum can hence factor out too. They are basic symmetries of an electronic state that are not always apparent.

Using primed coordinates in the laboratory-centered frame to help comparisons with Pack-Hirschfelder's work [8, 15], the molecular hamiltonian

$$H_m(\Gamma') = K_{N'} + H_e(\Gamma') \tag{6}$$

can be transformed with an origin shift given by $R' = \mathbf{C}n' + R$ and $\rho' = \mathbf{C}n' + \rho$ into:

$$H_m(\Gamma') \to K_N + H_e(\Gamma) + K(C_{n'}) + \hbar\,\nabla(C_{n'})\cdot\Sigma_k(\hbar/M_k)\nabla_k + \hbar\,\nabla(C_{n'})\cdot\Sigma_i(\hbar/m_e)\nabla_i \tag{7}$$

The spectra related to the stationary states is the same for both molecular hamiltonians. The information concerning the stationary frame is hidden in the electronic wave function. This is equivalent to saying that this theory makes transparent the symmetries of the otherwise opaque total hamiltonian in the laboratory frame. For all practical purposes, $H_m(\Gamma')$ transforms into $H_m(\Gamma)$ for momentum stationary states and the electronic stationary states are obtained from a set of equations such as eq. (5).

One can now write the total hamiltonian as

$$H_m = H_m^o + K(\mathbf{C}_{n'}) + W = H_o + W \tag{8}$$

and the operator coupling the global motion to the electro-nuclear system

$$W = \hbar\, \nabla(\mathbf{C}_{n'}) \cdot \Sigma_k(\hbar/M_k)\nabla_k + \hbar\, \nabla(\mathbf{C}_{n'}) \cdot \Sigma_i(\hbar/m_e)\nabla_i = \hbar\, \nabla(\mathbf{C}_{n'}) \cdot \mathbf{p} \tag{9}$$

The reference frame depends upon the stationary electronic state (if any) derived from a diagonalization of the electronic hamiltonian $H_e(\Gamma)$.

For non-stationary processes the motion of the center of mass produced by external sources (sound waves for instance) can be coupled to the electro-nuclear system via its total momentum operator **p** (cf. eq.(9)). Not only vibrational excitations but also electronic ones can be mediated by non-stationary motion of the center of mass. This is a feature related to the stationary frame determined by a particular stationary electronic state.

It remains now to find the stationary states related to the nuclear dynamics.

2.4. NUCLEAR DYNAMICS AND SPECTRAL REPRESENTATION

If quantum nuclear fluctuations are sought, from eq.(8) one only needs the first term, $H_m^o = K_N + H_e(\Gamma)$. These fluctuations include rotations as a whole of a frame rigidly bound to the stationary external Coulomb sources. The coupling terms in W, involving the momenta operators of electrons and nuclei, have no diagonal components; they will contribute to changes in the populations of stationary levels related to the hamiltonian H_m^0. Integrating over electronic coordinates with a particular $\Xi_n(\rho;\alpha^o_n)$, the collective nuclear motion can be obtained as a solution to

$$(K_N + E_{en}(R))\, \Lambda_{nk}(R) = \varepsilon_{nk}\, \Lambda_{nk}(R) \tag{10}$$

This is the exact solution to the problem in a fixed frame related to the stationary geometry of the n-th electronic state. The electronic quantum number in $E_{en}(R)$ is sufficient to remind of the existence of a priviledged frame (and possibly specific symmetry constraints). In the stationary case, the hamiltonians $H_e(\rho;\alpha)$ and

$H_e(\rho, R)$ can be taken as isomorphous as α takes on all accessible values. The total wave function acquires the form: $\Phi_{nm}(\rho, R) = \Lambda_{nm}(R)\ \Xi_n(\rho;\alpha^o)$ because the two sets of dynamical variables are separable in the sense discussed above. The set of product functions differ from the strictly diabatic electronic basis [16] in that the electronic wave functions are related to Coulomb attractors, via a potential energy function that is invariant under translations and rotations.

To construct a spectral representation, use is made of the set of orthogonal wave functions $\{\Xi_n(\rho;\alpha^o{}_n)\Lambda_{nk}(R) \leftrightarrow \langle\rho R|nk\rangle \leftrightarrow |nk\rangle\}$. The general kets $|nk\rangle$ would correspond to the molecular states associated with the real electro-nuclear spectra. Provided integration over electronic degrees of freedom is carried out before the nuclear ones one obtains:

$$H^o_m = \Sigma_n \Sigma_{(k(n))}\ |nk\rangle \varepsilon_{nk} \langle nk| \tag{11}$$

The index k(n) recalls that the nuclear fluctuation quantum states in eq.(11) are determined by the electronic quantum state via potential energy $E_{en}(R)$. Once the electronic problem is fully solved, via a complete set of eq.(5), it is not difficult to see that $\langle\rho R|nk\rangle$ multiplied by the box-normalized wave solutions (see p. 428, ref. [17] 2nd ed.) are eigenfunctions of the hamiltonian H_o and, for stationary global momentum solutions, the molecular hamiltonian is also diagonalized thereby solving eq. (2).

By forming translational wave packets, molecules can henceforth be localized in momentum space. This process may happen under non stationary conditions. This possibility results from the separation approach used here.

3. Chemical framework

The separability used here leads to a clear relationship between chemical species and ground state electronic wave functions. Each isomeric species is determined by its own stationary ground state electronic wave function. The latter determines a stationary arrangement of Coulomb sources which is different for the different isomers. The nuclei are then hold around a stationary configuration if eq.(10) has bound solutions. An interconversion between them would require a Franck-Condon process, as it is discussed in Section 4.

Schrödinger's eq.(5) looks pretty much the same as in standard BO theory as both schemes are based on a separability ansatz. They are, however, fundamentally different. While in the BO scheme the electronic wave functions $\{\Phi_n(\rho,R)\}$ are assumed to form complete orthonormal sets for all R-parameters (see, e.g., refs [2, 18]), the present theory may put constraint as a function of the electronic state, so that not all nuclear configurations may be accessible. The new theory requires the existence of a complete set of quantum states related to external stationary Coulomb sources $\{\alpha_n\}$.

In principle both approaches may differ in the index of the stationary geometries sensed by the eigenvalues of the Hessian matrix. The variational principle implies, in the present electronuclear approach, that any stationary conformation is a minimum since geometric variations must conserve symmetry; to do actual computing one should use variations in the α_n-space. On the other hand, in the BO-scheme, if no symmetry restrictions are used (α-space optimization) some states may display saddle point character with indices equal to or larger than 1. When such a situation is met the very fact that there are solutions with imaginary frequencies indicate they are not acceptable as physical stationary state solutions. For this reason, it is common practice, when calculating any property related to the molecular spectra, to discard these solutions as it is done in evaluating vibrational partition functions to get chemical rates [19].

3.1. POTENTIAL ENERGY HYPERSURFACES

The equality $E_{en}(\alpha;\alpha^o_n) = E_{en}(R)$ is not obvious as R stands for nuclear coordinates appearing in the dynamics of such species, while α represents a space which can be used to define gauge fields [8] for either classical or quantum mechanical models. This latter issue is not examined here. For the time being, let us take the equality and use it by relaxing symmetry constraints that may be related to α_n but not necessarily to α. Then, eq.(10) follows trivially from eq.(2) after averaging with the electronic function, $\Xi_n(\rho;\, \alpha^o_n)$.

Now the potential energy hypersurface, $E_{en}(R)$, does not change adiabatically its electronic state. The wave function $\Xi_n(\rho;\, \alpha^o_n)$ covers all the configurational space in a way compatible with the appropriate boundary and symmetry conditions. The variational principle implies that $E_{en}(\alpha;\alpha^o_n) > E_{en}(\alpha^o_n)$ for external Coulomb sources having configurations differing from the stationary one, $\alpha \neq \alpha^o_n$. The Hessian cannot have negative eigenvalues as this would imply that there are directions where the energy along directions on the hypersurface would be lower than the stationary value. This property differs from the one assumed in ref. [4].

Deforming the geometry of Coulomb sources adiabatically is an allowed process, since these are classical variables; the electronic energy, eq.(4), can only increase, but there cannot be a change of electronic state. The energy of two electronic states with the same space symmetry for a given α can be accidentally equal. Still, they cannot produce an "avoided crossing" in the sense of the BO picture. If each state represents, for instance, the reactants (|1R>) and products (|1P>), respectively, and the system is prepared as reactants, the matrix element

<1R|W|1P>, if different from zero, opens a probability to actually populate the product states, and vice versa. A coupling of this kind may create a linear superposition thereby producing finite probabilities to populate the quasi degenerate state. A finite life time may hence be related to such process. Theoretically, if population of product states is required, a decoherence process must take place [20]. Thus the electronic reaction dynamics happens in a neighborhood of the crossing region. If the crossing appears with coordinates that are different from the corresponding stationary coordinates (which would mean that the system is in one of the rovibrational excited states), once the coherence is lost, the system will evolve from the intramolecular excited state via nuclear dynamics relaxation to the ground state of the product or the reactant channel. Quantum mechanics can only deal with probabilities. But, because the chemical interconversion process is electronic, femtosecond time scales can be involved in producing it.

3.2. PAIR CLUSTERS AND ASYMPTOTIC STATES

To discuss chemical reactions of bimolecular kind, it is appropriate to define asymptotic systems. Let us consider a system clustered in two groups that can be decomposed into n_1 and n_2 electrons and m_1 and m_2 nuclei, with $n = n_1+n_2$ and $m = m_1+m_2$, respectively. Each cluster's ground electronic state is assumed to have particular stationary geometry and separability using them as local frames is considered now. Terms coupling nuclear and electronic dynamics cancel out rigorously for the stationary situation. Without loss of generality, let us put the origin of the frame at the center of mass in cluster (molecule) 1 and measure the relative position of cluster (molecule) 2 center of mass with the vector R_{12}; we take the z-axis along this direction. The asymptotic molecular hamiltonian is defined by the limit

$$Hm(|R_{12}| \rightarrow \infty) = Kcm1 + Hm1 + Kcm2 + Hm2 \quad (12)$$

H_{mi} is the i-th cluster hamiltonian referred to its space fixed frame with axes parallel to the laboratory frame. K_{cmi} is the i-th cluster kinetic energy operator with total mass M_i. The existence of electronic wave functions for each system is taken for granted as their spectra can be determined in isolation.

For finite intermolecular distances R_{12} the molecular hamiltonian in the cluster representation is given by

$$H_m(\Gamma ; R_{12}) = K_{cm1} + K_{cm2} + H_{m1} + H_{m2} + V_{12}(\Gamma_1,\Gamma_2 ; R_{12}) \quad (13)$$

or, introducing the relative motion coordinates with reduced mass, $1/M_{12} = 1/M_1 + 1/M_2$, and the combined centre of mass, with mass $M = M_1 + M_2$, one gets:

$$H_m(\Gamma ; R_{12}) = K_{cm} + K_{12} + H_{m1} + H_{m2} + V_{12}(\Gamma_1,\Gamma_2 ; R_{12}) \quad (14)$$

This transformed molecular hamiltonian is useful in discussing molecular events in the interaction region when the clusters are (adiabatically) brought into contact. The use of the intermolecular distance as a reaction coordinate transforms the problem into a semiclassical one. In fact, this distance relates two stationary sources of Coulomb potential incorporated in the hamiltonians H_{m1} and H_{m2}. Coordinates related to these frames can be used to define classical reaction coordinates that may help visualising the chemistry. In a full quantum dynamical description, all integrals cover all possible distances and the scheme corresponds to standard quantum scattering theory (see next section). Still one can use the semiclassical framework to discuss mechanistic issues. One should not forget the energy levels of the supermolecule which will only become accessible for definite ranges of R_{12}. Within those boundaries, the quantum states of the collision pair and the supermolecule should have the smallest energy gaps. As shown below, the operator W will mediate the interconversion among them.

In principle, once the electronic states have been chosen, the nuclear dynamics for each partner can be solved too. Averaging with a particular rovibrational quantum state, $\kappa1(\alpha1)$ or $\kappa2(\alpha2)$ respectively, one obtains

$$E_{\alpha1\kappa1,\alpha2\kappa2}(R_{12}) = \varepsilon_{\alpha1\kappa1} + \varepsilon_{\alpha2\kappa2} + \langle\kappa1,\kappa2|\ V_{\alpha1\alpha2}(R_1,R_2;R_{12})|\kappa1,\kappa2\rangle \quad (15)$$

The form of this equation makes explicit the fact that intermolecular forces do depend upon their vibrational states as well as on their electronic states. Due to the antisymmetrization of the global electronic wave function, $V_{\alpha1\alpha2}(R; R_{12})$ contains Coulomb exchange terms and a direct term formed by the Coulomb multipole interactions and the infinite order perturbation electrostatic effects embodied in the reaction field potential [21, 22].

To sense the effect one cluster would have on the other let us consider the effective hamiltonian approach. There, one cluster is subjected to the field of the other. With obvious notations one can define an effective parametric electronuclear problem for each partner via the functional:

$$E_{\alpha1\kappa1,\alpha2\kappa2}(R_{12}) = \varepsilon_{\alpha1\kappa1} + (1/2)\ \langle\kappa1|\langle\kappa2|\ V_{\alpha1\alpha2}(R;R_{12})|\kappa2\rangle|\kappa1\rangle, \quad (16)$$

with a similar equation for the other partner. These effective equations are useful in the context of a microscopic description of molecular mechanisms and the theory of surrounding medium effects. An important result is that they show a given molecule can be seen as geometrically deformed by the forces exerted by the partner. Displacement from the stationary geometry will be produced by terms linear in the nuclear fluctuation variables (see for instance [22]). The intermolecular forces produce geometric molding and shift of the energy levels for each partner. A collision pair remains a collision pair albeit a deformed one if the partners are brought together in an adiabatic sense. This is a fundamental difference of interpretation between equations of the present theory and the BO framework-based computer approaches.

4. Theory of chemical reactions

The quantum states of the molecular hamiltonian can be labelled with the linear momentum of each rigid configuration for the sources of static Coulomb potential, $\mathbf{P}_1$ and $\mathbf{P}_2$ or $\mathbf{k}_{o1} = \mathbf{P}_1/\hbar$ and $\mathbf{k}_{o2} = \mathbf{P}_2/\hbar$, respectively. The total angular momentum of each cluster, $\mathbf{J}_1$ and $\mathbf{J}_2$ and their projections $\mathbf{J}_{1z}$ and $\mathbf{J}_{2z}$ together with the vibrational and electronic quantum numbers provide a set of labels helping to characterize the quantum states of the system:

$$| \mathbf{k}_{o1}, \mathbf{k}_{o2}, \mathbf{J}_1, \mathbf{J}_{1z}, \mathbf{J}_2, \mathbf{J}_{2z}, \alpha 1\kappa 1, \alpha 2\kappa 2> \tag{17}$$

These states are normalized:

$$<\mathbf{k}_{o1}',\mathbf{k}_{o2}',J_1',m_1',J_2,m_2',\alpha 1'\kappa 1',\alpha 2'\kappa 2'|\mathbf{k}_{o1},\mathbf{k}_{o2},J_1,m_1,J_2,m_2,\alpha 1\kappa 1, \alpha 2\kappa 2> =$$
$$\delta(\mathbf{k}_{o1}'-\mathbf{k}_{o1})\, \delta(\mathbf{k}_{o2}'-\mathbf{k}_{o2})\, \delta J_1',J_1\, \delta m_1',m_1\, \delta J_2',J_2\, \delta m_2',m_2\, \delta\alpha 1',\alpha 2'\, \delta\kappa 1',\kappa 2', \tag{18}$$

where $\delta(\mathbf{k}'-\mathbf{k})$ is the three dimensional Dirac distribution and the other deltas are Kronecker symbols. These orthogonality conditions are more general than those used by Golden in his quantum theory of chemical reactions [23].

The construction of global wave functions using the cluster decomposition can be done by using appropriate angular momentum composition rules [24]. As reactants and products have different angular momentum eigenfunctions, the allowed combinations must fulfill angular momentum conservation rules. In what follows, we use simple labels to identify the system, thereby masking the complexity. For actual computing, one has to reintroduce a more detailed state labelling [25].

A chemical reaction can be seen as happening during a collision of two (molecular) beams. Each beam is made of fragments, and the two of them form an asymptotic reactant system [26]. Under well defined conditions, and after crossing the interaction region, the system that was in its ground electronic state may populate a different cluster, this latter in its ground electronic state too. To describe this type of situations, one has to find solutions to the Schrödinger eq.(2) for a given energy E:

$$(H_o + W - E) | \Phi > = 0 \tag{19}$$

where H_o is defined in eq.(8). One can now prepare the system in one eigenstate of H_m^o, create translational wave packets to represent the beams, and the system will evolve with eq.(1) populating different electronic and rovibrational accessible quantum states via the interaction W.

One selects the important quantum states of reactant and products to define the subspace P. All other states, including different types of supermolecules, are gathered in the orthogonal subspace Q. To simplify the analysis, consider a bimolecular reaction in the ground electronic state, the two molecules are described by the state $| 1R(P) \rangle$, and the products in the electronic ground state, are represented by $| 1P (P) \rangle$. Operationally, the spectra of reactants and products are used to identify them. The symbols 1R and 1P are there to indicate the quantum number for the linear and angular momentum for each fragment and the rovibrational state for he asymptotic fragments involved (cf. eqs.(17), (18)).

4.1. SCATTERING FORMALISM

The chemical reaction corresponds to a preparation-registration type of process. With the volume periodic boundary conditions for the momentum eigenfunction, the set of stationary wavefunctions form a Hilbert space for a system of n-electrons and m-nuclei. All states can be said to exist in the sense that, given the appropriate energy E, if they can be populated, they will be. Observe that the spectra contains all states of the supermolecule besides the colliding subsets. The initial conditions define the reactants, e.g. $| 1R(P) \rangle$. The problem boils down to solving eq.(19) under the boundary conditions defining the characteristics of the experiment.

Preparation-registration processes have been discussed by Arno Bohm and coworkers (see for instance [27, 28]). The issues developed in reference [27] are left out in the present analysis. We mention these references to indicate the existence of subtle and complex mathematical issues related to rigged Hilbert spaces. Here we follow the quantum scattering approach to an extent required to discuss specifically chemical features.

A physical process would correspond to the passage from a given population prepared far back in the past for instance ($t = -\infty$) to another (different) population to be measured in the future ($t = +\infty$). The initial quantum state ($t = -\infty$) is represented by the ket $|\chi_i^{(+)}\rangle$ and $|\chi_f^{(-)}\rangle$ stands for the final state ($t = +\infty$). A quantum state at an intermediate time t is obtained as usual with a time evolution operator: $U(t, t_o = -\infty)$, so that

$$|\chi(t)\rangle = U(t,t_o=-\infty)\,|\chi_i^{(+)}\rangle \tag{20}$$

with the boundary condition

$$|\chi_f^{(-)}\rangle = \lim_{t=+\infty} |\chi(t)\rangle = U(+\infty,-\infty)\,|\chi_i^{(+)}\rangle = S^{\dagger}\,|\chi_i^{(+)}\rangle \qquad (21)$$

where $S^{\dagger}$ is (by convention) the transpose complex conjugate scattering matrix.

The initial and final asymptotic states are always expanded in the time independent basis associated with the molecular hamiltonian; the scattering matrix is unitary. Note again that the basis contains all possible resonance and compound states. If there is no interaction, the scattering matrix is the unit matrix 1. Formally, one can write this matrix as S= 1+iT where T is an operator describing the non-zero scattering events including chemical reactions. Thus, for a system prepared in the initial state $|\Phi_i\rangle$, the probability amplitude to get the system in the final state $|\chi_f^{(-)}\rangle$ is given by $\langle\chi_f^{(-)}|T\ \Phi_i\rangle$. As this is an integral over the full configurational space, what one does is to look after those regions that may contribute most to the integral. This is akin to construct particular mechanistic pathways for chemical processes.

To construct a particular expression for the probability amplitude, the quantum mechanical problem in eq.(19) must be solved.

Consider the asymptotic system including quantum states for reactants, $r = r_1 + r_2$ and products, $p = p_1 + p_2$. For the given energy E, one selects all quantum states which can be measured at $t= +\infty$. These are gathered in a subspace P with projection operator P. These states are called open channels. All other states would form the orthogonal complement to P with projection operator Q. As usual, $P^2=P$, $Q^2=Q$, $PQ = QP = 0$ and $P+Q = 1$. The state vector can be decomposed as:

$$|\Psi\rangle = P\,|\Psi\rangle + Q\,|\Psi\rangle \qquad (22)$$

The events related to the suspace $P|\Psi\rangle$ would correspond to direct collisions between reactants that may either leave the system as it is or it may interconvert into products via the interaction hamiltonian W. The component $Q|\Psi\rangle$ contains events involving states in the orthogonal complement that includes supermolecule states. To populate these states require physical excitation processes; these states may have finite lifetimes. The Q-component is called the time-delaying component by Feshbach in the context of nuclear reactions [29]. The $Q\,|\Psi\rangle$ component is given by

$$Q\,|\Psi\rangle = (E+i\eta - QHQ)^{-1}\, QHP\,|\Psi\rangle = W_{QQ}\, QHP\,|\Psi\rangle \qquad (23)$$

where $H_{QQ} = QHQ$, $H_{QP} = QHP$, etc., are the projected hamiltonians, and we write H instead of H_0+W. The imaginary energy is included in case open channels are found in the Q-subspace; η tends to zero from positive values. $E+i\eta$ will be indicated as $E^{(+)}$ in what follows. The partitioning technique leads to an effective hamiltonian for the P-system:

$$H_{eff} = H_{PP} + H_{PQ}\,(E^{(+)} - H_{QQ})^{-1}\, H_{QP} = H_{PP} + H_{PQ}\, W_{QQ}\, H_{QP} \quad (24)$$

and the Schrödinger equation looks like

$$(E - H_{eff})\, P\,|\Psi\rangle = 0 \quad (25)$$

The effective hamiltonian is energy dependent, complex and nonlocal. To get a formal solution for $P|\Psi\rangle$, the Q-projected hamiltonian ($QHQ = H_{QQ}$) must be diagonalized. This amounts, in chemistry, to finding out stationary states for the supermolecule that might be relevant for describing different mechanistic pathways.

The orthogonal Q-space plays an important role in describing the mechanism with which to accomplish a given reaction. This is true when the electronic wave functions for reactants and products have equal parity, as discussed below.

The H_{QQ} operator may have discrete (e_s) or continuous (e) eigenvalues, with eigenfunctions $|\Phi_s\rangle$ and $|\Phi(e,\alpha)\rangle$, and normalization conditions:

$$\langle\Phi_s|\Phi_{s'}\rangle = \delta_{ss'};\ \langle\Phi_s|\Phi(e,\alpha)\rangle = 0\ ;\ \langle\Phi(e',\alpha')|\Phi(e,\alpha)\rangle = \delta(e' - e)\,\delta(\alpha-\alpha')$$

The effective hamiltonian can be finally written (see ref.[29]) as:

$$H_{eff} = H_{PP} + \sum_s \frac{H_{PQ}|\Phi_s\rangle\langle\Phi_s|H_{QP}}{E - e_s} + \int d\alpha \int de\, \frac{H_{PQ}|\Phi(e,\alpha)\rangle\langle\Phi(e,\alpha)|H_{QP}}{E^{(+)} - e} \quad (26)$$

In the state basis diagonalizing H_o the mixing of states is due to the coupling operator W. In the P-space one has the ingoing quantum state for the reactant (say, $|\,1R(P)\rangle$) and the product state (say, $|\,1P(P)\rangle$), and one is interested in getting the probability to populate the latter when the system was prepared in the former state. Two situations are found:

i) the two electronic states have different parity;

ii) the electronic states have the same parity.

In H_{PP} the important off-diagonal matrix elements are $\langle 1R(P)|W|1P(P)\rangle$, which can easily be factorized as the operator has the following product form: $\langle 1R(P)|\hbar\, \nabla(Cn').\mathbf{p}|1P(P)\rangle$. Let us consider electro-nuclear transitions controlled by $\mathbf{p}$. If the electronic wave functions have the same parity, the matrix element is globally zero. If they have different parity the electronic contribution can be different from zero. This process has a clear Franck-Condon nature.

In the second case, reading eq.(26), it is necessary that at least one of the Q-space states has the same parity as the operator **p**. Then, in (26) there may be a non-zero off-diagonal element connecting the ingoing to the outgoing channels. This state is called here a transition state (TS) and the coordinates of the stationary arrangement of external Coulomb sources α^{o}_{TS} (or α_{TS}) is defined as a transition state structure (TSS). The TSS is a fundamental electronic property, while the quantum states of the TS include translational and rovibrational states with their characteristic density of states.

From the structure of the matrix elements of W, one can see that the change in electronic state is of Franck-Condon type and, consequently, it takes place with frozen nuclei positions. For chemistry this fact has important consequences. The geometric arrangement at interconversion must be the same for both electronic quantum states. For each quantum state the electronic energy ($E_{en}(\alpha;\alpha^{o}_{n})$) depends upon the geometry of the external sources and on the specific n. The maximal probability for interconversion is attained when the energy gap is minimal between the different electronic states involved, and the external energy can be used to produce resonance conditions. For case i) this is simply a geometric activation process followed by geometric relaxation (similar to an inelastic scattering process), which will bring the system to the asymptotic products. This activation-relaxation involves changes in the rovibrational state populations of the colliding partners. One uses here the operator $H_m(\Gamma; R_{12})$ to help visualizing things on a more chemically flavored approach.

The novel situation is found for the case ii): reactant and product channel electronic states have the same parity. Now, if the electronic Q-state sustains bound vibrational states, the stationary external Coulomb sources define a geometric arrangement around which the nuclei would fluctuate, α^{o}_{TS}. These (normal) modes, at low energy, cover a configurational space region with well defined (symmetry) constraints. To prompt for a change of electronic structure, the reactants and products configurations must also populate such conformational regions. For reactants having an equilibrium external Coulomb sources, which are much different from the one related to the geometry of the Q-state compound, a molding process on the collision pair is required if a Franck-Condon transition is to take place with reasonable probability. This is a *necessary condition* that appears for the first time in a quantum mechanical description of chemical reactions. That it is also a *sufficient condition* derives from this formalism, as it has been documented by actual calculations of transition state structures. The highest probability would be attained for the minimal compatible energy gap. Once the Q-state (the one with a particular transition state structure) is populated, the probability to open the product channel becomes, in principle, different from zero.

Thus, the chemical interconversion for equal electronic parity channels has four separated aspects: i) activation via molding of reactants; ii) population of TS rovibrational quantum states; iii) population of reactants molded into configurations covered by the TS, and iv) relaxation towards products in their ground states. All such changes are submitted to energy and angular momentum conservation rules.

It may take a finite time to get nuclear configurations compatible with the molded products channel. Once the electronic state of the molded product becomes populated, its geometry has to relax in order to get the asymptotic species in their ground electronic structure. It follows that the interconversion process generates a finite lifetime for the transition state.

At this stage we leave the formal scattering theory. The calculation of rates amounts to determine the matrix elements of the T-operator of eq.(21) [29]. The situation in chemistry is complicated by the nuclear configuration activation-relaxation processes. They may range from acoustic (cf. eq.(9)) to thermal stochastic motion and up to coherent nuclear dynamics. The rate limiting step is then difficult to identify on a too general a base. Note, before leaving this formalism, that the interaction with one Q-space state can be seen as a resonance process (cf. Feshbach ref. [29]).

4.2. CHEMICAL EXAMPLES

To realize the import of this approach to chemical reaction descriptions, let us consider the elementary electron transfer process at an adiabatic level:

$$H(1s) + H^+ \rightarrow H^+ + H(1s) \tag{27}$$

The electronic quantum state of the pair $|H,1s\rangle|H^+\rangle = |in\rangle$ remains invariant at all distances. The electron transfer will not take place in a direct manner because the electronic parity is equal for both channels. The interconversion process requires a TS with parity -1. Among the states available to a system decomposable in one electron and two protons (or proton deuterium, etc) there are the hydrogen molecule ion species. The first electronic excited state ($1\sigma_u$) of the molecular ion H_2^+ provides an "intermediate" (Q-state) for the interconversion once angular momentum conservation rules are fulfilled. The state ($1\sigma_u$) is found above the $|in\rangle$ and $|out\rangle$ states leading to resonance in the cross section. This state may either relax to the ($1\sigma_g$) state yielding the hydrogen molecule ion and emitting a photon as this state is 2.8eV below dissociation, or it may take the product channels. This is a FC-like process. The reaction (27) is a prototype of electron transfer (ET). Thus, for any ET reaction whose $|in\rangle$ and $|out\rangle$ asymptotic electronic states share the same parity, the actual interconversion would require the mediation of a TS.

The quantum states relevant to the "harpoon mechanism" (see page 134 in ref. [26]) for the reaction of K + Br_2 producing KBr + Br are depicted as: |1> = |K + Br-Br >; |2> = | K^+ + Br-Br^- >; and |3>=| K-Br + Br >. Chemical graphs indicate where the main attractors (bonds) are. It is apparent that the stationary source for the negatively charged bromine molecule has a larger distance than the neutral species, and the rovibrational state may have larger nuclear fluctuations. To analyze the interconversion, one uses the asymptotic states gathered at times $t = -\infty$ and $+\infty$. The system is prepared as wave packets | K > and | Br-Br > each with their local symmetries (not shown to simplify notation) and momenta (cf. eq.(17)) so as to collide or interact within a bounded region, say between the parallel surfaces A and B oriented so as the particles approach together, R_{12} is at 90° from the normal, the advance of the beam particles as a whole is measured along the normal to those surfaces. The system prepared in state |1> will enter at A and, when measured beyond B, it may be either in the same electronic state (possibly different rovibrational states will be populated) or in state |3>. The system |1> is given an energy E along the intermolecular degree of freedom R_{12} which is now a "control coordinate" for the sake of description, the motion of the center of mass transports the system from A to B and onwards. The energy of |1> ($E_{e1}(\alpha;\alpha^{o}{}_1)$) as a function of the approach coordinate (measured as time using the relative speed of the beam particles) is almost constant, except at very short range (not considered here) where repulsion dominates. In the energy scale, the asymptotic charge transfer state (|2> with energy $E_{e2}(\alpha;\alpha^{o}{}_2)$) is above the asymptotic state |1> by about 418 - 325 KJ/mol = ΔE_{CT} which is the difference between the ionization potential of K and the electronaffinity of Br_2. The Coulomb interaction V12 (cf. eq.(14)) lowers the energy of |2> as the beams approach each other. At point A its energy equals E (this latter measured from the asymptotic state |1>) henceforth, as the intermolecular distance decreases it may "cross" the energy surface of |1>: a "crossing" distance can be defined $R^{\#}$(Å) = 14.4 / ΔE_{CT} (eV); cf. eq.(4.5) from ref.[26]. This latter distance is characteristic of the electronic system, it does not depend upon the impact parameter, b. If the experimental set up does not bring the fragments at distances smaller than $R^{\#}$ the energy of the charge transfer state will always be larger than E and no population of |2> can be expected (the crossing point A does depend upon b).

The initial conditions lead to an energy for the virtual product state |3> that is very high before the system comes into appropriate range. The direct reaction leading to population of state |3> can only happen in a region α where the K-Br distances (covered by the vibrational-rotational states) and the bromine atom position match possible K + Br-Br geometries with minimal energy gaps. This may happen in a very narrow region. The integral $\langle 1 | W | 3 \rangle$ can be different from zero by parity as the bromine atom can be in a P-state (L=1) and the parity of state |1> is positive. To estimate cross sections, it is a common procedure to approach it with a hard-sphere model. However, the cross section of the real reaction is much larger than the one expected from this model, in this case represented by the square modulus of $<1| W | 3 >$. Polanyi proposed an "harpoon mechanism" which in the present approach is simply translated into the population of the state |2> once conditions for energy, momentum and angular momentum conservation are fulfilled. In the first place, population of |2> will increase the life time of the transient complex (the system is actually a bounded ionic collision pair at appropriate energy E). To simplify the description we will assume the system has no coherent superposition of states (they might decohere fast, see for example [20]). This would mean that the electronic system is in one and only one electronic state between the Franck-Condon jumps. Now, as observed above $Br\text{-}Br^-$ has a fluctuation amplitude larger than the neutral molecule, and as the ionic pair is bounded, there will be a much large probability to match configurational space related to the product K + Br-Br increasing the probability to populate |3> molded into geometries of |2>. For, being now in channel |3> means that the nuclei are trapped by the electronic wave function for which K-Br has bounded rovibrational states and a free Br atom (most likely electronically excited). If the system leaves region B in state |3>, then the energies of states |1> and |2> go upwards steeply; the corresponding states cannot be populated, due to energy conservation rules. For those situations where the system is trapped by the charge transfer state there will be renewed chances to show a probability to populate the channels |1> and |3>. Finally, as the reaction is exoergic along channel |3>, the density of states at the region where a Franck-Condon process takes place is much larger than for channel |1>. This situation will enhance the cross section for reaction channel |3> beyond the value obtained for direct reaction.

The "harpoon" image is nice and can be incorporated in the present scheme. In fact, the important factor is the population of the charge transfer state. If this happen from a quantum state having a much different stationary configuration compared to the CT state, its time evolution will change the geometry and one can describe the process as if it were an "electron" transfer over very large distances provided you measure it from the relaxed CT state. The actual electronic process will take place in a femtosecond or subfemtosecond time scale, relaxation will take much longer time.

5. Discussion

5.1. THE BORN-OPPENHEIMER APPROACH

In the quantum mechanical description of molecules (atoms and clusters) one problem has been the identification and validity of adiabatic separations of electronic (ρ) and nuclear (R) coordinates [30] This problem has been with us ever since the Born-Oppenheimer (BO) theory was published in 1927 [1, 2]. But this approach, as implemented in quantum chemistry, has serious deceiving aspects.

Woolley and Sutcliffe [6] have pointed out serious riddles in the BO-theory, ranging from the arrangement of the nuclear particles as an "equilibrium configuration" for which the potential energy has an absolute minimum to states that do not lie in the Hilbert space of the molecular eigenstates [6, 31]. Goscinski *et al.* have carefully discussed issues related to adiabatic separations, broken symmetries and geometry optimization [10, 32]; see also Pack and Hirschfelder [15, 33] and more recently Littlejohn and Reinsh [8]. Most important to chemical theory, the possibility of justifying the existence of molecular shape and structure also creates problems in this context [6, 7]. The BO approach, however, is important as proven by the concept of potential energy function and its successful use in describing chemical dynamics [23, 26, 34], and in the determination of accurate molecular structures and electronic spectra [35, 36]. The crux of the problem resides in the parametric dependence of the electronic wave function on the instantaneous nuclear coordinates. The diabatic approach was meant to overcome this riddle [30]. However, conditions required for having strictly diabatic electronic basis have been examined by Mead and Truhlar [16], who showed that, except for diatomics, such bases do not exist. Although new and promising ideas are emerging, see for instance the END approach of Örhn and coworkers [37], a general quantum theoretical scheme having a chemical flavor is still needed. A step towards an answer is given here.

Note that in the present theory there is nothing like a nuclear equilibrium configuration, as normally used in the BO approximation [6]. It is the electronic wave function that provides the *attractor* towards the nuclei, as these have sources of Coulomb potential moving away from the geometric values defining the stationary sources. Then, as shown by eq.(9), if the attractor is sufficiently deep, bound nuclear states will develop. It is the electronic stationary state which responds for the stability of matter, the image of electrons as particles following the nuclei is not adequate. The concept of molecular structure appears then ingrained in the properties of the electronic stationary wave functions. As noted in our previous work [4], there is a deep relation with density functional theory (DFT) [38]. One can hence take advantage of important work done upon the stability of matter in this context, see for instance the paper by Lieb and Thirring [39] and references therein. Incidentally, the molecular vibrational problem in DFT can be handled in the way described above. The practical BO computing methods serve to generate models for the potential energy function.

A statistical relationship between the above description and the standard one can be obtained. In a molecular sample at time t, the nuclei are statistically distributed. Each molecule shows its own particular energy gap between the electronic states involved in the process. On the average, things may look like as an electro-nuclear adiabatic process which could be *modelled* as a wave packet propagating on an adiabatic potential energy surface. This is the point where standard BO simulations of reaction processes [40] and the present view can be tied together. Individual systems are sensing electronic processes while the molecular sample may display a sort of apparent nuclear adiabatic-like process in a vicinity of the "crossing point".

The theory discussed here gives a special role to the stationary states of the molecular hamiltonian. In particular, there are stationary electronic states, not a set of electrons. For example, the hydrogen atom cannot be seen as formed by one proton plus one electron. It is the electronic spectra which define it, not the model we use to calculate the energy levels and wave functions. This may sound strange but consider a thermal neutron. This system decomposes into one proton plus an electron and a neutrino. One cannot say that a neutron is made of such particles. Matter may exist in different kinds of stationary states: processes can be seen as changes among them.

5.2. THE HAMILTONIAN FOR SYSTEMS OF CHARGES

The analysis presented so far is based on the same premises used to discuss the BO methodology. We have tried to show that a chemical interconversion is a Franck-Condon-like proces. But one important aspect when studying chemical systems is that they are always exposed to electromagnetic fields.

In general, charged systems are not separable in a unique way from the electromagnetic field. In the Lorentz gauge, for instance, charges interact via scalar photons and the electromagnetic field has transverse and longitudinal modes [41]. Here we do not use such a gauge but retain the standard approach where charged particles interact via the transverse vector potential **A** and the instantaneous Coulomb potential V [42]. In this frame, the scalar photon field compensates the longitudinal electromagnetic field [41]. This approach is adequate for the range of energy involved in normal chemical reactions (see also [43]).

In molecular theory one cannot get away from the transverse field **A**. For a set of electrons and nuclei considered as massive point charges, the total hamiltonian in a fixed laboratory frame is cast into the form:

$$H = H_m + U_{mrad} + H_{rad} \tag{28}$$

U_{mrad} describes the coupling between the system of charges and the transverse field; it contains terms linear and quadratic in **A** and the spin coupling to the magnetic field **B** for a Breit-Pauli hamiltonian [42]. Note that antisymmetrization of cluster electronic functions can be thought of as induced by this perturbing hamiltonian. H_{rad} is the free radiation hamiltonian. The radiation field is taken as a *source* or *sink* of energy. Charged particles, gauge invariance and molecular electrodynamics are discussed by Woolley [43].

In the present context, population changes among stationary states of H_m can be produced via spontaneous and/or induced emission of photons, or absorption. As discussed above, the chemical interconversion has several separated aspects. In neighborhoods of the lowest energy gap or at the "crossing" point, the electronic states can also be coupled via the electromagnetic field and change state by exchanging photons. The operator W would include now terms depending on **A** as found in U_{mrad}. This coupling may produce coherent linear superpositions between electronic states. For practical reasons, it is convenient to adopt the laws of electrodynamics as spelled out by Feynman without further ado [42] so that only terms linear in the transverse field are retained in W.

With the above caveat in mind, it is not difficult to understand that unimolecular reactions can be activated by ambient blackbody radiation. This has been experimentally reported [44]. Moreover, in this perspective, a chemical change would be closely related to a change (increase) in entropy via population variations in an ensemble.

To close this section, it should be noted that some of the ideas underlying the theory have been implicitly used as working hypotheses to analyze theoretical results in recent years [4, 22, 45-48].

6. Final remarks

A quantum mechanical separation scheme is introduced between electronic and nuclear degrees of freedom. The electronic stationary wave function determines the stationary geometry of the external Coulomb source completing the characterization of the molecular species. The hypothesis reflected by the formal equality $E_{en}(\alpha;\alpha^o{}_n) = E_{en}(R)$ warrants a discussion of the standard electronic problem with our special electronic wave functions, that do not depend upon the instantaneous positions of the nuclei. Each electronic state generates a potential energy function given by the average value of the molecular hamiltonian taken with respect to the relevant electronic wave function. Pictorially speaking, it is the stationary electronic wave function which glues the nuclei to the stationary geometry. The nuclear dynamics problem is then solved, as usual, by using this potential energy function generated by this electronic wave function. It is shown that the molecular hamiltonian can be diagonalized in the electro-nuclear basis of stationary states for a system of n-electrons and m-nuclei in a volume V, namely, the molecular hamiltonians H_o and H_m have the same spectra. Conservation principles enter in a natural manner to define selection rules. In particular, electronic parity is fundamental. Chemical reactions are described as belonging to a particular type of Franck-Condon spectroscopy.

Reactions having Q-states provide mechanistic pathways to achieving the electronic interconversion which can be modulated by external electromagnetic fields such as, for instance, microwaves as well as thermal blackbody radiation, due to the quasi energy degeneracy. In non-stationary situations, the operator $\hbar\ \nabla(Cn')\cdot\mathbf{p}$ appearing in eq.(8) may couple the electro-nuclear system to sound waves as well. This conclusion follows from the nature of the present theory.

Important classes of chemical reactions in the ground electronic state have equal parity for the in- and out-going channels, *e.g.*, proton transfer and hydride transfer [47, 48]. To achieve finite rates, such processes require accessible electronic states with correct parity that play the role of transition structures. These latter acquire here the quality of true molecular species which, due to quantum mechanical couplings with asymptotic channel systems, will be endowed with finite life times. The elementary interconversion step in a chemical reaction is not a nuclear rearrangement associated with a smooth change in electronic structure, it is a Franck-Condon electronic process with timescales in the (sub)femtosecond range characteristic of femtochemistry [49].

Acknowledgments

The author thanks Drs G. A. Arteca and C. T. Reimann for many enlightening discussions, and Dr Jean Maruani for calling his attention to possible effects of sound on chemical (biological) processes. The Swedish Research Council (NFR) has given financial support to this work.

References

1. Born, M. and Huang, K.: *Dynamical theory of crystal lattices*, Clarendon, Oxford, 1954.
2. Aroeste, H.: Towards an analytical theory of chemical reactions, *Adv.Chem.Phys.*, **6** (1964) 1-83.
3. Child, M. S.: *Molecular collision theory*, Academic Press, London, 1974.
4. Tapia, O. (2000) Electro-nuclear quantum mechanics beyond de BO approximation: towards a quantum electronic theory of chemical reactions, in Hernandez-Laguna, A., Maruani, J., Mc Weeny, R. and Wilson, S. (eds), *Quantum systems in chemistry and physics, Vol II: Advanced problems and complex systems*, Kluwer, Dordrecht, pp.193-212.
5. Woolley, R. G.: Quantum theory and molecular structure, *Adv.Phys.*, **25** (1976) 27-52.
6. Woolley, R. G. and Sutcliffe, B. T.: Molecular structure and the Born-Oppenheimer approximation, *Chem.Phys.Lett.*, **45** (1977) 393-398.
7. Woolley, R. G.: Must a molecule have a shape ?, *J.Am.Chem.Soc.*, **100** (1978) 1073-1078.
8. Littlejohn, R. G. and Reinsch, M.: Gauge fields in the separation of rotations and internal motions in the n-body problem, *Rev.Mod.Phys.*, **69** (1997) 213-275.
9. Flurry, R. L. and Siddall, I. T. H.: On the group of the independent-particle hamiltonian, *Mol. Phys.*, **36** (1978) 1309-1320.
10. Goscinski, O. and Mujica, V. (1987) Adiabatic separation, broken symmetries and geometry optimization, in Erdahl, R. and Smith Jr, V. H. (eds), *Density matrices and density functionals*, Reidel, Dordrecht.
11. Witkowski, A.: Separation of electronic and nuclear motions and the dynamical Schrödinger group, *Phys.Rev.A*, **41** (1990) 3511-3517.
12. Bransden, B. H. and Joachain, C. J.: *Physics of atoms and molecules*, Longman Scientific and Technical, New York, 1983.
13. Gilbert, T. L.: Hohenberg-Kohn theorem for nonlocal external potentials, *Phys.Rev.B*, **12** (1975) 2111-2120.

14. Mezey, P. G. (1989) Topology of molecular shape and chirality, in Bertran, J. and Csizmadia, I. G. (eds), *New theoretical concepts for understanding organic reactions*, Kluwer, Dordrecht, pp. 55-100.
15. Pack, R. T. and Hirschfelder, J. O.: Separation of rotational coordinates from the N-electron diatomic Schrödinger equation, *J.Chem.Phys.*, **49** (1968) 4009-4020.
16. Mead, C. A. and Truhlar, D. G.: Conditions for a definition of a strictly diabatic electronic basis for molecular systems, *J.Chem.Phys.*, **77** (1982) 6090-6098.
17. Sakurai, J. J.: *Modern Quantum Mechanics*, Benjamin/Cummings, Menlo Park, 1985.
18. Smith, F. T.: Diabatic and adiabatic representations for atomic collision problems, *Phys.Rev.*, **179** (1969) 111-123.
19. Steinfeld, J. I., Francisco, J. S. and Hase, W. T.: *Chemical kinetics and dynamics*, Prentice-Hall, New Jersey, 1989.
20. Omnès, R.: *Understanding quantum mechanics*, Princeton University Press, Princeton, New Jersey, 1999.
21. Tapia, O. (1982) Quantum theories of solvent-effect representation: an overview of methods and results, in Ratajczack, H. and Orville-Thomas, W. J. (eds), *Molecular Interactions*, Wiley, Chichester, pp. 47-117.
22. Tapia, O., Andres, J. and Stamato, F. M. G. (1996) Quantum theory of solvent effects and chemical reactions, in Tapia, O. and Bertran, J. (eds), *Solvent effects and chemical reactivity*, Kluwer, Dordrecht.
23. Golden, S.: The quantum mechanics of chemical kinetics of homogeneous gas phase reactions, *J.Chem.Phys.*, **17** (1949) 620-643.
24. Kleiman, V. D., Park, H., Gordon, R. J. and Zare, R. N.: *Companion to Angular Momentum*, Wiley and Sons, 1998.
25. Tully, J. C.: Diatomics-in-molecules potential energy surfaces. II. Nonadiabatic and spin-orbit interactions, *J.Chem.Phys.*, **59** (1973) 5122-5134.
26. Levine, R. D. and Bernstein, R. B.: *Molecular reaction dynamics and chemical reactivity*, Oxford University Press, New York, 1987.
27. Bohm, A., Antoniou, I. and Kielanowski, P.: The preparation-registration arrow of time in quantum mechanics, *Phys.Lett.A*, **189** (1994) 442-448.
28. Bohm, A.: *Quantum mechanics*, Springer-Verlag, New York, 1993.
29. Feshbach, H.: *Theoretical nuclear physics. Nuclear reactions*, Wiley and Sons, New York, 1992.
30. Cimiraglia, R. (1992) Adiabatic and diabatic sets in molecular calculations, in Broeckhove, J. and Lathouwers, L. (eds), *Time-dependent quantum molecular dynamics*, Plenum Press, New York, pp. 11-26.
31. Sutcliffe, B. T.: The idea of a potential energy surface, *J. Mol. Struct (Theochem)*, **341** (1995) 217-235.
32. Goscinski, O. and Palma, A.: Electron and nuclear density matrices and the separation of electronic and nuclear motion, *Int.J.Quantum Chem.*, **15** (1979) 197-205.
33. Pack, R. T. and Hirschfelder, J. O.: Energy corrections to the Born-Oppenheimer approximation. The best adiabatic approximation, *J.Chem.Phys.*, **52** (1970) 521-534.
34. Levine, R. D.: *Quantum mechanics of molecular rate processes*, Clarendon Press, Oxford, 1969.
35. Roos, B. O. (edr), *Lecture Notes in Chemistry*, Springer-Verlag, Berlin, 1992.
36. Roos, B. O.: Theoretical studies of electronically excited states of molecular systems using multiconfigurational perturbation theory, *Acc.Chem.Res.*, **32** (1999) 137-144.
37. Deumens, E., Diz, A., Longo, R. and Öhrn, Y.: Time-dependent theoretical treatments of the dynamics of electrons and nuclei in molecular systems, *Rev.Mod.Phys.*, **66** (1994) 917-983.
38. Hohenberg, P. and Kohn, W.: Inhomogeneous electron gas, *Phys.Rev.*, **136** (1964) B864-B871.
39. Lieb, E. H. and Thirring, W. E.: Bound for the kinetic energy of fermions which proves the stability of matter, *Phys.Rev.Lett.*, **35** (1975) 687-689.

40. Zewail, A. H.: *Femtochemistry. Ultrafast dynamics of the chemical bond*, World Scientific, Singapore, 1994.
41. Cohen-Tannoudji, C., Dupont-Roc, J. and Grynberg, G.: *Photons and Atoms. Introduction to quantum electrodynamics*, Wiley and Sons, New York, 1989.
42. Feynman, R. P.: *Quantum electrodynamics*, Benjamin, New York, 1961.
43. Woolley, R. G.: Charged particles, gauge invariance and molecular electrodynamics, *Int.J. Quant.Chem.*, **74** (1999) 531-545.
44. Dunbar, R. C. and McMahon, T. B.: Activation of unimolecular reactions by ambient black-body radiation, *Science*, **279** (1998) 194-197.
45. Tapia, O. and Andres, J.: On a quantum theory of chemical reactions and the role of in vacuum transition structures. Primary and secondary sources of enzyme catalysis, *J.Mol.Str.-Theochem*, 335 (1995) 267-286.
46. Tapia, O., Andres, J., Moliner, V. and Stamato, F. L. M. G. (1997): Theory of solvent effects and the description of chemical reactions. Proton and hydride transfer processes, in Hadzi, D. (edr), *Theoretical treatments of hydrogen bonding*, John Wiley and Sons, New York, pp. 143-164.
47. Tapia, O., Moliner, V. and Andres, J.: A quantum electronic theory of chemical processes. The inverted energy profile case: $CH_3^+ + H_2$ reaction, *I. J. Quantum Chem.*, **63** (1997) 373-391.
48. Tapia, O.: Quantum mechanics and the theory of hydrogen bond and proton transfer. Beyond a Born-Oppenheimer description of chemical interconversions, *J.Mol.Str.-Theochem*, **433** (1998) 95-105.
49. Zewail, A. H.: Femtosecond transition-state dynamics, *Faraday Disc.Chem.Soc.*, **91** (1991) 207-237.

THEORETICAL STUDY OF REGULARITIES IN ATOMIC AND MOLECULAR SPECTRAL PROPERTIES

I. MARTIN, C. LAVIN, AND E. CHARRO
Departamento de Química Física, Facultad de Ciencias, Universidad de Valladolid, 47005 Valladolid, Spain

Abstract. Regularities in the transition intensities along isoelectronic sequences, indi-vidual spectral series, and analogous transitions in homologous atomic and molecular systems are proving to be very useful in theoretical and experimental investigations of spectra, as well as for supplying additional data. We have tested such regularities in cal-culations performed mostly with the Quantum Defect Orbital method, both in its non-relativistic (QDO) and relativistic (RQDO) formulations. A brief summary of the under-lying theory for such regularities (and deviations from them) is given, followed by a number of numerical examples presented in tabular or graph form.

1. Introduction

The interpretation of the intensities of lines observed in astrophysical sources requires a wide variety of reliable atomic and, to a lesser extent, molecular data [1]. Also, the steady development of high temperature plasmas, in relation to the fusion programmes ongoing in several countries, has given rise to a considerable interest in the spectroscopy of heavy and/or highly ionised atoms [2]. The spectacular advance of some experimental techniques has not diminished the need for reliable theoretical data. In the production of spectroscopic quantities such as oscillator strengths to fulfill the present demands of both the astrophysics and plasma physics communities, several authors [3-5] have emphasised the need for both experimentalists and theoreticians to self-assess the data they supply.

Studies of regularities in transition intensities along isoelectronic sequences, in individual spectral series, in analogous transitions in homologous atomic and molecular systems, ..., are proving to be very useful in theoretical and experimental investigations of spectra. They can serve the purpose of assessing the adequacy of a given theoretical or experimental procedure and can be exploited to interpolate or extrapolate non-available data, and often reveal peculiarities in the behaviour of many-electron systems.

J. Maruani et al. (eds.), New Trends in Quantum Systems in Chemistry and Physics, Volume 2, 49–63.

When studying the absorption of increasing photon energy by an atom or ion initially in a given bound state, to be gradually excited until it becomes ionised, and to have afterwards the free electron increase its kinetic energy, there is no discontinuity in the oscillator strength spectral density at the ionisation threshold. An adequate theoretical calculation must reproduce such continuity, which may also be exploited to interpolate a value for the threshold photoionisation cross section.

We shall next summarise the basic approaches that justify the above features, to then illustrate some of them on various atomic and molecular systems with the results of some calculations performed by us with either the Relativistic Quantum Defect Orbital (RQDO) formalism [6, 7] or the non-relativistic version of this method (QDO) [8, 9].

2. The *Z*-charge expansion theory

Studies of isoelectronic sequences of atomic systems (that is, a number of elements with gradually increasing atomic number and degree of ionisation such that they all possess the same number of electrons) are, as mentioned above, very fruitful in theoretical, semiempirical and experimental investigations of atomic spectra. They reveal important regularities and pecularities in the behaviour and physical characteristics of many-electron atoms and ions, help to identify and classify their energy levels within an optimal coupling scheme, find values of semiempirical parameters, etc.

Let us discuss briefly the structure of multiply charged ions. It follows from calculations that the mean distance of the electronic shells from the nucleus rapidly diminishes along the isoelectronic sequence with an increase in ionisation degree, i.e., the ion shrinks swiftly under the action of a growing (due to less screening) nuclear charge field. In contrast to the neutral atom, which in fact is a comparatively weakly bound system, a highly ionised atom is a compact strongly bound structure. Changing the electronic configuration of such an ion via excitation of the outer electrons does not alter significantly the distribution of charge density [10].

In a similar way, the peculiarities of *electronic transitions* may be investigated along isoelectronic sequences. Line and oscillator strengths as well as Einstein coefficients usually change fairly smoothly along them.

In non-relativistic perturbative atomic *Z-expansion theory*, as recently summarised [11], a new scaled length, $\rho = Z\, r$, and a scaled energy, $\varepsilon = Z^{2} E$, are introduced in the many-electron wave equation. That is, the units of length and energy are changed to $1/Z$ and Z^2 a.u., respectively. The Hamiltonian then takes the form

$$H = Z^2 (H_0 + Z^{1} V), \tag{1}$$

where

$$H_0 = -\tfrac{1}{2} \Sigma_i (\nabla_i^2 + 2/\rho) \tag{2}$$

and

$$V = \Sigma_{i>j} 1/\rho_{ij}. \tag{3}$$

Then, the Schrödinger equation adopts the form:

$$(H_0 + Z^{-1} V)\, \psi = (Z^{-2} E)\, \psi \tag{4}$$

with V regarded as a perturbation. The expansions of ψ and E in powers of the perturbation parameter, Z^{-1}, are:

$$\psi = \psi_0 + Z^{-1} \psi_1 + Z^{-2} \psi_2 + \dots \tag{5}$$

and

$$E = Z^2 (E_0 + Z^{-1} E_1 + Z^{-2} E_2 + \dots) \tag{6}$$

Accordingly, the line strength for an electronic transition between the states i and f:

$$S = S(i,f) = S(f,i) = <\Psi_f \,/ \mathbf{P}^{(k)} /\, \Psi_i>^2, \tag{7}$$

and its related quantity, the absorption oscillator strength:

$$f_{if} = 2/3\, (E_f - E_i)\, S(i,f)/g_i \tag{8}$$

vary with Z as follows

$$Z^2 S = S_0 + S_1 Z^{-1} + S_2 Z^{-2} + \dots \tag{9}$$

$$f = f_0 + f_1 Z^{-1} + f_2 Z^{-2} + \dots \tag{10}$$

These expressions can be applied, according to the theory, to an isoelectronic sequence.

From the *non-relativistic* Z-expansion theory for transitions that may occur through the electric dipole mechanism, or $E1$ transitions, it is inferred:

As $1/Z \to 0$, $f \to f_0$ (hydrogenic value) and thus:

For $\Delta n = 0$ transitions: $f_0 = 0$

For $\Delta n \neq 0$ transitions: $f_0 \neq 0$

but a limiting f-value at $1/Z = 0$ can also be set. This offers the great advantage that once the f vs. $1/Z$ systematic trend has been established, non- computed (or non-measured) f-values may be read off from the graph.

However, in some cases, these regularities are valid only approximately. Especially large deviations from them may occur when correlation or relativistic effects are large [10]. For instance, near the neutral end (large $1/Z$ value), strong perturbations due to configuration mixing may occur. Also, irregularities may be observed where cancellation in the transition integral takes place: A dip in the f vs. $1/Z$ curve will appear. In addition, is has been shown [12] that at sufficiently large Z and ionisation degree, *relativistic effects* set in, causing considerable deviations from the behaviour described by equations (5) to (10): The scaled line strength $Z^2 S$ is rather less affected by the contraction (or expansion) in the wavefunctions than the excitation energy which experiences sizable changes. In f, except at high Z-values, these effects are partly compensated.

In some of our calculations of oscillator strengths on different isoelectronic sequences we have sought the possible regularities complied with by different transitions. We next summarise the theoretical methodologies we have mostly employed to yield the results we show as examples.

3. The Relativistic Quantum Defect Orbital (RQDO) method

The RQDO radial, scalar, equation derives from a non-unitary decoupling of Dirac's second order radial equation. The analytical solutions, RQDO orbitals, are linear combinations of the large and small components of Dirac radial function [6,7]:

$$\left[-\frac{d^2}{dr^2}+\frac{\Lambda(\Lambda+1)}{r^2}-\frac{2Z'_{net}}{r}\right]\Psi_k^{RD}=2e^{RD}\Psi_k^{RD}, \tag{11}$$

where the model potential parameters are related to relativistic quantities:

$$\Lambda=\begin{cases} s-1-\delta'+c & if\ j=l+\frac{1}{2} \\ -s-\delta'+c & if\ j=l-\frac{1}{2} \end{cases}, \tag{12}$$

$$Z'_{net}=Z_{net}\left(1+\alpha^2 E^X\right). \tag{13}$$

The relativistic quantum defect, δ', is determined either empirically from the state energy value or by inter/extrapolation techniques.

$$\frac{E^X\left(1+\frac{1}{2\alpha^2E^X}\right)}{\left(1+\alpha^2E^X\right)^2}=-\frac{Z_{net}^2}{2(\tilde{n}-\delta')^2}\,. \tag{14}$$

In order to explicitly account for core-polarization effects in the calculation of oscillator strengths, a corrected transition operator is frequently employed in our calculations,

$$\mathbf{P}(r)=\mathbf{r}\left\{1-\frac{\alpha_c}{r^3}\left[1-\exp\left(-\frac{r}{r_c}\right)\right]^3\right\} \tag{15}$$

where α_c is the core dipole polarizability and r_c a cut-off radius.

The transition integrals derived from the above wavefunctions are closed-form analytical expressions and, thus, their calculation is free from convergence problems. Another advantage is that the computational cost does not increase with the size of the atomic system.

The non-relativistic version of this formalism (QDO), as originally proposed [8, 9] is recovered by taking the value of the fine-structure constant, α, in equations (11) to (14) to be equal to zero.

4. Numerical examples of systematic trends along isoelectronic sequences

In Figure 1 we illustrate the regularities in the fine-structure oscillator strengths for the *3s-3p* and *3p-3d* arrays, respectively, along the sodium isoelectronic sequence, covering a wide range of atomic species, from Z=11 to Z=99 [13]. The deviations from the behaviour predicted from the Z-expansion theory where relativistic effects become important can be observed. The results displayed in part (a) and (c) of the figure were obtained with the QDO method, in which relativistic effects are only included implicitly by using experimental energies, whilst in parts (b) and (d) the results of the relativistic formulation RQDO are shown. The change in the slope of some of the lines of both multiplets due to the explicit inclusion of relativistic effects is apparent. The RQDO systematic trends reproduce very closely the experimental ones for the *3s-3p* transitions, for which measurements are available [13].

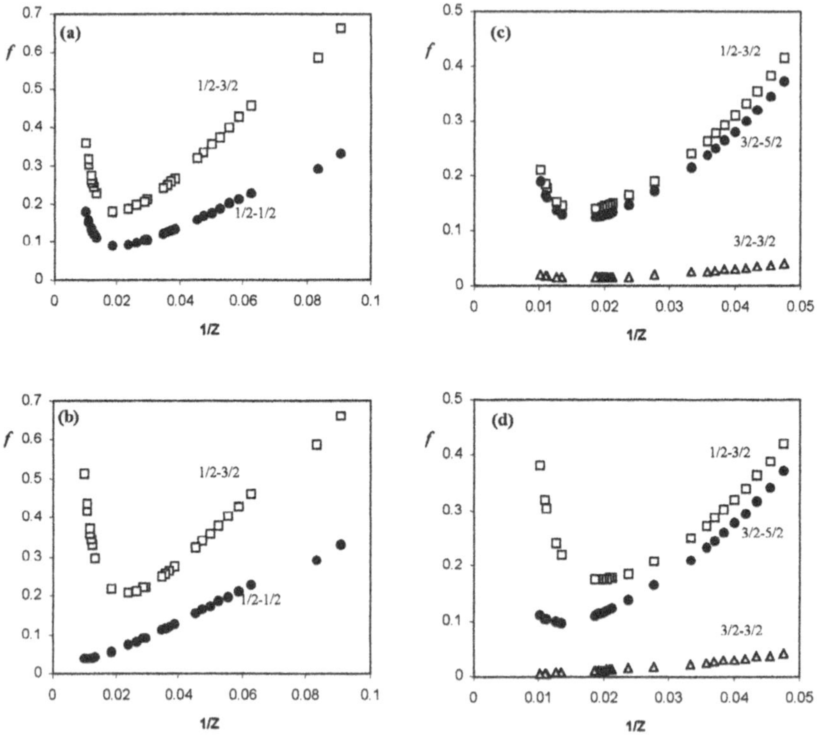

Figure 1. Systematic trends of the fine-structure QDO and RQDO, (a) and (b), respectively, oscillator strengths for the 3s ^2S- 3p $^2P^o$ transitions. Systematic trends of the fine-structure QDO and RQDO oscillator strengths for the 3p $^2P^o$- 3d ^2D transitions, (c) and (d) respectively, in the sodium isoelectronic sequence [13].

In Figure 2 the systematic trend of the oscillator strength for the resonance transition, $3s^2\ ^1S^0$- $3s\ 3p\ ^1P_o^1$, of the magnesium isoelectronic sequence (from Z=12 to Z=92) is shown. Here we have not only employed the RQDO formalism but also performed an extensive configuration interaction calculation with the CIV3 code [14, 15] in which part of the relativistic effects have been included, and a multiconfiguration Dirac-Fock calculation with the GRASP code [16]. The departure from the slope predicted by the non-relativistic *Z*-expansion theory at about Z=33 is apparent. We would like to point out the very good agreement between the RQDO f-values and those of the much more expensive CIV3 and GRASP calculations.

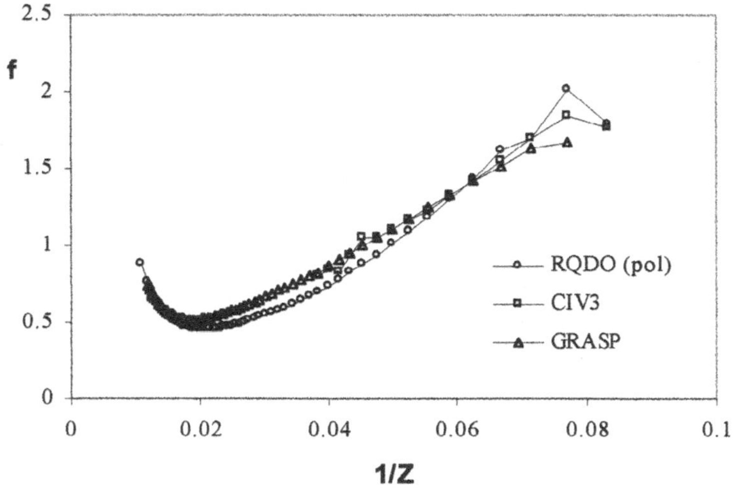

Figure 2. Systematic trends of the oscillator strengths of the $3s^2\ ^1S_0$- $3s3p\ ^1P_1^o$ transitions in the magnesium isoelectronic sequence. RQDO (pol) are the core-polarisation-corrected RQDO f-values. Both CIV3 and GRASP f-values correspond to the length form of the transition integral

We display in Tables 1 to 3 and in Figure 3 the multiplet oscillator strengths for the 3d- 4p and 3d- nf (n= 4-10) transition arrays of some of the potassium-like "iron group" ions Sc III to Cu XI, as they have been classified since the early days of solar spectroscopy. These ions have been found to be of generally high abundance in the solar photosphere and to be fairly well represented in the solar corona [17]. This astronomical object is the subject of future major space projects, where the suitability of some of the presently studied lines for temperature and density diagnostics is at stake. In Tables 1 to 4 in addition to the very recently calculated RQDO f-values, the Opacity Project (OP) data reported by Verner et al. [18] and those of an earlier Critical Compilation (CC) by Mori et al. [19], on Fe VIII, have been included. A good general agreement between the RQDO and OP oscillator strengths is found, no matter the large differences in the complexities of the two calculations.

TABLE 1. Oscillator strength for the 3d-4p transitions

Ion	Method	$^2D_{3/2}-$	$^2D_{3/2}-$	$^2D_{5/2}-{}^2P_{3/2}$	$^2D-{}^2P$
Sc III	RQDO[a]	0.1830	0.3658(-1)	0.2196	0.2196
Ti IV	RQDO[a]	0.1715	0.3411(-1)	0.2050	0.2052
V V	RQDO[a]	0.1575	0.3121(-1)	0.1878	0.1882
Cr VI	RQDO[a]	0.1463	0.2891(-1)	0.1740	0.1745
Mn VII	RQDO[a]	-	0.2721(-1)	0.1639	-
Fe VIII	RQDO[a]	0.1310	0.2549(-1)	0.1536	0.1548
	OP[b]	0.1200	0.2420(-1)	0.1450	0.1450
	CC[c]	0.5144(-1)	0.1036(-1)	0.6190(-1)	-
Co IX	RQDO[a]	0.1306	0.2566(-1)	0.1548	0.1553
Ni X	RQDO[a]	0.1193	0.2327(-1)	0.1405	0.1413
Cu XI	RQDO[a]	0.1161	0.2259(-1)	0.1365	0.1374

In this and the remaining tables, *A* (*B*) denotes $A.10^{(B)}$

[a] Relativistic Quantum Defect Orbital Method, this work

[b] Verner et al. [18]

[c] Critical Compilation, Nori et al. [19]

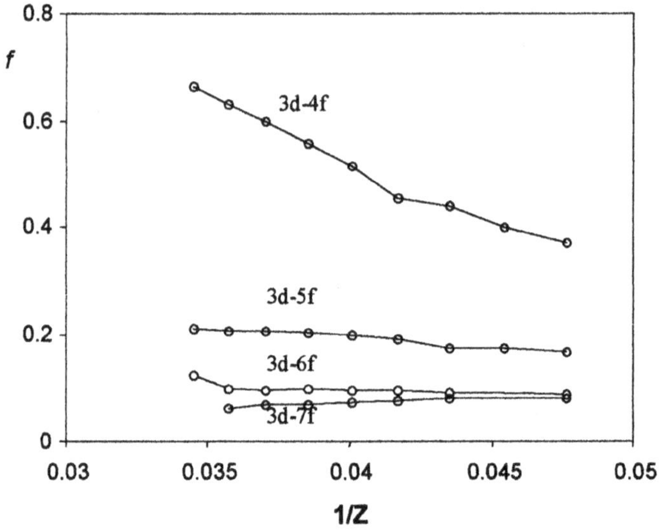

Figure 3. Systematic trends of the RQDO multiplet oscillator strengths of the 3d ^2D- nf $^2F^o$ transitions in the "iron group" potassium-like ions.

The RQDO f-values for individual lines follow a smooth systematic trend. The ions studied in this group do not seem to be sufficiently heavy for the oscillator strengths to present substantial deviations from the slope predicted by the non-relativistic Z-expansion theory.

TABLE 2. Oscillator strength for the 3d-nf (n=4,5) transitions

Ion	Method	$^2D_{3/2} - {}^2F_{5/2}$	$^2D_{5/2} - {}^2F_{5/2}$	$^2D_{5/2} - {}^2F_{7/2}$	$^2D - {}^2F$
3d – 4f					
Sc III	RQDO[a]	0.3715	0.1773(-1)	0.3547	0.3720
Ti IV	RQDO[a]	0.3992	0.1906(-1)	0.3811	0.3998
V V	RQDO[a]	0.4385	0.2094(-1)	0.4213	0.4407
Cr VI	RQDO[a]	0.4535	0.2166(-1)	0.4309	0.4529
Mn VII	RQDO[a]	0.5142	0.2456(-1)	0.4908	0.5149
Fe VIII	RQDO[a]	0.5568	0.2660(-1)	0.5323	0.5580
	OP[b]	0.5690	0.2700(-1)	0.5400	0.5680
	CC[c]	0.4349	-	0.4132	-
Co IX	RQDO[a]	0.5954	0.2844(-1)	0.5687	0.5965
Ni X	RQDO[a]	0.6305	0.3012(-1)	0.6024	0.6317
Cu XI	RQDO[a]	0.6594	0.3151(-1)	0.6301	0.6617
3d – 5f					
Sc III	RQDO[a]	0.1652	0.7881(-2)	0.1576	0.1654
Ti IV	RQDO[a]	0.1741	0.8305(-2)	0.1661	0.1743
V V	RQDO[a]	0.1738	0.8293(-2)	0.1649	0.1734
Cr VI	RQDO[a]	0.1890	0.9014(-2)	0.1803	0.1892
Mn VII	RQDO[a]	0.1969	0.9387(-2)	0.1876	0.1970
Fe VIII	RQDO[a]	0.2022	0.9637(-2)	0.1928	0.2023
	OP[b]	0.1980	0.9400(-2)	0.1880	0.1970
	CC[c]	0.1727	-	0.1641	-
Co IX	RQDO[a]	0.2056	0.9798(-2)	0.1960	0.2057
Ni X	RQDO[a]	0.2064	0.9832(-2)	0.1967	0.2065
Cu XI	RQDO[a]	0.2075	0.9883(-2)	0.1978	0.2076

See footnotes to Table 1

5. Regularities in intensities of analogous transitions in homologous atoms

Another interesting type of regularities in the intensity of lines concerns homologous atoms, that is, atoms with the same outer electron structure. The partial sum rules that apply to the oscillator strength suggests themselves as starting points to look for possible regularities. In particular, the Wigner-Kirkwood sum rule [20, 21] for transitions of the type $n\,\ell \rightarrow n'\,(\ell+1)$ states that the summation of the corresponding oscillator strengths over all values of n', including the continuum as well as all possible lower states (which may be filled shells), is equal to:

$$1/3(\ell+1)\ (2\ell+3)\ (2\ell+1)^{-1}. \qquad (16)$$

TABLE 3. Oscillator strength for the 3d-nf (n = 6,10) transitions

Ion	Method	$^2D_{3/2} - {}^2F_{5/2}$	$^2D_{5/2} - {}^2F_{5/2}$	$^2D_{5/2} - {}^2F_{7/2}$	$^2D - {}^2F$
3d –6f					
Sc III	RQDO[a]	0.8642(-1)	0.4120(-2)	0.8240(-1)	0.8648(-1)
V V	RQDO[a]	0.9148(-1)	0.4361(-2)	0.8723(-1)	0.9154(-1)
Cr VI	RQDO[a]	0.9408(-1)	0.4483(-2)	0.8979(-1)	0.9420(-1)
Mn VII	RQDO[a]	0.9524(-1)	0.4358(-2)	0.9077(-1)	0.9528(-1)
Fe VIII	RQDO[a]	0.9583(-1)	0.4564(-2)	0.9130(-1)	0.9585(-1)
	OP[b]	0.1060	0.5020(-2)	0.1010	0.1050
	CC[c]	0.8422(-1)	-	0.8025(-1)	-
Co IX	RQDO[a]	0.9471(-1)	0.4509(-2)	0.9022(-1)	0.9472(-1)
Ni X	RQDO[a]	0.9640(-1)	0.4618(-2)	0.9247(-1)	0.9681(-1)
Cu XI	RQDO[a]	0.1208	0.5787	0.1158	0.1213
3d – 7f					
Sc III	RQDO[a]	0.7874(-1)	0.3751(-2)	0.7502(-1)	0.7876(-1)
V V	RQDO[a]	0.7955(-1)	0.3788(-2)	0.7581(-1)	0.7958(-1)
Cr VI	RQDO[a]	0.7684(-1)	0.3657(-2)	0.7317(-1)	0.7684(-1)
Mn VII	RQDO[a]	0.7286(-1)	0.3465(-2)	0.6932(-1)	0.7281(-1)
Fe VIII	RQDO[a]	0.6802(-1)	0.3231(-2)	0.6502(-1)	0.6815(-1)
	OP[b]	0.7300(-1)	0.3470(-2)	0.7300(-1)	0.7300(-1)
	CC[c]	0.4771(-1)	-	0.4537(-1)	-
Co IX	RQDO[a]	0.6651(-1)	0.3183(-2)	0.6367(-1)	0.6670(-1)
Ni X	RQDO[a]	0.6056(-1)	0.2892(-2)	0.5766(-1)	0.6056(-1)
3d-8f					
V V	RQDO[a]	0.3381(-1)	0.1611(-2)	0.3224(-1)	0.3384(-1)
Cr VI	RQDO[a]	0.3387(-1)	0.1613(-2)	0.3227(-1)	0.3388(-1)
Mn VII	RQDO[a]	0.3333(-1)	0.1587(-2)	0.3174(-1)	0.3333(-1)
Co IX	RQDO[a]	0.3174(-1)	0.151(-2)	0.3023(-1)	0.3174(-1)
3d-9f					
V V	RQDO[a]	0.2281(-1)	0.1087(-2)	0.2182(-1)	0.2287(-1)
Cr VI	RQDO[a]	0.2279(-1)	0.1085(-2)	0.2171(-1)	0.2279(-1)
Co IX	RQDO[a]	0.2087(-1)	0.9926(-2)	0.1985(-1)	0.2086(-1)
3d-10f					
Cr VI	RQDO[a]	0.1604(-1)	0.7640(-3)	0.1528(-1)	0.1604(-1)
Co IX	RQDO[a]	0.1469(-1)	0.6984(-3)	0.1397(-1)	0.1468(-1)

See footnotes to Table 1

One finds for the most comon type of transitions:

$$\begin{aligned} n\,s \rightarrow n'p:&\ \ \Sigma_{n'} f = 1 \\ n\,p \rightarrow n'd:&\ \ \Sigma_{n'} f = 10/9 \\ n\,d \rightarrow n'f:&\ \ \Sigma_{n'} f = 7/5 \end{aligned} \tag{17}$$

If it happens that a single transition overwhelmingly dominates the spectral series then the f-value for this transition will be approximately given by the sum rule, equation (16). Such a situation may be expected to prevail when the ℓ and $(\ell+1)$ wavefunctions overlap very strongly and are radially separated from the core, e.g., in $s \rightarrow p$ transitions. Transitions to be compared in the case of the alkali elements are [22]

$$[n\,\ell \rightarrow n'\ell']_{\rm Li} \rightarrow\ [(n+1)\,\ell \rightarrow (n'+1)\,\ell']_{\rm Na} \rightarrow\ [(n+2)\,\ell \rightarrow (n'+2)\,\ell']_{\rm K} \rightarrow \ldots$$

That is, if the corresponding transition dominates the sum (16), as one proceeds along columns of the periodic table, the sume rule then provides a method for estimating f-values for homologous atoms. It is true that complex atomic structures, as well as cases involving strong cancellation in the integrand of the transition integral, generally do not adhere to this regular behaviour [23].

We have studied this type of regularities in the group of the halogen atoms. Their electronic structure is complex enough for not all of their energy levels to belong to the same spin-orbit coupling scheme. We have found that most levels of F and Cl comply with the *LSJ* scheme whilst no much degree of excitation is needed for the levels of Br and I to comply with the pair-coupling scheme $J_c\ell$. We have performed RQDO calculations on a number of fine-structure and multiplet transitions involving levels in different degrees of excitation in the first four halogen atoms, for many of which comparative data from either experimental or theoretical sources are very scarce.

TABLE 4. RQDO oscillator strengths for multiplet transitions in the halogen atoms. All calculated in the LSJ coupling scheme

Transition	F I (n=3)	Cl I (n=4)	Br I (n=5)	I I (n=6)
		ns-np transitions		
^{4}P – ^{4}P	0.2740	0.2942	0.3050	0.3135
^{4}P – ^{4}D	0.4679	0.5073	0.5107	0.5309
		np-(n+1)s		
^{4}P – ^{4}P	0.2588	0.2521	0.2743	0.2705
^{4}D – ^{4}P	0.2806	0.2796	0.2914	0.2900

Therefore, we have attempted to assess the correctness of our results by analysing their possible regular behaviour along the group. As an example, we have collected in Table 4 the RQDO oscillator strengths for a few multiplets that correspond to $nl \rightarrow n'l'$ transition arrays (n=3, 4, 5, 6, for F, Cl, Br and I, respectively). Inspection of the table reveals that the intensities of the multiplets expressed as f-values are very similar in the different halogens, as long as the coupling scheme remains the same.

6. Regularities in analogous transitions in molecules having the same united atom limit

During the last few years we have applied a molecular-adapted version of the QDO formalism, as proposed by Martín et al. [24], to the calculation and analysis of intensities in the electronic spectra of Rydberg molecules. This classification can be attributed to Herzberg [25] and corresponds to certain molecules or radicals for which the ground state is either unstable or dissociative and all the excited electronic states possess Rydberg character, having long lifetimes as compared to rotational or vibrational periods. The geometry of the Rydberg states usually conforms very well with that of the rather stable core ion, in the absence of level mixings. Several Rydberg molecules have been observed, or predicted to exist, in regions of astrophysical importance. They also offer a more general chemical interest, as some of them have been detected as short-lived intermediates in chemical reactions. In spite of the very recent spectacular advances in the experimental techniques for the detection and isolation of these species, such as the zero kinetic energy (ZEKE) spectroscopic technique [26], there still is a strong demand of theoretical calculations on the structure and spectra of such molecules. Some rather expensive multi-configurational calculations (MRD-CI) [27] as well as multichannel quantum defect theory (MQDT) [28] have been applied to the prediction of the structure and spectra of a number of Ryberg molecules. Nevertheless, given that a, highly excited, molecular Rydberg electron behaves in many ways as an atomic Rydberg electron, we have modified the QDO method as formulated to treat atomic transitions [8, 9, 29] in order to give account of the splittings originated in the molecular Rydberg states by the specific symmetry of the core [30] and applied it to prediction of spectral intensities in Rydberg radicals. In various Rydberg radicals for which the electronic spectra had been measured, our calculations have been able to reproduce and assign the main spectral features. Also, our QDO oscillator strengths and Einstein coefficients have been found to conform well with those of much more complex calculations, in particular for the strongest transitions [24, 31-34].

In treating molecular Rydberg states we have found rather important remarks in Herzberg's review [25], which refer to similarities in the intensities of bands observed in some rare-gas hydrides and their corresponding united-atom limits. In our view similarities of the same sort should, with more reason, occur in the spectra of isoelectronic Rydberg radicals, which are also "homologous" in the sense of all having the same outer shell structure, determined by the behaviour of the Rydberg electron.

In Tables 5 and 6 we compare the spectral intensities, in the form of oscillator strengths, for 3s-3p and 3p-3d transitions (in the united atom notation) between states of the same and different molecular symmetries within each corresponding molecular point group, in the Rydberg radicals CH_5, H_3O, H_2F, and NeH, all of which have sodium as the united atom limit.

TABLE 5. Oscillator strengths for the 3s - 3p electronic transitions between molecular states of the same and different symmetry

[a] NeH Transition	f	[b] H_2F Transition	f	[c] H3O Transition	f	[d] CH_5 Transition	f
$^2\Sigma^+(3s)$-$^2\Sigma^+(3p)$	0.255	$^2A_1(3s)$-$^2A_1(3p)$	0.266	$^2A_1(3s)$-$^2A_1(3p)$	0.297	$1^2A'(3s)2^2A'(3p)$	0.2(
	0.105[e]		0.259[e]		0.293[e]		0.306[e]
			0.212[f]				
						$1^2A'(3s)$-$1^2A'(3p)$	0.283
							0.279[e]
		$^2A_1(3s)$-$^2B_1(3p)$	0.267			$1^2A'(3s)$-$1^2A''(3p)$	0.298
			0.258[e]				0.313[e]
			0.244[f]				
$^2\Sigma^+(3s)$-$^2\Pi(3p)$	0.510			$^2A_1(3s)$-$^2E(3p)$	0.523		
	0.414[e]				0.739[e]		

[a] QDO, Martín et al. [33]
[b] QDO, Martín et al. [30]
[c] QDO, Lavín and Martín [31]
[d] QDO, Martín et al. [32]
[e] FSSO, Raynor and Herschbach [35]
[f] MRD-CI, Petsalakis et al. [27]

TABLE 6. Oscillator strengths for the 3p - 3d electronic transition between molecular states of the same and different symmetry

[a]NeH Transition	f	[b] H_2F Transition	f	[c]H_3O Transition	f	[d]CH_5 Transition	f
$^2\Pi(3p)$-$^2\Pi(3d)$	0.237	$^2B_1(3p)$-$^2B_1$ (3d)	0.246	$^2E(3p)$-$^2E(3d)$	0.222	$^2A''(3p)$-$^2A''(3d)$	0.229
	0.227[e]		0.236[e]		0.483[e]		0.242[e]
			0.298[f]				
$^2\Sigma^+(3p)$-$^2\Sigma^+$ (3d)	0.312	$^2A_1(3p)$-$^2A_1$ (3d)	0.335	$^2A_1(3p)$-$^2A_1$ (3d)	0.325		
	0.251[e]		0.265[e]		0.248[e]		
			0.325[f]				
$^2\Sigma^+(3p)$-$^2\Pi$ (3d)	0.467			$^2A_1(3p)$-^2E (3d)	0.503		
	0.438[e]				0.509[e]		
$^2\Pi(3p)$-$^2\Sigma^+$ (3d)	0.079			$^2E(3p)$-$^2A_1$ (3d)	0.080		
	0.063[e]				0.106[e]		
$^2\Pi(3p)$-$^2\Delta$ (3d)	0.474						
	0.553[e]						

See footnotes to Table 5

In addition to our QDO values [31-33], the oscillator strenghs derived by us from Einstein coefficients of a self-consistent-field procedure with a basis of floating spherical Slater orbitals (FSSO) by Raynor and Herschbach [35] and those of a MRD-CI calculation by Petsalakis et al. [27] have been included, for the transitions for which they are reported. Inspection of Tables 5 and 6 reveals the expected similarities for the "homologous" Rydberg radicals in the intensities of transitions which are "analogous" in the sense that, apart from corresponding to the same variation in the angular momentum of the Rydberg electron, take place between molecular states for which the difference in degeneracy, as indicated by that of the irreducible representations they belong to, is the same. Further work on molecular Rydberg transitions is presently in progress.

Acknowledgments

This work was supported by DGICYT (Project nbs PB88-0343, PB91-0207-C02-01 and PB94-1314-C03-03) and DGES (Project nb PB97-0399-C03-01) of the Spanish Ministry of Education and by JCL (Project nb VA21/97). E.C. and I.M. are grateful to A. Hibbert for his kind hospitality at the Department of Applied Mathematics and Theoretical Physics of Queen's University, Belfast, where part of the work was done.

References

1. Fournier, K.B., Goldstein, W.H., Finkenthal, M., Bell, R.E., and Terry, J.L., *J. Electron Spectroscopy and Related Phenomena* **80** (1996) 283-295.
2. Träbert, E., *Nuclear Instrumentation and Methods in Physics Research* **B 98** (1995) 10-16.
3. Hibbert, A., *Physica Scripta* **T65** (1996) 104-109.
4. Wiese, W. L., *Physica Scripta* **T65** (1996) 188-191.
5. Wiese, W.L., and Kelleher, D. E., in P.J. Mohr and W.L. Wiese (eds.), *Atomic and Molecular Data and Their Applications*, NIST Special Publication 926, US Government Printing Office, Washington, DC, pp. 105-116, 1998.
6. Martín, I., and Karwowski, J., *J. Physics B: Atomic, Molecular and Optical Physics* **24** (1991) 1539-1545.
7. Karwowski, J., and Martín, I., *Physical Review A* **43** (1991) 4832-4839.
8. Simons, G., *J. Chemical Physics* **60** (1974) 645-651.
9. Martín, I., and Simons, G., *J. Chemical Physics* **62** (1975) 4799-4811.
10. Rudzikas, Z.: *Theoretical Atomic Spectroscopy*, Cambridge University Press, Cambridge, 1997.
11. Froese Ficher, C., in G.W.F. Drake (ed.), *Atomic, Molecular, and Optical Physics Handbook*, American Institute of Physics, Woodbury, New York, pp. 243-257, 1996.
12. Martin, W. C., and Wiese, W.L., *Physical Review A* **13** (1976) 699-711.
13. Martín, I., Karwowski, J., Lavín, C., and Diercksen, G.H.F., *Physica Scripta* **44** (1991) 567-573.
14. Hibbert, A., *Computational Physics Communications* **9** (1975) 141-165.
15. Hibbert, A., unpublished (1999).
16. Dyall, K.G., Grant, I.P., Johnson, C.T., Parpia, F.A., and Plummer, E.P., *Computational Physics Communications* **55** (1989) 425-442.

17. Anders, E., and Grevesse, F., *Geochimica and Cosmochimica Acta* **53** (1989) 197-205.
18. Verner, D.A., Verner, E.M., and Ferland, G.J., *Atomic Data and Nuclear Data Tables* **64** (1996) 1-165.
19. Mori, K., Otsuka, M., and Kato, T., *Atomic Data and Nuclear Data Tables* **23** (1979) 1-172.
20. Wigner, E., *Physik Zeistchrift* **32** (1931) 450-456.
21. Kirkwood, J.G., *Physik Zeistchrift* **33** (1932) 521-534.
22. Weiss, A.W., *J. Quantitative Spectroscopy and Radiative Transfer* **18**, (1977) 481-496.
23. Martin, W. C., and Wiese, W.L., in G.W.F. Drake (ed.), *Atomic, Molecular, and Optical Physics Handbook*, American Institute of Physics, Woodbury, New York, pp. 135-153, (1996).
24. Martín, I., Lavín, C., and Karwowski, J., *Chemical Physics Letters* **255** (1996) 89-93.
25. Herzberg, G., *Ann. Rev. Physics. Chem.* **38** (1987) 27.
26. Merkt, F., *Annals of Review of Physical Chemistry* **48** (1997) 675-694.
27. Petsalakis, I.D., Theodorakopoulos, G., Wright, J.S., and Hamilton, I.P., J. *Chemical Physics* **88** (1988) 7633-7638.
28. Jungen, Ch, Roche, A.L., and Arif, M., *Philosophical Transactions from the Royal Society London A* **355** (1977) 1475-1490.
29. Martín, I., and Simons, G., *Molecular Physics* **32** (1976) 1017-1022.
30. Martín, I, Lavín, C., Velasco, A.M., Martin, M.O., Karwowski, J., and Diercksen, G.H.F., *Chemical Physics* **202** (1996) 307-315.
31. Lavín, C., and Martín, I., *Advances in Quantum Chemistry* **28** (1997) 205-218.
32. Martín, I, Lavín, C., Pérez-Delgado, Y., Karwowski, J., and Diercksen, G.H.F., *Advances in Quantum Chemistry* **32** (1999) 181-196.
33. Martín, I., Pérez-Delgado, Y., and Lavín, C., *Chemical Physics Letters* **305** (1999) 178-186.
34. Velasco, A.M., Martín, I., and Lavín, C., *Chemical Physics Letters* **264** (1997) 579-83.
35. Raynor, S., and Herschbach, D.R., *J. Physical Chemistry* **86** (1982) 3592-3598.

EXCITED STATES OF HYDROGEN PEROXIDE: AN OVERVIEW

P. K. MUKHERJEE
Department of Spectroscopy, Indian Association for the Cultivation of Science, Jadavpur, Calcutta 700 032, India

AND

M. LUISA SENENT AND YVES G. SMEYERS
Instituto de Estructura de la Materia, C.S.I.C, Serrano 113 bis, 28006 Madrid, Spain

Abstract. The current status of excited states of hydrogen peroxide and related properties is briefly presented. This chemically smallest nonrigid molecule displays internal rotation in its ground and lowest excited states. Large changes in the preferred conformation are observed in the excited states. A short discussion on the ground-state structural parameters is given. Ab initio and other model calculations are discussed with special emphasis on the explanation of the experimental VUV spectrum, including vibrational structure. Results of different levels of approximation are compared. Frequencies of different vibrational modes and spectroscopic parameters are presented, and a critical analysis is given towards possible calculations for the excitation properties of this molecule.

1. Introduction

Hydrogen peroxide is an interesting molecule from both structural and chemical point of view. It is chemically the smallest molecule showing internal rotation. It is an important constituent of troposphere and stratosphere, the recombination of the two HO_2 radicals being the main cause of H_2O_2 formation in atmosphere. It is related to acid rain formation by the oxidation of SO_2 by H_2O_2 either in gas phase or in a water droplet [1-4]. Techniques for the detection of H_2O_2 can be developed using the emission line of OH $(A^2\,\Sigma^+ \rightarrow X^2\,\Pi)$ from the photodissociation of H_2O_2. Currently the molecule is being utilized for the development of amperometric biosensors and also in a plasma catalytic reactor [5, 6]. The ground state of hydrogen peroxide is most stable in skew conformation having a point group symmetry C_2 with the spectroscopic term 1^1A. The most accurate data for the dihedral angle was predicted from infrared spectroscopic

J. Maruani et al. (eds.), New Trends in Quantum Systems in Chemistry and Physics, Volume 2, 65–83.

observations to be $111.9^\circ \pm 0.4^\circ$ [7]. This is corroborated by the accurate theoretical estimates of Thakkar et al [8] and Senent et al [9]. A number of experimental and theoretical investigations have been performed for this molecule to study the structure and properties in the ground state, the vibrational minimum and excited states from its electronic ground state [7-28], its response properties under static electric and magnetic fields [29, 30] and parity violating energy states [31].

The photo-dissociation of hydrogen peroxide into OH fragments is an interesting dynamical process from several point of view. Because of the importance of the OH radical, it is useful to have a well characterized photochemical source. The state distribution of the OH radical can in principle be determined theoretically from an assumed or ab-initio calculated upper state potential. This can be obtained as the tetra-atomic molecule is relatively simple. Finally as the absorption spectra of H_2O_2 is mostly a succession of broad continua, photo-fragmentation spectroscopy is one of the few techniques which can provide information about the mostly repulsive excited states. Specifically photo-dissociation products are mainly vibrational unexcited OH radicals with rotationally excited channels. This is originated from the torsional motion connected with the dihedral angle $\angle HOOH$ and bending modes connected with $\angle OOH$ angles. The photo-dissociation process at different wavelengths ranging from 193 nm to 266 nm was the one mostly studied by different experimental groups earlier [32-40]. The experimental photoelectron spectra of hydrogen peroxide is known for a long time [41-43] and theoretical interpretations provided [44, 45]. Experimental investigations on the absorption cross section in the UV and VUV are rather scanty. The absorption cross section in the UV region (190-400 nm) has been measured by some authors [46-48], while in the VUV region (106-190 nm) only two experiments are available [49,50] of which that of Suto and Lee [50] is the most convincing one. The absorption cross section as measured by Suto and Lee [50] at 135-170 nm region shows unambiguously several small vibrational structures superimposed on the continuum indicating the existence of quasi-bound excited states. In the same region, vibrational structures appear in the fluorescence cross section. Theoretical calculations, quite small in number, have been performed to interpret the experimental observations [51-56]. In the following sections details of the works done on experimental spectrum and its theoretical interpretation are given. To have consistency a brief outline is provided in the next section on the structure and properties of the ground state.

2. The ground state

Hydrogen peroxide has the electronic configuration ------- $4a^2 5a^2 4b^2$: ^1A in the ground state with symmetry C_2. Because of the interesting structure, a good number of experimental [6, 11-21] and theoretical [8, 9, 22-28, 57] investigations at different levels of sophistication was done earlier. It was noted that a successful theoretical calculation must use elaborate basis functions which incorporate polarization functions to account for accurate structural parameters and level staggering. Cremer [57] first performed a critical analysis of the basis set effects and correlation corrections with fairly extended basis sets, using Moller-Plesset perturbation theory vis-à-vis existing experimental data. As the molecule is twisted, definite rotational barriers exist for the cis and trans conformations of the ground state. Penny and Soutterland [60] first proposed an explanation of the stability of skew form based on valence bond theory. According to Penny and Southerland [60] a suitable interplay between the repulsion of oxygen lone pairs and the interactions of OH bond dipole moments is responsible for the stability of the skew structure. The argument was corroborated in a different language by Radom, Hehre and Pople [61] interpreting this stabilization as due to lone pair delocalisation in the polar bond. The interpretation given by Veillard [62] is due to hyperconjugative interaction between the OH bond and the vicinal lone pair. Table 1 displays the optimized geometry parameters as calculated by different theoretical methods of which the fairly recent ones are done with high level of sophistication [8, 9, 28, 29]. Available experimental values have been listed for comparison. It is clearly observed that theoretical calculations using multireference CI, Coupled Cluster or Moller-Plesset perturbations of second and fourth order with extended basis sets produce structural parameters in very good agreement with the experimental data. The rotational barriers have also been theoretically estimated and compared favourably with experimental values. Highly accurate calculation on the ground state rototorsional levels has been performed by Senent et al [9] using non-rigid molecular group theoretical framework which gives much more insight in the detailed rototorsional structure of this molecule.

3. Photo-dissociation

The photo-dissociation process of hydrogen peroxide has been well studied and the process

$$H_2O_2(^1A) + h\nu \rightarrow H_2O_2^* \rightarrow OH(^2\Pi) + OH(^2\Pi)$$

was investigated by different workers [32-40, 53] at excitation wavelengths ranging from 193 nm to 266 nm using Doppler and polarization spectroscopy with laser. Laser induced fluorescence (LIF) technique was used as a probe for the detection of the translational photo-fragment distribution. The photo-fragments are found to be vibrationally unexcited [34] but with relatively high rotational excitations. The O-H bond lengths for the present molecule and the isolated fragments are found to be roughly identical indicating weak vibrational-translational coupling in the exit channel. Single photon photolysis at 193 nm, 248 nm and 266 nm shows no generation of electronically excited OH radical, and both fragments are found in their ground $^2\Pi$ state. The available energy is mostly transferred to fragment recoil and only a small part can be found in OH product rotation. Experiment with synchrotron radiation for the absorption cross section of H_2O_2 in the VUV region (106-193 nm) was performed by Suto and Lee [50]. The fluorescence cross section of the OH photo-fragments was measured indicating electronically excited OH radicals. This suggests a second reaction channel:

$$H_2O_2 + h\nu \rightarrow H_2O_2^* \rightarrow OH(^2\Sigma) + OH(^2\Pi)$$

The $OH(^2\Sigma) \rightarrow OH(^2\Pi)$ resonant emission was observed from the photo-dissociation process at Kr (123.6 nm) and Xe (147 nm) radiations [40]. Most of the released energy is transferred into OH rotation during excitation. Two photon process also take place at $\lambda \sim 193$ nm producing OH photo-fragments in $^2\Sigma$ state.

The resonant radiation from OH photo-dissociation can be used to develop techniques for the detection of H_2O_2.

Photo-dissociation processes yield accurate knowledge about the entire internal-state distribution, along with Λ and spin doubling. The translational motion and molecular alignment can be detected. Evaluation of observed Doppler profiles of the recoiling OH photo-fragments indicates the anisotropic character of the dissociation process. Such studies yield scalar and vector properties of the products, namely, the internal state distribution, alignment, translation and correlation between translational and internal motions of OH fragments and the translational dipole moment of the parent molecule which can indicate the nature of the intermediate electronically excited state of the molecule.

Excitations above 193 nm generate $\tilde{A}\,^1A$ excited state predominantly but at $\lambda \leq 193$ nm about two thirds of the OH photo-fragments are formed via $\tilde{A}\,^1A$ excited state and roughly one third via the $\tilde{B}\,^1B$ state of the parent molecule. In principle, the existence of still higher excited states is possible. Further at or below 157 nm electronically excited photo-fragments indicating new dissociation channels were observed with an inverted rotational state distribution [34]. The OH product rotation mainly originates from the torque provided by the strong dihedral angle dependence of the $\tilde{A}\,^1A$ or $\tilde{B}\,^1B$ excited state potential. Bershon and Shapiro [63] analyzed the dissociation of H_2O_2 from low lying electronic states by the classical trajectory method, where the parameters of the excited state functions were chosen to fit the observed rotational state distribution and to be consistent with the electronic spectrum.

TABLE 1. Ground state energy, geometry and rotational barriers of hydrogen peroxide

	E(a.u.)	R(O-O) A°	r(O-H) A°	∠ O-O-H Deg.	∠ H-O-O-H Deg.	Barrier Heights (Kcal/mole) CIS	TRANS
SCF[a]/[4s3p1d/2s1p]	-151.26844	1.456	0.970	99.5	118.8	8.50	0.87
SCF[b]/MIDI+POL+DIF	-150.8264	1.390	0.947	102.6	118.4		
SCF[c]/QPOL.Fn+POL		1.397	0.943	102.7	111.2		
SCF[d]/MIDI+POL+DIF		1.391	0.948	102.5	115.5		
MP2[a]/[4s3p1d/2s1p]	-151.28031	1.451	0.967	99.3	119.3	8.13	0.94
MP4[c]/QPOL.Fn+POL		1.465	0.963	99.8	113.8		
MP2[e]+MP4/Aug-cc-pVTZ	-151.38472	1.448	0.964	99.8	112.6	7.51	1.16
CCSD(T)/cc-pvDZ[f]	-151.19402	1.470	0.971	98.9	118.0	7.90	0.68
CCSD(T)/cc-pvTZ[f]	-151.35861	1.458	0.964	99.6	113.9	8.17	1.05
CCSD(T)/cc-pvQZ[f]	-151.40822	1.453	0.963	99.9	112.5	7.28	1.08
MP2/611+G(3df, 2p)[g]	-151.37429	1.441	0.964	100.1	111.9	7.79	1.20
Expt.[h]		1.464	0.965	99.4	111.8		
Expt.[i]					111.9	7.33	1.11
Expt.[j]		1.467	0.965	98.5	120.0		
Expt.[k]		1.452	0.965	100.0	119.1		
Expt.[l]					112.8	7.57	1.10

a. Ref. [57]; b. Ref. [45]; c. Ref. [8]; d. Ref. [55]; e. Ref. [9]
f. Ref. [28]; g. Ref. [26]; h. J. Koput, J. Mol. Spectros. **115**, 438(1986)
i. Ref. [7]; j. P. A. Giguere and T. K. K. Srinivasan, J. Mol. Spectros. **66**, 168(1977)
k. G. A. Khachkuruzov and I. N. Przhevalskii, Opt. Spectros. **36**, 172 (1974); l. C. S. Ewig and D. O. Harris, J. Chem. Phys. **52**, 6268 (1970)

The photo-dissociation dynamics at 193 nm was analyzed in detail and the observed rotational state distribution was obtained by using the "rotation reflection principle" by Schinke and Stasemler [53]. All rotational state distributions depend sensitively on the anisotropy of the dissociative potential energy surface. These are interpreted as a mapping of the bound state wave function onto the quantum number axis. The mapping is mediated by the classical excitation function determined by running classical trajectories onto the potential energy surface within the dissociative state. This so-called rotation-reflection principle establishes a direct relation among the bound states of the parent molecule, the anisotropy of the dissociative potential and the final rotational state distribution.

4. Ionic states

Photoelectron spectra (PE) of H_2O_2 were known for a long time in the region between 10-20 eV [41-43]. The PE spectrum (Figure 1) as obtained by Osafune and Kimura [41] between 10 to 14 eV appears to originate from two distinct ionization processes, with a clear vibrational structure on the rise of the first band progression of frequency 1050 cm^{-1}. More detailed structure was provided by the PE spectrum of Ashmore and Burgess [43]. This strongly suggests a non-dissociative nature of the first two ionic states of H_2O_2 falling in this region. While ab-initio calculations of the ionic states of H_2O_2 are rather scanty [45], the vertical ionization potentials (VIP) have been estimated by different theoretical methods [42, 44, 45, 54].

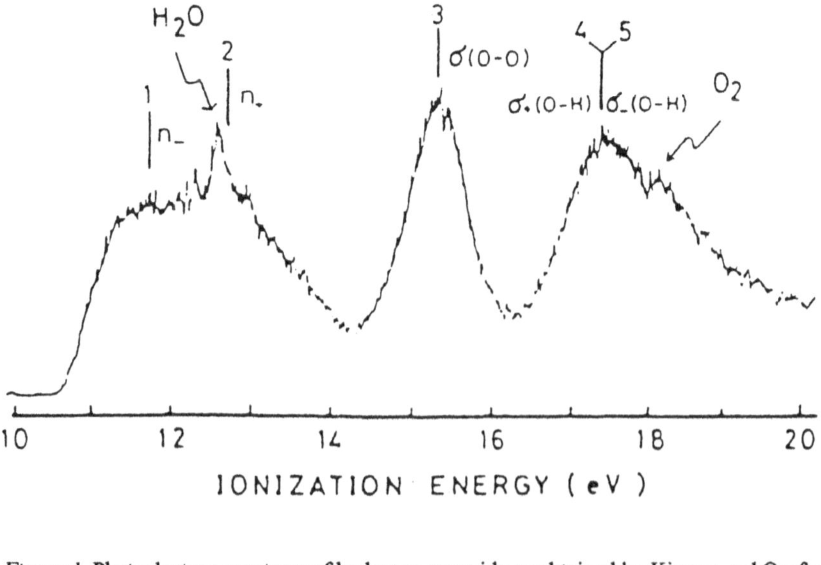

Figure 1. Photoelectron spectrum of hydrogen peroxide as obtained by Kimura and Osafune.

Table 2 displays the VIP's for several levels of H_2O_2.

TABLE 2. Vertical ionization potentials of hydrogen peroxide

Meth.	Vertical ionization potential (eV)				
	$(4b)^{-1}$	$(5a)^{-1}$	$(4a)^{-1}$	$(3b)^{-1}$	$(3a)^{-1}$
Expt.[a]	11.69	12.69	15.33	17.40	17.40
Expt.[b]	11.51	12.56	15.26	17.35	17.35
Expt.[c]	11.7	13.0	15.4	17.5	18.5
Calc[d]	12.8	13.4	14.8	18.0	19.0
Calc[e]	11.15	12.34	14.98	17.08	17.46
50%	11.69	12.28	15.93	18.04	18.65
Calc[g]	11.7	12.8	15.2		
Calc[h]	11.33	12.40			
Calc[i]	11.23	12.66			

a. Ref. [41]; b. Ref. [42]; c. Ref. [43]; d. CNDO results reported in [42]
e. Results with third order RSPT and geometric approximation using DZ+Pol basis [44]; f. CI results reported in [42]; g. SDCI results with DZ+Pol basis [54]
h. SCF results with extended basis + Pol. Functions [45]
i. SDCI results reported in [45]

The molecular orbitals from which the ionization takes place are assigned. The first one $(4b)^{-1}$ is due to the out of phase combination of oxygen lone pairs while the second one $(5a)^{-1}$ is due to the in phase combination of the lone pairs and the σ (O-O) bond. Detailed investigations of the potential energy surfaces of the first two ionic states, namely, ^2B and ^2A have been performed by Takeshita and Mukherjee [45] with SCF and SDCI methods using extended basis set (11s7p) → [7s4p] due to Salez and Veillard [64] for oxygen and Tatewaki-Huzinaga [65] basis for H with proper scale factor [66]. Appropriate polarization functions [65] have been incorporated (MIDI+POL+DIF). The SCF method was used to optimize geometries for respective states with respect to all the parameters to find the potential energy surface and to calculate the harmonic force constants. A sketch of the PE diagram with respect to the dihedral angle obtained by Takeshita and Mukherjee [45] is given in Figure 2.

The PE diagram for the ground state is also shown for continuity. It has been noted that, while the ground state ^1A has a skew symmetry at its minimum, the ^2A state is most stable in the cis conformation, with $\angle$HOOH = 0°, and the ^2B state is stable in the trans conformation, corresponding to an angle of $180°$. The calculated barrier height for the ^2B state is 3.69 eV and that for ^2A is 2.24 eV. The cis and trans barrier heights for the ground ^1A state are approximately 0.35 eV and 0.04 eV, respectively. Thus the rotational barriers are much larger in the ionic states. Table 3 displays the optimized geometry parameters for the ionic states at the SCF level along with the energy values at SCF and SDCI levels due to Takeshita and Mukherjee [45]. The authors have also estimated the vibrational frequencies for four symmetric modes of the ionic and ground states using a harmonic force field. The calculated O-O stretching frequency and H-O-O-H torsional frequency become considerably larger than in the ground state. This is due to shortening of the O-O bond distance accompanying ionization and large rotational barriers for the torsional oscillation modes in the ionic states. The vibrational frequencies will be displayed in a subsequent table. The Frank-Condon factors have also been calculated with a view to compare with the PE profile. Figure 3 shows striking similarity of the calculated PE profile with that due to Osafune and Kimura [41] indicating validity of harmonic approximation.

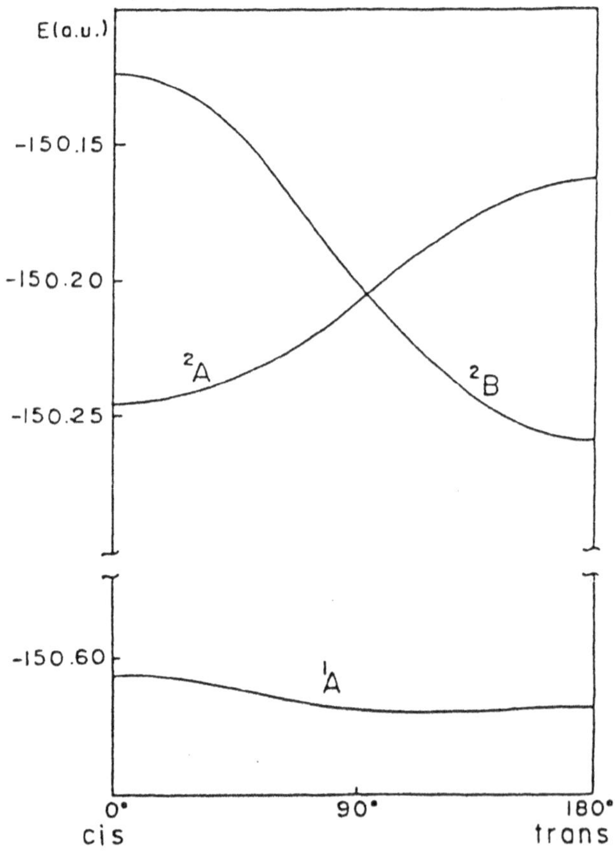

Figure 2. Potential energy vs. dihedral angle for the ionic states of hydrogen peroxide.

TABLE 3. Energy and optimized geometry for the ^2A and ^2B states of hydrogen peroxide

State	Energy (a.u.)	Geometry parameters O-O A°	O-H A°	∠O-O-H Deg.	∠H-O-O-H Deg.
^2A	-150.4443[a] -150.8190[b]	1.249	0.975	113.2	0
^2B	-150.4596[a] -150.8346[b]	1.250	0.976	116.8	180

a. SCF/MIDI+POL+DIF energy of Ref. [45]
b. SCF/MIDI+POL+DIF energy of Ref. [45]

5. Excited states

The absorption cross section of hydrogen peroxide has been measured by a few authors in the 190-400 nm region [46–48], which shows more or less featureless characteristics. The somewhat old experiment in the VUV region ($\lambda \langle$ 190 nm) [49] is not very consistent in the overlapping region (185-200 nm) and the need for consistent data in the VUV region was felt [67]. The most convincing experiment in the VUV region (106-193 nm) for absorption cross section was performed by Suto and Lee [50] using synchrotron radiation. In addition to absorption cross section, they measured the fluorescence cross section from OH (A$\rightarrow$X) transition. The absorption cross section shows several small vibrational structures superimposed on the continuum in the 135-165 nm region. This is displayed in figure 4. The fluorescence cross section corresponding to the same wavelength region has also been measured and shown in figure 4. In the region 155-167 nm the detailed structure shows a vibrational progress with frequency 1350 cm^{-1} In the 143-153 nm region an intense progression with vibrational frequency 1140 cm^{-1} was noted. Each member has two sub-bands with vibrational frequencies 770 cm^{-1} and 260 cm^{-1}. Suto and Lee [50] assigned tentatively the respective frequencies in terms of vibrational frequencies of selected ground state modes on the assumption that vibrational modes for the ground and excited states are more or less similar. This assumption has been found later incorrect by ab-initio calculations for the excited states. In the fluorescence cross section, weak vibrational structures with frequency ~1000 cm^{-1} have been observed in the 137-150 nm region, which is in conformity with the absorption structures. The main conclusion of this experiment is the existence of quasi-bound electronic excited states for this molecule, a prediction which follows from the study of photo-dissociation products [53] discussed earlier.

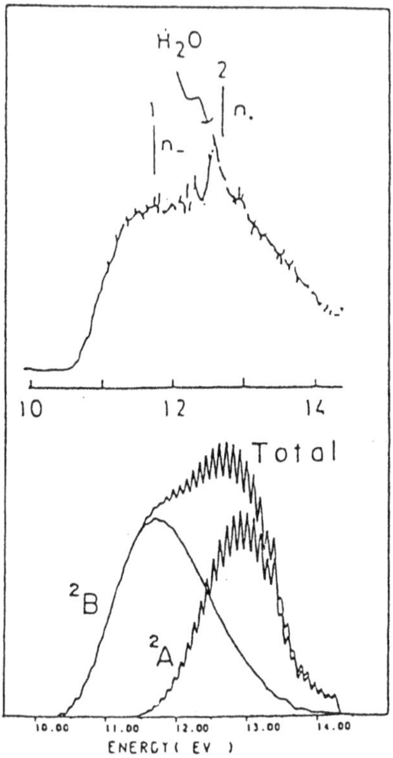

Figure 3. Comparison of experimental and theoretical ionization profiles for the lowest two ionic states of hydrogen peroxide.

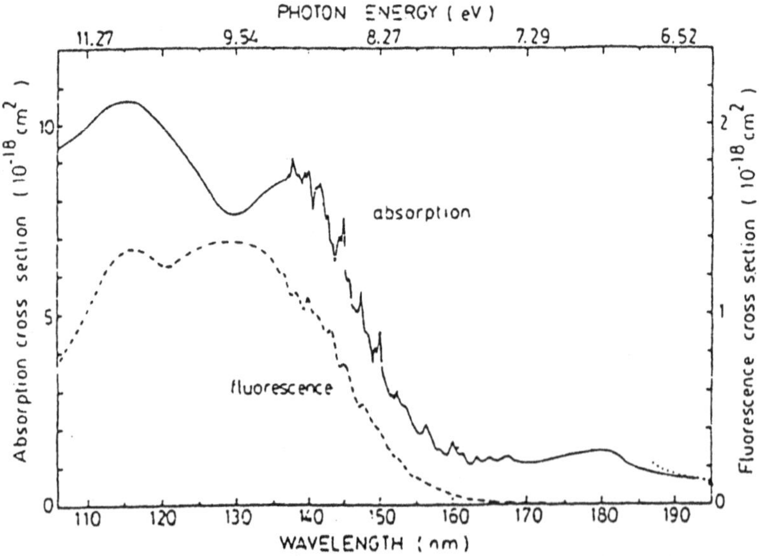

Figure 4. Absorption and fluorescence cross sections of hydrogen peroxide in the 106-193 nm region.

Theoretical studies on the electronic excited states of H_2O_2 are rather scanty. Semi-empirical calculations for ground and excited state correlation diagrams were performed by Evleth and Kassab [51] with a view to study the photochemical cleavage of H_2O_2. Rauk and Barriel [52] used perturbation theory to generate ground and excited state wave functions equivalent to singles CI, and calculated the transition energies, oscillator strengths and optical rotatory power of H_2O_2 with DZ basis augmented with diffuse functions. An empirical correction equal to the difference between the experimental vertical ionization potential and that estimated by Koopman's theorem has been added to the Rydberg like first two excited state energies. The interpretation of the VUV spectrum is, however, not satisfactory. The first accurate ab-initio calculation in this regard was performed by Cevaldonnet et al [54] using Huzinaga basis [68] with Dunning [69] contraction including polarization and diffuse functions. Variation perturbation technique was adopted to account for electron correlation effects. Multi-reference CI calculation was performed with single and double substitutions for several excited states and transition moments evaluated. However, the ground state geometry due to Cremer [57] was used to calculate the excited state energies and properties. According to Cevaldonnet et al [54], the first two vertical excited states are of ^1A and ^1B symmetry. This was corroborated by the photodissociation studies by Schinke and Staemler [53] using classical dynamics and ab-initio potential energy surfaces of the first two lowest excited states which have been found to be in the ratio ~0.7/0.3. The calculation of Cevaldonnet et al [54], although accurate has the drawback of the choice of incorrect geometry for the excited states. It was explicitly shown by Takeshita and Mukherjee [55] that the excitation process is accompanied by large change of geometry, particularly the dihedral angle. Using the basis set of Tatewaki & Huzinaga [65] augmented with polarization and diffuse functions the optimum molecular structure for the 1^1B excited state was determined by the RHF method with gradient techniques and the SCI method for the 2^1A excited state. The optimum geometry structure for the first two excited states have also been determined recently by Mukherjee et al [56] using a much more extended quadrupole zeta basis due to Thakkar et al [8] augmented with adequate polarization and diffuse functions. The configuration interaction with single substitutions method was used to find geometry parameters. It was established that the excited ^1A state has a stable structure at cis conformation while the ^1B state has its minimum at trans conformation. This fully corroborates the findings by Takeshita and Mukherjee [55]. Figure 5 displays the general nature of the dependence of potential energy with respect to dihedral angle for the 2^1A and 1^1B states.

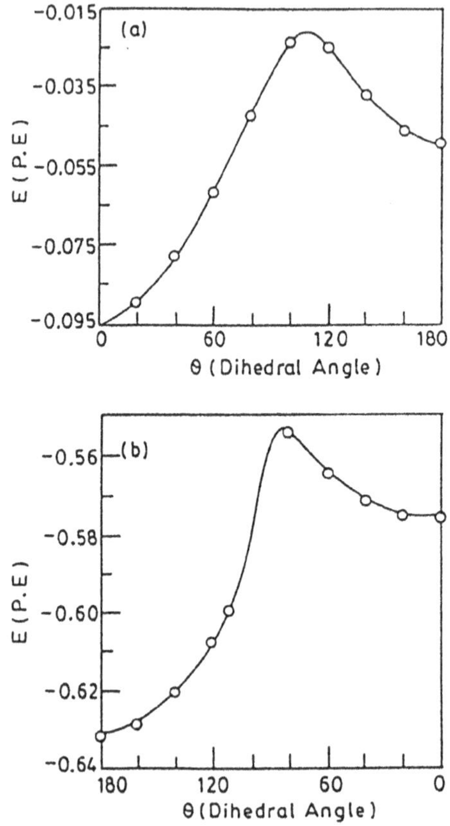

Figure 5. Potential energy vs dihedral angle for the first two excited states of hydrogen peroxide. a. SCI calculation for 2^1A state; b. HPHF calculation for 1^1B state. Energy scales are chosen for convenience.

Both excited states are found to posses Rydberg-like character. Table 4 contains the optimized geometry parameters for the first two excited sates.

TABLE 4. Optimized geometry of the 2^1A and 1^1B states of hydrogen peroxide

State	Geometry parameters			
	O-O	O-H	∠H-O-O	∠H-O-O-H
	A°	A°	Deg.	Deg.
2^1A	1.263[a]	1.064[a]	102.0[a]	0[a]
	1.253[b]	1.069[b]	102.5[b]	0[b]
1^1B	1.277[a]	1.003[a]	101.98[a]	180
	1.248[b]	0.989[b]	104.7[b]	180

a. Ref. [56]
b. Ref. [55]

Mukherjee et al [56] also used the half projected Hartee-Fock (HPHF) method due to Smeyers and co-workers [70-74], which is based on variational formulation with two DODS determinant for the representation of the wave function and is proved to be very effective in calculating singly excited states of molecules. Table 5 displays the energy values only in the cis and trans conformations of the two excited states.

TABLE 5. Energy values for 2^1A and 1^1B states of hydrogen peroxide in cis and trans conformations

State	∠H-O-O-H Deg.	Energy$^+$ (a.u.) SCI(G-94)	HPHF
2^1A	0	-150.59493*	-150.65552*
	180	-150.54886	-150.60670
1^1B	0	-150.51766	-150.57519
	180	-150.57329*	-150.63065*

+ Ref. [56]
* Minimum energy configurations.

Table 5 indicates that HPHF method yields slightly better results than single CI with same basis functions. The adiabatic excitation energies calculated for the respective states by Mukherjee et al [56] and by Takeshita and Mukherjee [55] are displayed in Table 6.

TABLE 6. Adiabatic excitation energies for the 2^1A and 1^1B states of hydrogen peroxide

State	Adiabatic excitation energy (eV) SCI (G-94)[a]	HPHF[a]	SDCI[b]
2^1A	6.84	5.19	5.78
1^1B	7.43	5.78	6.09

a. Ref. [56], basis set QPOL+POL.Fn+DIF
b. Ref. [55], basis set MIDI+POL+DIF

It is interesting to observe the behavior of the 2^1A and 1^1B potential energies with respect to O-O stretching. The general observation is that close to the minimum conformation for both states one gets an equilibrium O-O bond length which increases continuously with change of dihedral angle. The two PE diagrams cross near $\angle$HOOH~100° and the PE becomes dissociative with respect to O-O stretching and remains dissociative afterwards. The level crossing can only be avoided by an elaborate CI calculation [55]. Table 7 shows the relative values of equilibrium O-O bond length with respect to change of dihedral angle for both states using SCI method due to Mukherjee et al [56]. The assignment of the vibrational spectrum is discussed in the next section.

TABLE 7. Equilibrium bond length for O-O against dihedral angle

$\angle$H-O-O-H (Deg.)		0	20	40	60	80	100	120	140	160	180
Equilibrium bond length	a.b	1.2631	1.2631	1.2731	1.2831	1.3031	*	*	*	*	*
O-O A°		*	*	*	*	*	1.2971	1.2871	1.2771	1.2771	1.2771

* Dissociative

a. For 2^1A state Ref. [56]

b. For 1^1B state Ref. [56]

6. Assignment of vibrational spectrum

H_2O_2 has six fundamental vibrational modes corresponding to ν_1 (O-H stretching), ν_2 (O-H symmetric bending), ν_3 (O-O stretching), ν_4 (H-O-O-H torsion), ν_5 (O-H asymmetric oscillation and ν_6 (O-H asymmetric bending). Suto and Lee [50] assumed the excited state molecular geometry is approximately the same as that of the ground state and compared their experimental vibrational frequencies with respect to those of the ground state which were obtained by Giguere [75]. The geometry, however, changes considerably in the excited states and the frequencies were calculated by Takeshita and Mukherjee [55] using harmonic approximation and the Franck Condon factors evaluated. An appreciable change is observed for the frequencies of the excited state with respect to those of the ground state.

In fact, the frequency of the torsional oscillation mode ν_4 is found to be more than double that of the ground state. The frequency of the torsional oscillation mode was reevaluated by Mukherjee et al [56], using a very accurate representation of the one-dimensional vibrational Hamiltonian of the non-rigid rotor in terms of a Fourier series [76-78], and other spectroscopic parameters calculated for the first time taking care of anharmonicity. A new assignment of the experimental spectrum was given. The results are displayed in Table 8. For reference purpose the vibrational frequencies of the ionic states are also listed along with those of the ground state. Suto and Lee [50] assigned their vibrational progression frequencies of 1140 cm^{-1} and 1350 cm^{-1} as due to different electronic states of O-H asymmetric bending modes (ν_6). The theoretical FCF's of Takeshita and Mukherjee [55] showed explicitly that for the nontotally symmetric ν_6 modes, these are rather small while those of totally symmetric ν_4 modes are large. For the ^1B state the intensity calculations of Takeshita and Mukherje [55] compares favorably with that due to Suto and Lee [50]. Figure 6 explicitly demonstrates the overall agreement.

TABLE 8. Vibrational frequencies of ground, excited and ionic states of hydrogen peroxide

State	Vibrational frequencies (cm^{-1})			
	ν_1	ν_2	ν_3	ν_4
1^1A	4144[a]	1611[a]	1172	386[a]
	4126[b]	1615[b]	1134[b]	373[b]
	3426[c]	2647[c]	890[c]	465[c]
			880[d]	
	2374[b]	1297[b]	1515[b]	1083[b]
2^1A				1043[e]
				1345[f]
	3438[b]	1694[b]	1559[b]	931[b]
1^1B				827[e]
				1140[f]
				849[g]
	3826[a]	1499[a]	1575[a]	749[a]
^2A	3792[a]	1720[a]	1568[a]	951[a]
^2B			1080 ± 50[h]	

a. Ref. [45]; b. Ref. [55]; c. Expt. Ref. [73]
d. R. M. Miller and D. F. Horning, J. Chem. Phys. 34, 265 (1961)
e. SCI Calculations of Ref. [56]; f. Vibrational assignment of Suto and Lee [50]
g. HPHF Calculation of Ref. [56]; h. Expt. Ref. [42]

For the 2^1A state the FCFs are rather small and therefore a definite assignment cannot be given, although the calculated torsional frequency is similar to the experimental value. The extensive calculation of Mukherjee et al [56] including intensity evaluation supports the previous theoretical results [55]. Spectroscopic parameters such as $\omega_0, \omega_e, x_e\omega_e$ [79] have been estimated and are listed in Table 9.

TABLE 9. Spectroscopic parameters for the 2^1A and 1^1B states of hydrogen peroxide

State	Method*	Spectroscopic parameters (cm)		
		ω_0	ω_e	$x_e\omega_e$
2^1A	SCI	531.0	1060	-10.2
1^1B	SCI	416.1	835	-4.1
	HPHF	425.5	853	-2.1

*Ref. [56]

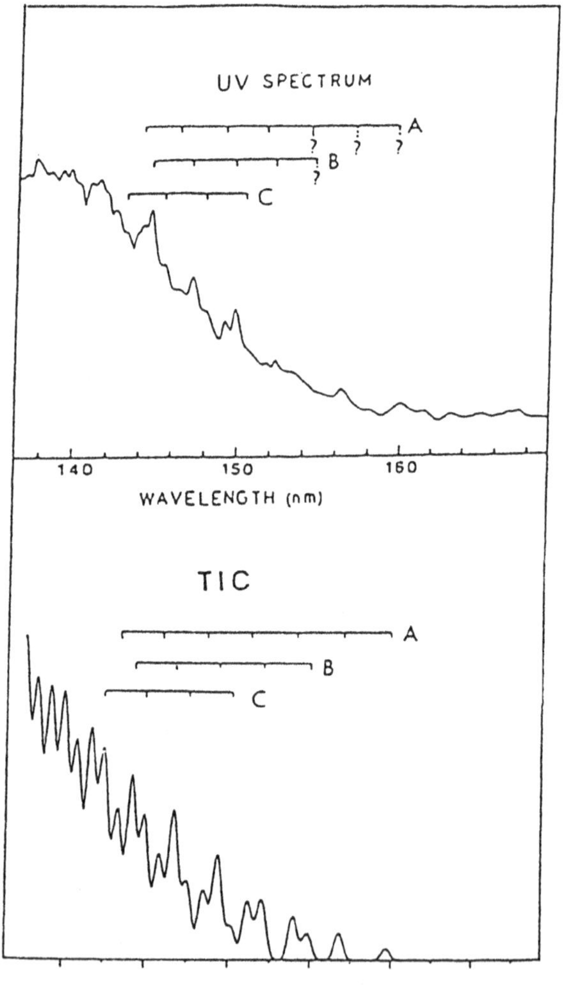

Figure 6. Comparison of experimental vibration spectrum and theoretical intensity pattern for the ^{1}B state of hydrogen peroxide.

7. Conclusions

Excited states of hydrogen peroxide and related properties remain a challenging area in theoretical chemistry. Although some of the general features relating to the experimental VUV spectrum have been correctly explained, accurate quantitative agreement relating to assignments in the vibrational spectrum is still not satisfactory. Elaborate multi-reference CI calculations with optimized geometry for each excited state, with sufficiently flexible basis functions, are necessary.

Acknowledgments

Author PKM greatly acknowledges financial assistance from the Spanish Ministry and CSIC, Madrid, where part of the work was conducted. PKM also thanks the Indian National Science Academy and the Indian Association for the Cultivation of Science for financial assistance to attend QSCP-IV at Marly le Roi in April 1999. Finally, PKM acknowledges a research grant from the Indian Council of Scientific and Industrial Research (CSIR) nb 03(0888)/99/EMR-II.

References

1. P. J. Cruitzen and J. Fishman, Geophys. Res. Lett. **4**, 321 (1977).
2. S. A. Penkett, B. M. R. Jones, K. A. Brice and A. E. J. Eggleton, Atmos. Environ. **13**, 123 (1979).
3. P. K. Dasgupta, Atmos. Environ. **14**, 272 (1980).
4. L. R. Martin and D. E. Damschen, Atmos. Enviorn. **15**, 1615 (1981); G. A. Takacs and C. J. Howard, J. Phys. Chem. **88**, 2110 (1984); S. P. Sander, J. Phys. Chem. **88**, 6018 (1984).
5. T. Lotzbeyer, W. Schumann and H. L. Schmidt, Sens. Actuators B, Chem. **B33**, 50 (1996).
6. S. Gamburzev, P. Atanasav, A. L. Grundilis, E. Wilkins, A. Kaisheva and I. Iliev, Sens. Actuators B, Chem. **B43**, 70 (1997).
7. J. M. Flaud, C. Camey-Peyert, J. W. C. Jones and B. Carli, J. Chem. Phys. **91**, 1504 (1989).
8. A. J. Thakkar, T. Koga, M. Saito and R. E. Hoffmeyer, Int. J. Quantum Chem. S **27**, 343 (1993).
9. M. L. Senent, S. Fernandez-Herrera and Y. G. Smeyers, Spectrochim. Acta A, in press.
10. D. Cremer, J. Chem. Phys. **69**, 4440 (1978).
11. P. Helminger, W. C. Bowman and F. C. De Lucia, J. Mol. Spectros. **85**, 120 (1981); ibid. **87**, 571 (1981).
12. J. J. Hillman, J. Mol. Spectros. **95**, 236 (1982).
13. W. B. Olson, R. H. Hunt, B. W. Yong, A. G. Maki and J. W. Broult, J. Mol. Spectreos. **127**, 12 (1988).
14. C. Camey-Peyret, J. M. Flaud, J. W. J. Johns and M. Noel, J. Mol. Spectros. **155**, 84 (1992).
15. W. B. Cook, R. H. Hunt, W. N. Shelton and F. A. Flaherty, J. Mol. Spectros. **171**, 91 (1995).
16. D. T. Petkie, T. M. Gogette, J. J. Holton, F. C. De Lucia and P. Helminger, J. Mol. Spectros. **171**, 145 (1995).
17. A. Perrin, A. Valentin, J. M. Flaud, C. Camey Peyret, L. Schriver, A. Schriver and P. Arcus, J. Mol. Spectros. **171**, 358 (1995).

18. J. M. Flaud, A. Perrin and C. Camey-Peyret, Spectrochim. Acta A, Mol. Spectros. **51A**, 1217 (1995).
19. A. Perrin, J. M. Flaud, C. Camey-Peyret, R. Shermoul, M. Winnewissr, J. Y. Mandin, V. Dana, M. Badaoui and J. Koput, J. Mol. Spectros. **176**, 287 (1996).
20. M. Bellini, E. Cataechini, P. Natale, G. Dilonardo, L. Fusina, M. Ingucio and E. Venuti, J. Mol. Spectros. **177**, 115 (1996).
21. B. Rebeurs, Int. J. Infrared. Millim. Waves, **16**, 1465 (1995).
22. J. E. Carpenter and F. Weinhold, J. Php. Chem. **92**, 4295 (1988); ibid. **92**, 4306 (1988).
23. A. W. Willets, J. F. Gaw, N. H. Handy and S. Carter, J. Mol. Spectros. **135**, 370 (1989).
24. S. Samdal, V. S. Mastryukov and J. E. Boggs, J. Mol. Struct. **346**, 35 (1995).
25. J. M. Seminario, Int. J. Quantum. Chem. (Symp.) **28**, 655 (1994).
26. A. Chung-Phillips and K. A. Jebber, J. Chem. Phys. **102**, 7080 (1995).
27. J. Antikainen, R. Friesner and C. Lefrestier, J. Chem. Phys. **102**, 1270 (1975).
28. J. Koput, Chem. Phys. Lett. **236**, 516 (1995).
29. J. Koput, Chem. Phys. Lett. **257**, 36 (1996).
30. S. M. Cybulski and D. M. Bishop, Mol. Phys. **93**, 739 (1998).
31. P. Lazaretti and R. Zanasi, Chem. Phys. Lett. **279**, 349 (1997).
32. S. Klee, K. H. Gericke, J. and F. J. Comes, J. Chem. Phys. **85**, 40 (1986).
33. K. H. Gericke, S. Klee and F. J. Comes, J. Chem. Phys. **85**, 4463 (1986).
34. A. U. Grunewaid, K. H. Gericke, J. and F. J. Comes, Chem. Phys. Lett. **132**, 121 (1986); J. Chem. Phys. **87**, 5709 (1987).
35. K. H. Gericke, S. Klee and F. J. Comes, Chem. Phys. Lett. **137**, 510 (1987).
36. M. P. Docker, A. Hodgson and J. P. Simons, Chem. Phys. Lett. **128**, 264 (1986).
37. G. Ondrey, N. Van Veen and R. Bershon, J. Chem. Phys. **78**, 3732 (1983).
38. K. H. Welge, Cac. J. Chem. **52**, 1424 (1974).
39. A. Jacobs, K. Klinermanns, H. Kugl and J. Wulfrum, J. Chem. Phys. **79**, 3162 (1983).
40. K. H. Becker, W. Groth and D. Kley, Z. Naturforsch. **20a**, 748 (1965).
41. K. Osafune and K. Kimura, Chem. Phys. Lett. **25**, 47 (1974).
42. R. S. Brown, Can. J. Chem. **53**, 3439 (1975).
43. F. S. Ashmore and A. R. Burgess, J. Chem. Soc. Faraday Trans II **73**, 1247 (1977).
44. T. Minato and D. P. Chong, Can. J. Chem. **61**, 550 (1983).
45. K. Takeshita and P. K. Mukherjee, Chem. Phys. Lett. **160**, 193 (1989).
46. R. B. Holt, C. K. Mc Lane and O. Oldenburg, J. Chem. Phys. **16**, 225 (1948).
47. C. L. Lin, N. K. Rohatgi and W. B. De More, Geophys. Res. Lett. **5**, 113 (1973).
48. L. T. Molina, S. D. Schinke and M. J. Molina, Geophys. Res. Lett. **4**, 580 (1977).
49. M. Schurgers and K. H. Welge, Z. Naturfosch. **23a**, 1508 (1968).
50. M. Suto and L. C. Lee, Chem. Phys. Lett. **98**, 152 (1983).
51. E. M. Evleth and E. Kassab, J. Am. Chem. Soc. **100**, 7859 (1978).
52. A. Rauk and J. M. Barriel, Chem. Phys. **25**, 409 (1977).
53. R. Schinke and V. Staemmler, Chem. Phys. Lett. **145**, 486 (1988).
54. C. Cevaldonnet, H. Cardy and A. Dargelos, Chem. Phys. **102**, 55 (1986).
55. K. Takeshita and P. K. Mukherjee, Chem. Phys. **182**, 195 (1994).
56. P. K. Mukherjee, M. L. Senent and Y. G. Smeyers, Int. J. Quantum. Chem. S (1999).
57. D. Cremer, J. Chem. Phys. **69**, 4440 (1978).
58. J. Koput, S. Carter, N. C. Handy, J. Phys. Chem. **A102**, 6325 (1998).
59. B. Kuhn, T. R. Rizzo, D. Luckhaus, M. Quack, M. A. Suhm, J. Chem. Phys. **111**, 2565 (1999).
60. W. G. Penny and G. B. B. M. Southerland, J. Chem. Phys. **2**, 492 (1934).
61. L. Radom, W. J. Hehre and J. A. Popple, J. Am. Chem. Soc. **93**, 281 (191); ibid. **94**, 2371 (1972).
62. A. Veillard, Theo. Chem. Acta, **18**, 21 (1970).
63. R. Bersohn and M. Sapiro, J. Chem. Phys. **85**, 1396 (1976).
64. C. Salez and A. Veillard, Theo. Chim. Acta. **11**, 441 (1968).

65. H. Tatewaki and S. Huzinaga, J. Compnt. Chem. **1**, 205 (1980).
66. P. K. Mukherjee and D. P. Chong, Chem. Phys. Lett. **86**, 190 (1982).
67. K. H. Welge, Can. J. Chem. **52**, 1424 (1974).
68. S. Huzinaga, J. Chem. Phys. **42**, 1293 (1965).
69. T. H. Dunning Jr., J. Chem. Phys. **53**, 2823 (1970).
70. Y. G. Smeyers, An. Fis. (Madrid) **67**, 17 (1971).
71. Y. G. Smeyers and L. Doreste-Suarez, Int. J. Quantum. Chem. **7**, 687 (1973).
72. Y. G. Smeyers and G. Delgado Barrio, Int. J. Quantum. Chem. **8**, 773 (1973); ibid. **10**, 461 (1976); Phys. Rev. **A15**, 135 (1977).
73. P. Fernandez Serra, V. Boltella and Y. G. Smeyers, A. Galano and G. Delgado Barrio, Int. J. Quantum. Chem. **54**, 305 (1995).
74. Y. G. Smeyers, M. B. Ruiz and P. Otto, J. Mol. Struct.(Theochem) **390**, 91 (1997).
75. P. A. Giguere, J. Chem. Phys. **18**, 88 (1950); O. Bain and P. A. Giguere, Can. J. Chem. **33**, 527 (1955).
76. Y. G. Smeyers, A. Nino and D. C. Moule, J. Chem. Phys. **93**, 5786 (1990).
77. Y. G. Smeyers, M. L. Senent, V. Botella and D. C. Moule, J. Chem. Phys. **98**, 2754 (1993).
78. A. Viver, V. H. Uc and Y. G. Smeyers, J. Chem. Phys. **109**, 2279 (1998).
79. G. Herzberg, Spectra of Diatomic Molecules, Van Nostrand, Princeton, N. J. 1950.

ON ELECTRON DYNAMICS IN VIOLENT CLUSTER EXCITATIONS

P. G. REINHARD
Institut für Theoretische Physik, Universität Erlangen
Staudtstr. 7, D-91058 Erlangen, Germany

AND

E. SURAUD
Laboratoire de Physique Quantique, Université Paul Sabatier
118 route de Narbonne, F-31062 Toulouse Cedex, France
Membre de l'Institut Universitaire de France

Abstract. We present a brief overview of recent progress in the theoretical description of electron and ion dynamics following strong excitations of metal clusters. We consider in particular the excitation by intense femtosecond laser pulses and by collisions with highly charged ions. We use the Time-Dependent Local-Density Approximation (TDLDA) complemented by local pseudopotentials and occasionally explicit ionic motion. After presenting the general framework of these studies, within emphasizing the complexity of this multi scale dynamics we show a few examples of applications.

The dynamical response of metal clusters to electro-magnetic excitations is a key feature for the understanding of these objects. These processes have been extensively studied over the past two decades in particular in the linear regime [1, 2]. Nevertheless, interest is now extending to more energetic excitations reaching dynamical regimes allowing acccess to strongly out of equilibrium situations. A typical nonlinear phenomenon is cluster fragmentation which has been intensively studied, in particular from the experimental side, and for some years [3, 4, 5]. It can for example be accessed by a successive stacking of photons into a cluster with help of a low fluency nanosecond laser. In such a way both, electronic and ionic, degrees of freedom are simultaneously excited, so that the cluster is already at thermal equilibrium before its deexcitation. Relaxation then proceeds by evaporation, fission or fragmentation, and can be described by statistical

J. Maruani et al. (eds.), New Trends in Quantum Systems in Chemistry and Physics, Volume 2, 85–105.

concepts. However, the recent experimental developments have meanwhile opened the path to almost instantaneous excitation processes, running at a time scale of the order of characteristic electronic times. The situation is here different from the previous experiments in that excitation and initial ionization proceed simultaneously, thus placing the charged cluster in a state far from equilibrium and with initially little intrinsic heat. Of course this initial electronic excitation will subsequently be transferred to ionic degrees of freedom. Still, the relaxation process, which may also be quite violent, does not a priori occur in a thermally equilibrated system. It is the aim of this contribution to briefly review recent theoretical work on the nonlinear electronic and ionic dynamics following such energetic electromagnetic excitations. We also refer the reader to [6], where can be found more extensive discussions of these questions.

The paper is organized as follows. Section 1 gives an overview of the physical situations studied, in terms of reaction mechanisms and time scales. The various models developed to deal with such situations are also briefly discussed. Section 2 provides a survey of the theoretical tools used in our calculations. Section 3 is devoted to a discussion of typical excitation processes by ion collisions or laser irradiations. Conclusions and perspectives are given in section 4.

1. Physical context

1.1. ON REACTION MECHANISMS

In practice, we shall consider here two classes of rapid, intense, excitations: collisions with highly charged energetic ions [7] and irradiation by intense femtosecond laser pulses [8, 9]. Both mechanisms take place between tens of fs (a typical pulse length for nowadays intense lasers) down to below 1 fs (which can be attained by means of an ionic projectile at a velocity comparable to the electronic velocity inside the cluster). The excitation time thus turns out to be directly comparable to characteristic time scales of the valence electron cloud. In figure 1, we have schematically represented the various processes following such an excitation. The upper parts symbolize the excitation mechanisms (time scale τ_{exc}), namely laser excitation by a weavy line extending from tens up to beyond this plot for nanosecond lasers, and fast ion collisions by a short peak. The cluster response can be grouped in different stages, and the figure is accordingly splitted into three panels (from left to right) corresponding to three successive phases of the excitation and relaxation. The first phase of the "reaction" is fully electronic and characterized by a direct electron emission (time scale τ_{esc}) and a collective oscillation of the Mie plasmon (time scale τ_{pl}). During the second stage, still of electronic nature, damping of the collective Mie plasmon occurs (Landau

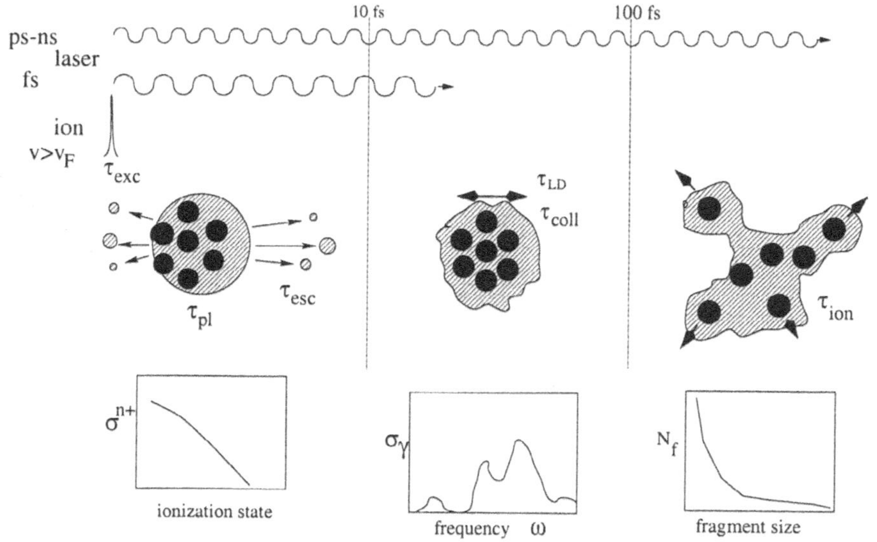

Figure 1. Schematic view of dynamical processes, associated time scales (see text), and observables. The first bottom panel displays a typical ionization cross-section, the second one a typical optical spectrum associated to Mie plasmon and the third one sketches a possible distribution of fragments. From [6].

damping with time scale τ_{LD}) and electron-electron collisions (time scale τ_{coll}) take a leading role. On longer times (right panel) electronic degrees of freedom will slowly couple to ionic motion (time scale τ_{ion}) to lead to the final deexcitation of the cluster by ion, electron or monomer evaporation, fission, or cluster fragmentation. This final phase of deexcitation is dominated by ionic effects. Ionic motion stems here from two sources: i) Coulomb repulsion due to the net charge of the cluster which was accumulated by ionization; ii) energy exchanges between the now "hot" electron cloud and the still "cold" ions. Note finally that the thermal evaporation of electrons proceeds on a very long time scale, in most cases slower than ionic processes.

A complete theoretical description of the whole dynamical process, from details of the excitation phase to the actual fragmentation, thus represents a formidable task, which requires to account for both electron dynamics (to be resolved at the sub-fs level) and ion dynamics (in the ps time scales). Before discussing how one can treat such a problem, it is interesting to quantify a bit more our above discussion of time scales, and to see how the various involved times do compete with each other.

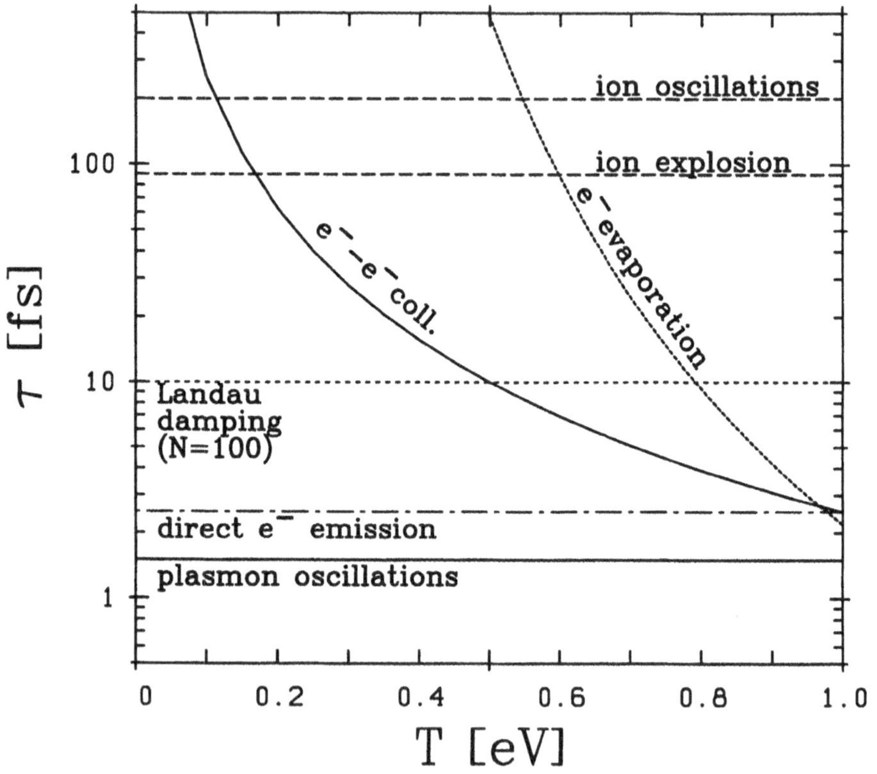

Figure 2. Illustration of the various time scales in cluster dynamics. Times are drawn versus temperature T to accomodate the two processes which strongly depend on temperature. From [6].

1.2. ON THE VARIOUS TIME SCALES

In Figure 2 we try to summarize the various time scales competing in the dynamics of metal clusters. We focus here on the case of Na clusters. The results are drawn versus electronic temperature T in the cluster, because at least two of these times strongly depend on T. The fastest time is given by the period of the plasmon oscillations, immediatley followed by the time needed for direct electron emission. Direct emission is induced by the force pulse from the external field (as delivered by the ionic projectile or the exciting laser) with some amplification through the plasmon response. It is to be distinguished from thermal electron evaporation, which corresponds to a statistical process in a thermally equilibrated system, and which can thus be evaluated following Weisskopf's formula [10]. The value obtained for such a time is usually far beyond any other scale, as can be seen from the figure. The evaporation, however strongly depends on temperature and could come into play for very highly excited systems. A while after plasmon oscillations and direct emission comes Landau damping, caused by coupling of the Mie plasmon mode to nearby particle-hole ($1ph$) states. The Landau damping time very much depends on the actual density of $1ph$ states which, in turn, strongly depends on the size of the system. There is actually little Landau damping for very small clusters ($N \approx 8$) and for $N \longrightarrow \infty$. In the case of singly charged Na clusters, a maximum is found around $N \approx 300$ [11]. In the typical case of a cluster with $N \approx 100$ one can deduce from the spectral width a relaxation time of about 10 fs. These three times, plasmon period, emission time, and Landau damping, only weakly depend on the temperature of the system. We have thus neglected in Figure 2 any temperature dependence and drawn straight horizontal lines.

Electron-electron collisions induce further thermalization and transport excitation energy from the $1ph$ modes to $2ph$ and higher configurations. The collision rate is however much suppressed at low excitation energies due to Pauli blocking of the final scattering states. The relaxation times from these electron-electron collisions thus strongly depend on cluster's temperature. The evaluation of these times in metal clusters is not yet standard. We rely here on a first attempt in this direction, performed at a semi-classical level [12]. In such a model, one obtains for Na clusters $\tau_{\mathrm{col}}^{-1} = 0.4\,T^2 \mathrm{fs}^{-1} \mathrm{eV}^{-2}$ which has been drawn in figure 2.

The time scales for ionic motion somewhat depend on the violence of the excitation. Ionic vibrations are nearly always present and can be associated to a period of about 200 fs. In turn, very energetic excitations can ionize the cluster quickly into a high charge state. The strong Coulomb force then drives the system to fragmentation and the speed of this process depends, of course, on the degree of ionization, but also on the amount of deposited excitation energy (primarily in the electron cloud, and later on as ionic temperature). For example, we have investigated Na_{12} charged to $Na_{12}{}^{+++}$ with a short laser pulse and found that significant effects from ionic motion arise already at around 100 fs [13]. Other cases reacted even a bit more quickly (section 3.2).

Finally, one should relate the times shown in figure 2 to the time scale of the excitation process. The situation is rather simple for fast ionic collisions, as they are completed before any one of the constituents of the cluster can react. Laser excitations, ranging nowadays from 10 – 20 fs to anything longer, provide however a much wider choice of time profiles, which all interfere with several of the internal cluster processes. This gives rise to a rich variety of dynamical processes, which, to a large extent, have yet to be explored.

Altogether, Figure 2 emphasizes the complexity of the situations we aim at describing. This complexity is particularly reflected in the fact that one has to deal with several possibly competing time scales. How to treat such a complex situation requires some discussion. We now briefly consider this aspect.

1.3. ON THEORETICAL APPROACHES

There are basically three ways of treating such a problem:

The first class of approaches could be labelled as "exact". A complete diagonalization of the full static electrons + ions hamiltonian in principle allows to access any dynamical process. It should thus provide a fully detailed description of the dynamics of the system. However, up to now, such calculations have focused on structural properties rather than on dynamical ones. Furthermore, even today's computer capabilities barely allow such calculations for clusters of more than very few atoms [14]. Cluster's size limitations are comparable for molecular physics's technics, based on the time propagation of quantal wavepackets [15]. These "exact' approaches hence mainly provide benchmarks for the other theories, but do not really allow a full exploration of the various facets of the physics involved.

In a second class of approaches, the problem is solved with an adiabatic approximation, either by using model hamiltonians, or with an effective treatment of the electrons by means of the Density Functional Theory (DFT) [16]. Electrons are here assumed to relax very fast towards the ground state potential energy surface, for a given ionic configuration (Born-Oppenheimer approximation) [17, 18]. In this sense, electrons are thus only playing a passive role in the dynamics. Such approaches allow to consider long time dynamics and/or ensembles of trajectories. But, by construction, they miss most of the crucial non-equilibrium features of the electronic response following a strong perturbation. They are thus restricted to situations with weak to moderate initial electronic excitations.

In the third class of approaches one relies on an effective propagation of the most relevant degrees-of-freedom, by means of the Density Functional Theory (DFT) [16], but without any adiabatic approximation. It should be noted that such DFT based approaches have been used since long in metal clusters, in particular for structure or optical response problems [2]. But their use in truly dynamical problems (Time Dependent DFT, TDDFT), beyond the linear regime, and possibly with coupling to ions, is very recent [13, 19, 20]. The DFT approaches turn out to provide an efficient and flexible access to truly non linear dynamics of the electrons. This is the approach which we have been using in our calculations. In the following we shall thus show how this method provides interesting information on the microscopic mechanisms of metal cluster response to strong perturbations.

2. Theoretical framework

2.1. DESCRIPTION OF ELECTRONS AND IONS

2.1.1. *Density Functional Theory for electrons*

In the Time Dependent Density Functional Theory (TDDFT) [16], the correlated many-electron problem is mapped into a set of coupled Schrödinger equations for each single electronic wavefunctions ($\phi_j(\mathbf{r},t)$, $j = 1, n$), which yields the so-called Kohn-Sham equations (in atomic units)

$$i\frac{\partial}{\partial t}\phi_j(\mathbf{r},t) = \left(-\frac{\nabla^2}{2} + v_{KS}[\rho](\mathbf{r},t)\right)\phi_j(\mathbf{r},t) \quad , \tag{1}$$

$$v_{KS}[\rho](\mathbf{r},t) = v_{\rm ext}(\mathbf{r},t) + v_{\rm ion}(\mathbf{r},t) + \int d^3r' \frac{\rho(\mathbf{r}',t)}{|\mathbf{r}-\mathbf{r}'|} + v_{\rm xc}[\rho](\mathbf{r},t) \tag{2}$$

$$\rho(\mathbf{r},t) = \sum_{j=1}^{n} |\phi_j(\mathbf{r},t)|^2 \quad . \tag{3}$$

The effective Time Dependent Kohn-Sham (TDKS) potential $v_{KS}[\rho](\mathbf{r},t)$ is decomposed into several pieces. The external source field $v_{\rm ext}(\mathbf{r},t)$ characterizes the excitation mechanism, namely the electromagentic pulse as delivered by a by passing ion or a laser pulse. The term $v_{\rm ion}(\mathbf{r},t)$ accounts for the effect of ions on electrons (the time dependence reflects here the fact that ions are allowed to move). Finally, appear the Coulomb (direct part) potential of the total electron density ρ, and the exchange correlation potential $v_{\rm xc}[\rho](\mathbf{r},t)$. The latter xc potential is expressed as a functional of the electronic density, which is at the heart of the DFT description. In practice, the functional form of the potential has to be approximated. The simplest choice consists in the Time Dependent Local Density Approximation (TDLDA). This latter approximation amounts to express $v_{\rm xc}[\rho(\mathbf{r},t)]$ as $v_{\rm xc}^{\rm TDLDA}(\mathbf{r},t) = \delta E_{\rm xc}^{\rm LDA}/\delta\rho\Big|_{\rho=\rho(\mathbf{r},t)}$, where $E_{\rm xc}^{\rm LDA} = \int d^3r\, e_{\rm xc}^{hom}[\rho(\mathbf{r})]$ is the static xc energy and $e_{\rm xc}^{hom}$ is the xc energy of the homogeneous electron gas. We use here the simple parametrization of [21] (figures 3, 5 and 7) or [22] (figures 4 and 6).

It should be noted that the above TDLDA picture a priori involves two "touchy" approximations. The first one consists in using the LDA, which basically relies on the assumption of weakly varying (in space) electron density. This LDA approximation has been widely used in metal clusters and does not raise problems with respect to the observables we are interested in. The second approximation is to use in a dynamical context a functional which has been tuned to static problems. The extension of LDA to TDLDA is thus a further approximation which can be considered as "adiabatic", in the sense that we are using, at each instant, the energy density as expressed from the actual density $\rho(\mathbf{r},t)$. In practice, TDLDA gives good results even in the case of rather rapid time dependence, as can be checked by considering more elaborate versions of the TDDFT, such as the ones based on the Self Interaction Correction (SIC). As shown in [23] the gross characteristics of the electron response are very similar in both (TDLDA and Time Dependent SIC) approaches. The TDLDA thus represents an efficient, reliable and simple scheme for describing the electronic response.

In our calculations we solve, directly in time, the TDKS equations, either in 2D cylindrical or in full 3D geometry (no symmetry restriction). Absorbing boundary conditions have been used in all our calculations, and their impact has been carefully checked, see [24] for details. We use grid techniques for representing the electronic wavefunctions. In the 2D case we use finite difference formulae for kinetic energy and a Crank-Nicholson time propagation. The 2D geometry allows to treat large clusters (up to 100 atoms and more). The 3D simulations use fast Fourier techniques for the kinetic energy and the Coulomb solver [25]. The time step relies on an interlaced local and kinetic propagation with complex exponentials [26]. The 3D simulations are best suited for intermediate clusters (10-40 atoms).

2.1.2. *Description of the positive ionic background*

The interaction between ions and valence electrons is described in a standard manner by pseudopotentials. Sodium clusters allow to use local pseudopotentials. We employ here a soft shape in terms of error functions

$$v_{\mathrm{Ps}}(\mathbf{r}-\mathbf{R}_I) = \sum_{i=1,2} v_i \frac{\mathrm{erf}\left(|(\mathbf{r}-\mathbf{R}_I)|/\sigma_i\right)}{|\vec{r}-\vec{R}|} \tag{4}$$

with

$$\mathrm{erf}(r) = \int_0^r dr' \exp\left(-\frac{r'^2}{2\sigma_i^2}\right) \tag{5}$$

with $\sigma_2 = 2\sigma_1$, $\sigma_1 = 0.8\mathrm{a}_0/(2\sqrt{\ln 2})$ ($a_0 \simeq 0.529$Å), $v_1 = 4.59\sigma_1^2$ and $v_2 = -1.21\sigma_2^2$, so that $v_{ion}(\mathbf{r},t) = \sum_I v_{\mathrm{Ps}}(\mathbf{r}-\mathbf{R}_I)$ in Eq.(2). The form is taken over from the local part of the pseudopotentials of [27]. For the actual parameters entering Eq.(4) we use the choice of [26] (figures 3 and 5) or the more recent fit of [28] (figures 4 and 6). Both provide appropriate binding together with plasmon position. And altogether, one obtains an accuracy of our calculations comparable to the one attained in quantum chemistry [14].

The time scales of excitation characterizing the violent cluster perturbations which we want to describe are so short that, to a very good approximation, ions can be considered as fixed during the excitation process and for an early stage of the electronic relaxation (typically up to 100 fs).

On longer times one nevertheless needs an explicit account of ionic motion. We then treat the ions as classical particles described within the standard framework of molecular dynamics. The force acting on the ions originates from the electrons (through pseudopotentials), from ion-ion interactions (treated as point charges) and from the external field (laser, projectile). For ion number I, it thus reads

$$\frac{d}{dt}\mathbf{P}_I = -\nabla_{\mathbf{R}_I}\left[v_{\text{ext}}(\mathbf{R}_I,t) - \int d\mathbf{r}\,\rho(\mathbf{r},t)v_{\text{Ps}}(|\mathbf{R}_I-\mathbf{r}|)\right. \tag{6}$$

$$\left. + e^2\sum_{J\neq I}\frac{1}{|\mathbf{R}_I-\mathbf{R}_J|}\right] \tag{7}$$

$$\frac{d}{dt}\mathbf{R}_I = \mathbf{P}_I/M_{\text{ion}} \quad . \tag{8}$$

The above molecular dynamics equations are then solved using the the standard and robust leap-frog algorithm [29].

When ions can be considered as frozen, one can simplify the treatment of the ionic background even further. Previous investigations of the plasmon response have shown that details of the ionic structure seem to play only a minor role for the electronic response, and become less and less important, the stronger the excitation [24]. The ionic background of the cluster (nuclei plus core electrons) can thus be simplified to the jellium approximation. In the following we shall actually also use the so called "soft" version of the jellium model, in which the (originally) sharp ionic distribution is folded with an Ashcroft pseudoptential. Practically, this is equivalent to using a Woods-Saxon type distribution. The surface width of this soft jellium is directly related to the position of the plasmon, and one can show that the soft jellium model in fact provides the right plasmon frequencies [30].

2.2. OBSERVABLES

As mentioned above, the electronic response of the cluster may be analyzed, at least during the early phases of the process, in terms of the collective Mie plasmon, ionization and energy deposit. These quantities can easily be accessed in the TDLDA.

2.2.1. *Plasmon response*

The cluster's collective response is dominated by the dipole mode (Mie plasmon) and can thus be characterized by the dipole moment in real time

$$\mathbf{D}(t) = \int_{\mathcal{V}} d^3r\,\mathbf{r}\rho(\mathbf{r},t) \quad . \tag{9}$$

The dipole is, in principle, a space vector but the response is usually concentrated along a prefered axis. In the case of a laser excitation, for example, the dipole motion is concentrated along the laser polarization axis (when a linearly polarized laser is used) which we will then take as z axis. Note that in the definition Eq. (9), the space integration is limited to an analyzing volume $\mathcal{V}$. We take in practice a region with a broad margin of 2-3 Wigner-Seitz radii r_s around the cluster volume [24]. For details on the interest and impact (or rather the lack of impact) on the spectral analysis, as well as on the ionization (section 2.2.2), see [24].

The dipole response in real time gives access to the response in frequency domain by Fourier transfrom $\tilde{D}(\omega)$, from which one can extract the strength function $S(\omega) = \Im\{\tilde{D}(\omega)\}$ and the power spectrum $P(\omega) = |\tilde{D}(\omega)|^2$. The strength function is the more suited quantity in the linear regime, where it can be related to the photoabsorption cross section [31], while the power spectrum better applies for spectral analysis in the non linear regime [24].

2.2.2. *Analysis of ionization*

The average signal of ionization is provided by the number of escaped electrons [32]

$$N_{\text{esc}}(t) = N(t{=}0) - N(t) \quad , \quad N(t) = \int_{\mathcal{V}} d^3r\, \rho(\mathbf{r},t) \quad , \qquad (10)$$

where $\mathcal{V}$ is again the finite analyzing volume introduced above. Going slightly beyond the strict limits of DFT, one can extract more information when analyzing single particle densities. This allows to establish an important link with experiments by evaluating approximate expressions for the ionization probabilities $P^k(t)$'s, namely the probabilities of finding the clusters in one of the possible final charge states k to which they can ionize. Explicit expressions for the $P^k(t)$'s can be obtained in terms of bound-state occupation probabilities $N_j(t) = \int_{\mathcal{V}} d^3r\, |\phi_j(\mathbf{r},t)|^2 = \int_{\mathcal{V}} d^3r\, \rho_j(\mathbf{r},t)$ associated with the single-particle KS densities ρ_j. The evaluation of these ionization probabilities rely on simple combinatorial arguments, which we do not repeat here. See [32] for details. Note however that, in a mean field description such as provided by the TDLDA, one should only expect reliable estimates for degrees of ionization much smaller than the total electron number.

From the ionization probablities P^k's one can recover the average number of escaped electrons $N_{esc} = \sum_k kP_k$ and an estimate of the corresponding variance $\Delta N_{esc}^2 = \sum_k k^2 P_k - (\sum_k kP_k)^2$. Ionization cross-sections can also been evaluated from the knowledge of the P^k's [32].

2.2.3. *Energy deposit*

Energy flow is another crucial observable in dynamics and the branching into its various components contains most interesting information. During the first phase of the process, the cluster excitation manifests itself preferably in terms of electronic degrees of freedom. As already mentioned, the initial excitation energy thus primarily goes into collective (dipole) motion and into direct electron emission. But in subsequent stages, collectivity runs out of phase and collective energy is converted to intrinsic motion. The associated thermal excitation energy can be measured as

$$E^* = E_{kin} - E_{coll} - E_{Fermi} \tag{11}$$

where E_{kin} is the total kinetic energy, E_{coll} the kinetic energy stored in collective motion, and E_{Fermi} labels the kinetic energy of Fermi motion for which we actually use an estimate from an extended Thomas-Fermi approximation.

The thermal excitation energy deposited in the electron cloud will later on be transferred to ionic degrees of freedom. The ionization yields a repulsive Coulomb force amongst the ions and the thermalized electrons will enhance this 'destabilizing effect' by heating up the ions. This coupling between electronic and ionic degrees of freedom takes place on a time scale beyond about 100 fs. A measure of the energy transferred to the ions will then simply be obtained by computing the ionic kinetic energy.

3. Examples of applications

3.1. EXCITATION BY AN IONIC PROJECTILE

We first focus on the case of excitations by a charged projectile. Such an excitation mechanism is essentially instantaneous (at least for most of the ionic velocities considered here) and we discuss the subsequent electronic response.

In figure 3 we compare the time evolution of various electronic observables: the dipole signal along the axis joining the projectile to the cluster's center of mass, the number of escaping electrons $N_{\rm esc}(t)$, and ionization probabilities. Results are shown for a typical test case of a Ar^{8+} projectile on a Na_9^+ cluster. The projectile velocity is $v = 50\ a_0/\text{fs}$, about twice the electronic Fermi velocity v_F and impact parameter is $b = 22\ a_0$. The collision is thus quite peripheral, as the radius of Na_9^+ is about $8\ a_0$. The Coulomb field of the ion acts for an extremely short time, actually less than 1 fs. At the instant of closest approach, here set at $t = 0$, one can see a sudden change of the dynamical state of the system. Large amplitude dipole oscillations are excited almost instantaneously, as can be seen in the uppermost panel of figure 3. Electron emission follows immediately and is

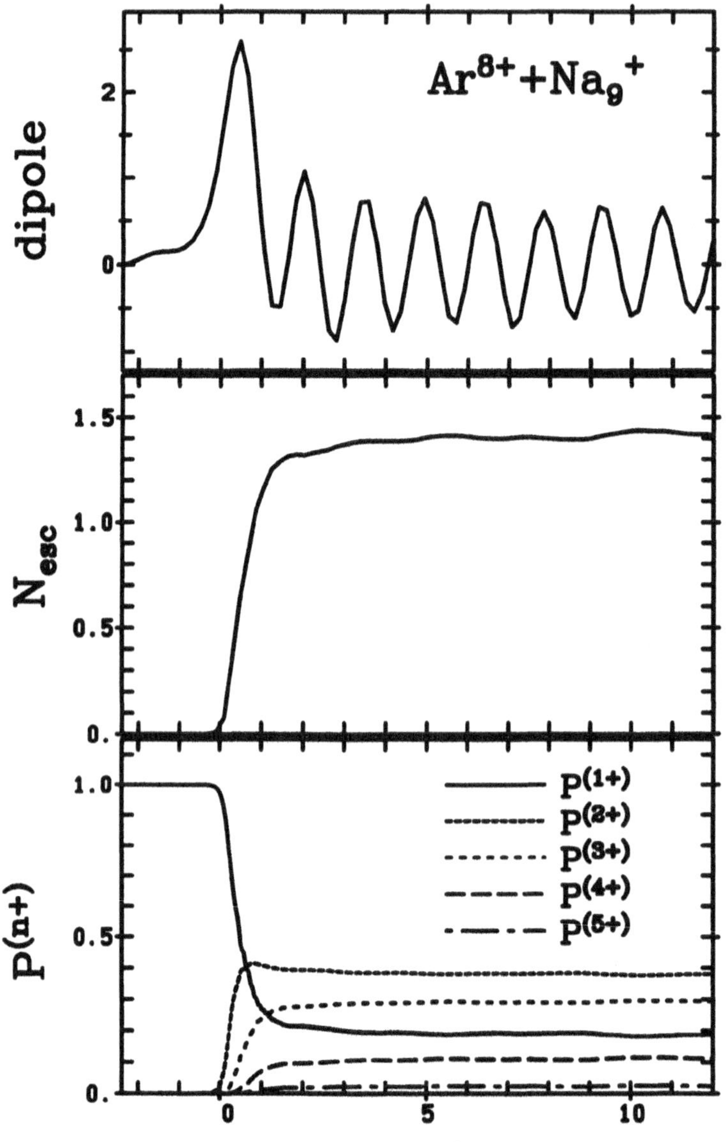

Figure 3. Example of a typical response of the electron cloud for the collision of Ar^{8+} with velocity $v = 50\,a_0/\mathrm{fs} = 2\,v_F f$ on Na_9^+ at impact parameter $b = 22a_0$. We show the dipole moment (upper part), the total number of escaped electrons (middle part) and ionization probabilities $P^{(n+)}$ (lower part) as a function of time. Ionic background is treated in the soft jellium approximation. From [32].

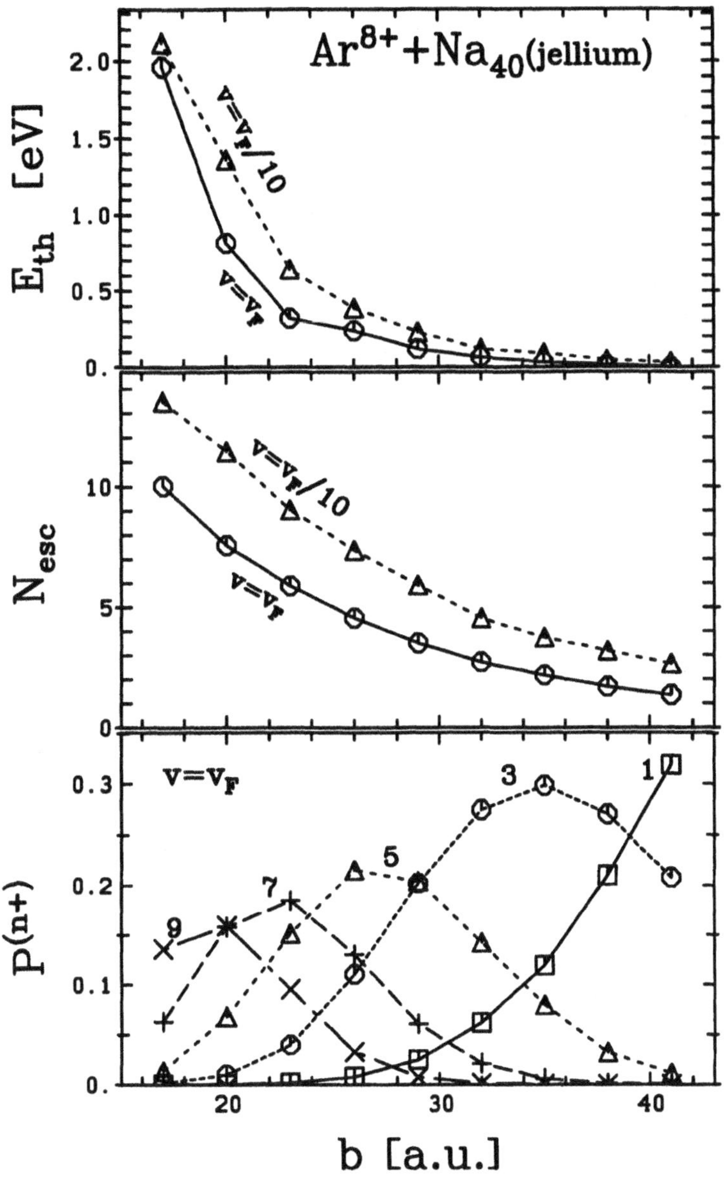

Figure 4. Trends with impact parameter for the asymptotic values of basic observables for a collision of an Ar^{8+} projectile on a Na_{40} cluster at two projectile velocities, as indicated. The upper panel shows the excitation energy, the middle panel the number of escaping electrons and the lower panel dispalys ionization probabilities in the case of $v = v_F$ velocity. All quantities are plotted as a function of impact parameters. Ionic background is treated in the soft jellium approximation.

accomplished within the time of about one plasmon cycle (less than 2 fs), as can be seen from the number of escaped electrons N_{esc} in the middle panel. The ionization probabilities (lowest panel) are built at the same fast time scale. Note also that the rapid loss of electrons causes a pronounced damping of the dipole oscillation in the earliest stage of the reaction. After this fast initial damping of the electron collective motion, the electron cloud oscillates steadily, while the ionization probabilities have stabilized at their asymptotic values.

The situation is somewhat different at lower impact velocity. In this case the electron cloud does not undergo plasmon oscillations but rather an 'adiabatic' polarization and depolarization. In turn, both energy deposit and ionization sensibly differ from the high velocity case. In order to quantify these effects we have made systematic calculations in the case of a Ar^{8+} projectile on a Na_{40} cluster at impact velocities v_F and $v_F/10$ and compared the outcoming results. This comparison is displayed in Figure 4 in which is plotted the thermal excitation energy E^* and ionization as a function of impact parameters. One can see from this figure that while the general tendencies are the same in both cases (both E^* and N_{esc} decrease with impact parameter, for the two velocities), quantitative differences are visible bewteen the low and high velocity cases. Generally speaking the low velocity case leads to more energy deposit and more ionization at a given impact parameter. But the higher the ionic velocity, the more volatile the encounter with the cluster", a trend expected from experimental investigations led with highly charged ions on clusters [5, 7].

3.2. COUPLING OF IONIC DEGREES OF FREEDOM IN LASER IRRADIATIONS

Beyond about t $\sim$ 100 fs the excitation stored in the electronic degrees of freedom is transferred to ionic motion. As already mentioned, this effect has a twofold origin: potential and thermal. The high net charge of the cluster following excitation will trigger Coulomb explosion. But this effect is enhanced by heating of the ions by the now hot electron cloud. The coupling between electrons and ions is thus beyond an adiabatic time evolution and has to be analyzed in a treatment where ions and electrons are propagated simultaneously. To this end, we complement the TDLDA equations by the ionic propagation (see Eq.(8)).

Such a non adiabatic coupled electrons+ions dynamics requires considerable numerical effort. Still our TDLDA scheme allows to perform such calculations, if not systematically, at least at a satisfactory level for the exploratory goal which we presently have. We have thus for example analysed this electrons-ions coupling in detail for a Na_{12}, with full ionic structure, irradiated by a gaussian laser pulse of 18 fs FHWM and at various frequencies

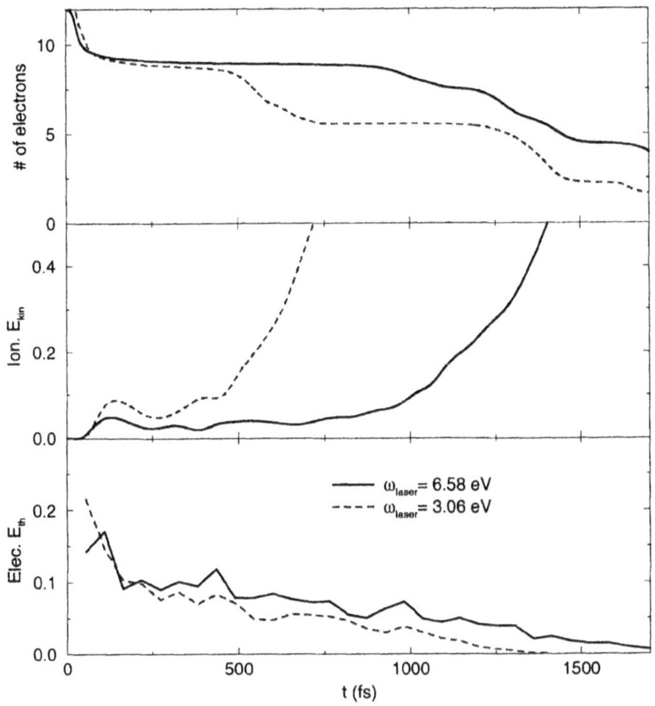

Figure 5. Time evolution of electron number $N_{\rm el}$ (upper panel), ionic kinetic energy $E_{\rm kin,ion}$ (middle, in Ry), and thermal energy of the electrons $E_{\rm th,el}$ (lower panel, in Ry). See [13] for the definition of the electronic excitation energy (lower panel) used in this calculation. From [13].

and peak intensities [13]. The laser parameters have actually been tuned to lead, at the end of the excitation phase, to the same ionization state: $\mathrm{Na}_{12} + h\nu \longrightarrow \mathrm{Na}_{12}^{3+*} + 3$ e, which allows a better comparison between the two cases. Note that the star reminds here that a non vanishing excitation energy is stored in the system (at that stage purely in the electronic degrees of freedom). The long time evolution of the system and in particular the coupling between electronic and ionic degrees of freedom is illustrated in Figure 5 [13]. From this figure one can make the following observations: i) ions are indeed frozen below about t $\sim$ 100 fs, which provides an a posteriori justification of several former calculations with frozen ions; ii) the switching on of ionic motion takes place, nevertheless, on an unexpected short time scale, between 100 and 200 fs, in any case at a time at which the electronic heat is still sizable; iii) the comparison of the 2 cases shows that while "initially" in the same 3+ charge state, the further evolution of the system strongly depends on laser parameters (upper and lower panels). This latter conclusion is confirmed by looking at the fragmentation pattern

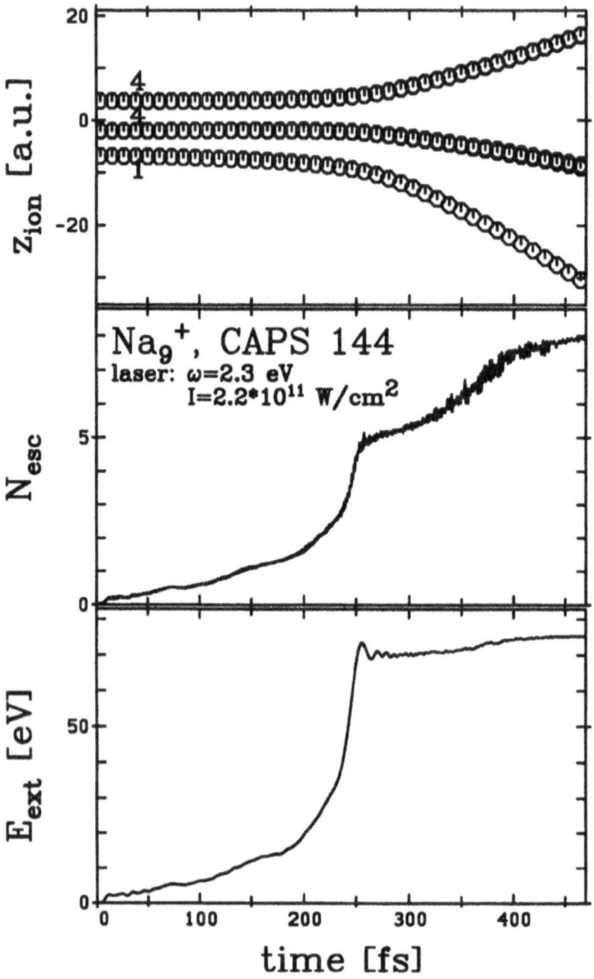

Figure 6. Time evolution of ionic structure and electronic key observables for Na_9^+ in a laser field with intensity $I = 2.2 * 10^{11}\,\mathrm{W/cm^2}$ and frequency $\omega = 2.3\,\mathrm{eV}$. Upper: The z-positions of the 9 ions with multiplicity (arrangement in rings according to CAPS) as indicated. Middle: Number of escaped electrons N_{esc}. Lower: Energy E_{ext} absorbed by the cluster from the external laser field. From [35].

in terms of ionic positions [13].

The case discussed in Figure 5 corresponds to a situation in which remains a decoupling between the excitation and the relaxation, because of the short duration of the laser pulse. The fact that we find a quick coupling to ionic motion however suggests that if using an only slightly longer laser pulse one might even spot a coupling between ionic motion and the excitation field. Ionic motion changes the resonance conditions which, in turn, has a large impact on the dynamical evolution. Experimental hints on such effects can be found in [33] for Pt. We give give here an analogous example for the simpler case of Na clusters. These effects are illustrated in Figure 6. We have chosen an illustrative case of a low frequency laser for this purpose. We start from a $Na_9{}^+$, excited by a "subcritical" laser field below resonance ($I = 2.2 * 10^{11}$ W/cm^2 and $\omega = 2.3\,\text{eV}$) with a never ending pulse. Initially, very little happens, but for a slow growth of ionization (middle panel) together with a slow absorption of external energy E_{ext} from the beam (lower panel). This continuous ionization has two competing effects: an electronic one and a ionic one. Ionization leads to a blue shift of the plasmon at fixed ionic configuration: this is a typical electronic effect [34]. In turn, ionization, on this long time scale, leads to a slow elongation of the ionic configuration, which lowers the plasmon peak. Finally, at around 250 fs the electron cloud comes into resonance with the laser beam. The effect is dramatic. One sees a sudden increase in absorbed energy (lower panel), from which a large portion is converted into direct electron emission. This enhances ionization and boosts Coulomb explosion. Note the short time interval during which all this happens (typically less than 10 fs). After that, energy absorption levels off, there is some delayed electron emission (probably thermal evaporation from the very hot cluster, figure 2), and the ions separate at a rather fast scale. This example proceeds, of course, in a particularly dramatic fashion by virtue of the chosen conditions. Other cases may evolve more gently, but in any case it is striking to see the role played by the ionic degrees of freedom in this situation.

This analysis of the coupling between electronic and ionic degrees of freedom has shown that this coupling takes place on a surprisingly short time scale of order 100 fs, at a time at which the electron cloud remains far from being relaxed on the Born Oppenheimer surface. This shows the importance of non adiabatic effects. We have even seen that ionic degrees of freedom may play a dominant role during the excitation phase itself, provided it is long enough. Furthermore, we have seen that the details of the excitation process have a strong influence on the long-time evolution, in particular on the fragmentation pattern. This is caused by a subtle interplay of various dynamical mechanisms (electron, monomer or ion emission versus fragmentation).

3.3. ELECTRON-ELECTRON DISSIPATION BEYOND MEAN-FIELD

The TDLDA approach used here for describing electronic dynamics is basically an effective mean-field theory. In particular, the possibly crucial dynamical correlations are missing. While the latters are expected to play a minor role at low excitations, they may become dominant in far from

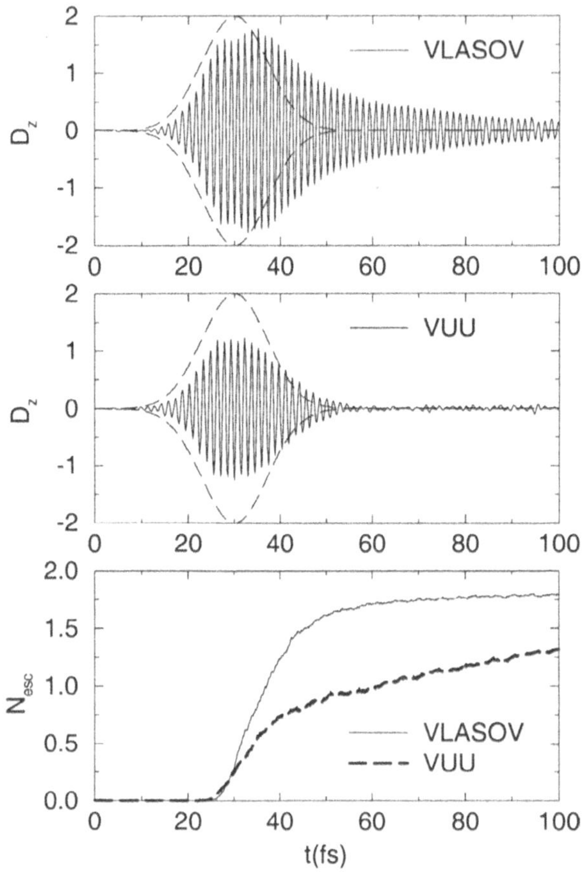

Figure 7. Dipole signal along the direction of laser polarization (D_z, upper part) and number of emitted electrons (N_{esc}, lower part) as a function of time during the interaction of a femtosecond laser pulse with Na_9^+. The laser frequency and fluency are $\hbar\omega_{\rm las} = 2.75$eV and $J = 6\,10^{11}$W/cm^2, and the pulse envelope is indicated in dashed line. Both Vlasov and VUU results are plotted for comparison. Ionic background is treated in the soft jellium approximation. From [44].

equilibrium situations (see Figure 2). It is hence of importance to evaluate the impact of these effects in realistic examples.

As known from nuclear physics, a direct extension of quantal mean-field is delicate [36]. Fortunately enough, in the high excitation regime, where semi-classical approximations are likely to become acceptable, Boltzmann-like kinetic equations, offer an efficient alternative. They thus have been extensively used in nuclear physics for describing heavy-ion collisions (VUU,

[37, 38]) in the Fermi energy domain in particular.

As a first shot to evaluate the impact of dynamical correlations on electronic dynamics, we have thus extended the semi-classical limit of TDLDA, namely the so-called Vlasov-LDA equation (already used in clusters [39, 40]) by adjoining it a VUU collision term. The cross section, which represents a key ingredient of the whole picture, has to be evaluated from a screened Coulomb potential, following the techniques used in similar cases in plasma physics [41]. We have obtained a screening length of $r_0 \sim 3\ a_0$, in agreement with bulk values for sodium [42]. The resulting VUU equation is then solved using the full ensemble technique [43].

An example of the impact of the VUU collision term on electron dynamics is shown in Figure 7. We have plotted here the dipole response of a Na_9^+ cluster irradiated by a laser pulse with peak intensity $6 \times 10^{11} W/cm^2$ and a frequency matching the plasmon resonance. This figure displays also the number of escaping electrons N_{esc} as a function of time. The comparison of VUU and pure Vlasov calculations shows a significantly stronger damping of the dipole in the case of VUU. This actually corresponds to the fact that more excitation energy has been stored in the VUU case (8 versus 4 eV). The behavior of N_{esc} is also worth a comment. One observes a reduced ionization in the case of VUU, at least on short times, which is another manifestation of the strong dipole damping in VUU. On longer times, while ionization levels off in the Vlasov case, it continues steadily in the VUU case, but its origin is now thermal. Altogether, these various observations hint that dynamical correlations might play an important role in the further global evolution of the system (see section 3.2).

The example of Figure 7 is essentially one illustration. We have performed a systematic analysis of the impact of the VUU collision integral on electron dynamics under various conditions, mainly as a function of the nature and duration of the excitation process [44]. The general findings are that the effect on ionization strongly depends on the duration of the excitation, but that one always finds an enhanced energy deposit due to electron-electron collisions. The effect on the dipole signal, although qualitatively important (see Figure 7) usually leads to little consequences in terms of stored energy. In any case, it is clear that dynamical correlations have non-negligible effects in the regime of very high excitations.

4. Conclusions and Perspectives

In this contribution, we have reviewed some patterns of the nonlinear dynamical response of metal clusters following strong excitations by ion collisions or femtosecond laser beams. We summarize quickly some of the most prominent features. Both, fast-ion collisions as well as femtosecond laser excitations, leave the cluster in about the same dynamical state having produced a comparatively large amount of ionization with minimum internal heating and thus constitute very efficient means of ionization. The excitation dynamics and subsequent evolution is far off equilibrium and cannot be treated by adiabatic methods (Born-Oppenheimer or related). The TDLDA approach offers here a very powerful tool, which can even be extended to account for dynamical correlations.

The few examples of this paper have shown many interesting effects and hinted further variants of the excitation mechanisms to be explored. It is to be remarked that these investigations are yet in an early stage and that we need many more systematic explorations of the richness of phenomea in this field, both at theoretical and experimental levels. It should finally be noted that numerous potential applications of these questions are showing up in related fields of physics, in particular in connection with surface and material science. Relations with these other fields are also currently being explored.

Acknowledgements

The results of this paper have been obtained in various collaborations. We thank M. Brack, F. Calvayarac, A. Domps, C. Kohl, S. Kümmel, and C. Ullrich for contributing to this work. We thank, furthermore, for many inspiring and motivating discussions with G. Bertsch, C. Guet, G. Gerber, H. Haberland, K.H. Meiwes-Broer, L. Schweikardt, and S. Voll. We also thank exchange program PROCOPE number 95073 and Institut Universitaire de France for financial support during the realization of this work.

References

1. W. DeHeer, Rev. Mod. Phys. **65** (1993) 611.
2. M. Brack, Rev. Mod. Phys. **65** (1993) 677.
3. *Large clusters of atoms and molecules*, NATO ASI **E313**, ed. by T. P. Martin, Kluwer, Dordrecht, 1996.
4. *Small particles and inorganic clusters*, ISSPIC8, ed. by H. H. Andersen, Zeit. Phys. **D40** (1997) 1-578.
5. U. Näher et al, Phys. Rep. **285** (1997) 245.
6. P. G. Reinhard and E. Suraud, J. Clust. Sci. (1999).
7. F. Chandezon et al, Phys. Rev. Lett. **74** (1995) 3784.
8. T. Baumert and G. Gerber, Adv. Atom. Molec. Opt. Phys. **35** (1995) 163.
9. R. Schlipper et al, Phys. Rev. Lett. **80** (1998) 1194.
10. V. Weisskopf, Phys. Rev. **52** (1937) 295.
11. J. Babst and P. G. Reinhard, Z. Phys. **D42** (1997) 209.
12. A. Domps, E. Suraud and P. G. Reinhard, Phys. Rev. Lett. **81** (1998) 15524.
13. F. Calvayrac, P. G. Reinhard and E. Suraud, J. Phys. **B31** (1998) 5023.
14. V. Bonačić-Koutecký, P. Fantucci and J.Koutecký, Chem. Rev. **91** (1991) 1035.
15. *Time dependent molecular dynamics*, ed. by J. Broeckhove and L. Lathouwers, Plenum, New York, 1992.

16. E. K. U. Gross et al, Top. Curr. Chem. **181** (1996) 81.
17. R. N. Barnett et al, J. Chem. Phys. **94** (1991) 608.
18. R. Car and M. Parinello, Phys. Rev. Lett. **55** (1985) 2471.
19. K. Yabana et al, Phys. Rev. **B54** (1996) 4484.
20. U. Saalmann and R. Schmidt, Z. Phys. **D38** (1996) 153.
21. O. Gunnarsson and B. I. Lundqvist, Phys. Rev. **B13** (1976) 4274.
22. J. P. Perdew and Y. Wang, Phys. Rev. **B45** (1992) 13244.
23. C. Ullrich, P. G.Reinhard and E. Suraud, J. Phys. **B31** (1998) 1871.
24. F. Calvayrac, P. G. Reinhard and E.Suraud, Ann. Phys. (N.Y.) **255** (1997) 125.
25. G. Lauritsch and P. G. Reinhard, Int. J. Mod. Phys. **C5** (1994) 65.
26. F. Calvayrac, P. G. Reinhard and E. Suraud, J. Phys. **B31** (1998) 1367.
27. G. B. Bachelet, D. R. Hamann and M. Schlüter, Phys. Rev. **B26** (1982) 4199.
28. S. Kümmel, M. Brack and P. G. Reinhard, *Proceedings of ISSPIC9*, Lausanne, Switzerland, September 1998, in press.
29. L. Verlet, Phys. Rev. **159** (1967) 98.
30. O. Genzken, P. G. Reinhard and M. Brack, Ann. Phys. (Leipzig) **5** (1996) 576.
31. J. M. Eisenberg and W. Greiner, *Nuclear Theory vol. 2*, North-Holland, Amsterdam, 1970.
32. P. G. Reinhard, E. Suraud and C. Ullrich, Euro. Phys. J. **D1** (1998) 303.
33. L. Köller, M. Schumacher, J. Köhn, S. Teuber, J. Tiggesbäumker and K.-H. Meiwes-Broer, Preprint, Rostock, 1998.
34. C. A. Ullrich, P. G. Reinhard and E. Suraud, J. Phys. **B30** (1997) 5043.
35. M. Brack et al, *Proceedings of ISSPIC9*, Lausanne, Switzerland, September 1998, in press.
36. K. Goeke and P. G. Reinhard (eds.), *TDHF and beyond*, Lect. Notes in Phys. **171**, Springer, Berlin, 1982.
37. G. F. Bertsch et al, Phys. Rep. **160** (1988) 189.
38. A. Bonasera, F. Gulminelli and J. Molitoris, Phys. Rep. **243** (1994) 1.
39. M. Gross and C. Guet, Z. Phys. **D33** (1995) 289.
40. L. Féret et al, J. Phys. **B29** (1996) 4477.
41. S. Ichimaru et al, Phys. Rep. **149** (1987) 91.
42. D. Pines and P. Nozieres, *The theory of quantum liquids* (Addison-Wesley).
43. G. Welke et al, Phys. Rev. **C40** (1989) 2611.
44. A. Domps, P. G. Reinhard and E. Suraud, Ann. Phys. (N.Y.) (1999), in press.

RELATIVISTIC EFFECTS IN NON-LINEAR ATOM-LASER INTERACTIONS AT ULTRAHIGH INTENSITIES

V. VÉNIARD, R. TAÏEB, C. SZYMANOWSKI AND A. MAQUET
Laboratoire de Chimie Physique-Matière et Rayonnement
Université Pierre et Marie Curie
11, rue Pierre et Marie Curie - 75231 Paris Cedex 05, France

Abstract. We address the question of the importance of relativity in selected non-linear radiative processes to be observed when an atomic system is submitted to an intense radiation field. To this end, we report on recent results obtained in the theoretical analysis of the following processes: i) two-photon transitions in high-Z hydrogenic systems, ii) laser-assisted Mott scattering of fast electrons, and iii) two-colour photoionisation spectra of atoms in the simultaneous presence of two ultra-intense laser fields. In each case, the signature of relativistic effects is evidenced.

1. Introduction

Relativity is expected to play an important role in several types of radiative processes in atoms. Its influence on the atomic levels fine structure has been most thoroughly investigated as its signature is easily evidenced in atomic x-ray spectra, [1], [2], [3]. It manifests itself also in some delicate aspects of the chemical reactivity of the elements, [4]. These effects arise from both the standard Dirac-like properties of electrons and from more sophisticated QED corrections. One of the major objective of the present paper is to show that the overall picture has dramatically changed recently, as a consequence of the considerable advances made in the design of ultra intense laser sources operated at intensities well beyond the so-called atomic unit of intensity $I_0 = 3.5 \times 10^{16}\, W/cm^2$, [5].

These advances have opened new fields of relativity-related research in two complementary directions. One is related to the advent of laser-based sources of coherent radiation in the X-UV domain, either from high harmonic generation [6], [7] or from X-ray-laser devices, [8]. The imple-

J. Maruani et al. (eds.), New Trends in Quantum Systems in Chemistry and Physics, Volume 2, 107–118.

mentation of these sources will make feasible, for the first time, to realize non-linear (multiphoton) optics experiments in the X-UV and X-ray ranges. Considerable advances are expected in the field, much in the spirit of those made in the IR-Visible ranges with standard laser sources. The question then arises of the importance of relativistic and of the (closely related) retardation effects on multiphoton transition amplitudes in high-Z atoms. This question has been the object of active investigations in the field of standard (one-photon) X-ray spectroscopies, [9]. As we will show below, for two-photon bound-bound transitions, see (Sec. 2), the magnitudes of the corrections differ significantly from what one could naively infer from the analysis of the one-photon results.

Another kind of processes in which intense lasers have brought relativity into play in atomic physics, results from the fact that even an electron initially at rest can acquire, within the laser field, a ponderomotive energy (i.e. its averaged oscillating energy) comparable to its rest mass energy. In addition, as a result of the dressing by the field, its mass can be significantly shifted from its bare value, [10]. As we shall show below these two combined effects can considerably modify the dynamics of typical atomic processes such as electron-atom collisions, (Sec. 3), and stronf field photoionization, (Sec. 4).

2. Two-photon bound-bound transitions in hydrogenic atoms

An intriguing observation made in several areas of X-ray spectroscopy (XAS, XES and even RIXS) is that the Non-Relativistic Dipole (NRD) approximation provides fairly accurate values for the oscillator strengths, well beyond its expected range of validity, [9]. For instance, for the element Sn, ($Z = 50, \alpha Z \approx .365$), the value of the K-shell photoionisation cross section deduced from a fully relativistic MCDF computation differs from the Non-Relativistic Dipole (NRD) value by less than 5% at photon energies up to $100\,keV$!, [9]. A commonly advanced explanation for this observation is that two distinct (though deeply entangled) effects may partially compensate each other: one arises from the contribution of retardation which becomes important at higher frequencies, i.e. when the momentum carried by the photon cannot be neglected. The other effect is a consequence of the modifications of the atomic structure under the combined effects of spin and nuclear charge. It seems that in most one-photon processes these two classes of contributions do partially compensate each other. The question is then of determining if whether this situation holds also for two-photon transitions in high-Z elements.

In order to address this question we have considered the simpler case of two-photon bound-bound transitions originating from the ground state of point-nucleus hydrogenic systems. For such systems it is possible to perform an *exact* calculation of the relevant second-order T-matrix element :

$$T^{(2)}_{f,i} = \langle\psi_f|VG(\mathbf{r},\mathbf{r}';\Omega)V|\psi_i\rangle, \quad (1)$$

where $V = \alpha.\mathbf{A}$, α,$\mathbf{A}$ denoting Dirac matrices and the potential vector associated to the electric field, respectively, and $G(\mathbf{r},\mathbf{r}';\Omega)$ is the Dirac Coulomb Green's function. Such second-order matrix elements can be computed conveniently to within a given accuracy with the help of a Sturmian expansion of $G(\mathbf{r},\mathbf{r}';\Omega)$, [11], [12]. When relativity is taken into account, the Z-dependence of the transition amplitude is intricate, as the nuclear charge enters in an complicated way into the expression of the relevant wave functions and Green's function. In contrast, in the Non-Relativistic Dipole (NRD) limit, it can be shown that the second-order matrix elements scale as Z^{-4}.

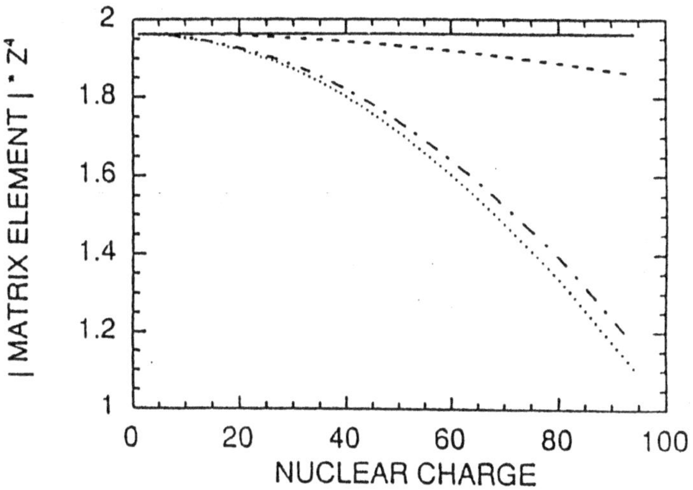

Figure 1. Modulus of the matrix elements M for the transition $|1\,^2S_{1/2}\rangle \rightarrow |2\,^2S_{1/2}\rangle$ in atomic units multiplied by a factor Z^4 as a function of nuclear charge Z. The solid line represents the nonrelativistic dipole (NRD), the dashed line the nonrelativistic retarded (NRR), the dot-dashed line the relativistic dipole (RD), and the dotted line the relativistic retarded (RR) result.

Accordingly, in order to clearly display the difference between the NRD approximate results and those deduced from more refined treatments, we have reported the scaled modulus $|T^{(2)}_{f/i}| \times Z^4$ in terms of Z. One of the main outcomes of the analysis, when performed for the most typical two-photon $|1^2S_{1/2}\rangle \rightarrow |2^2S_{1/2}\rangle$ transition, is that, as compared to the NRD approximation, relativistic contributions amount to well over 10% to the scaled second-order T-matrix element for $Z = 50$, i. e. for photon energies $\hbar\omega \approx 12,75\,keV$, see Figure 1. Moreover, one observes that the respective contributions of retardation and of the relativistic effects do not compensate each other. We have checked that this observation holds also for other two-photon bound-bound transitions, see [11]. To conclude, our model computations, performed for a well defined test-system, clearly indicate that the conclusions drawn from the analysis of one-photon transition amplitudes cannot be transposed to the case of multiphoton processes.

3. Laser-assisted Mott scattering

When a relatively dense medium (atomic clusters, for instance) is irradiated by a strong IR laser, fast electrons are produced very early during the rise of the pulse. Later in the pulse, these electrons can be scattered by ionic species while the laser has a very high intensity. It is thus of interest to determine to which extent relativistic effects will affect the dynamics of the laser-assisted collisions and, more generally, the exchange of energy between the electrons and the field. In order to address these questions, we have considered the simpler case of Coulomb scattering for a Dirac electron (the so-called Mott scattering) in the presence of a circularly polarized, single-mode, laser field. The main advantage of the model is that, though much simplified, its limitations are well delineated and, as a first-Born treatment of the collision stage can be performed exactlly, it leads to analytical expressions which can be analysed and discussed, [13]. The solutions of the Dirac equation for a free electron in the presence of a circularly polarized classical field, are the so-called Volkov wave functions:

$$\psi_q = \langle \vec{x}|q\rangle = \left[1 + \frac{\not{k}\,\not{A}}{2c(kp)}\right] \times \frac{u}{\sqrt{2QV}}\ \exp[-i\,(qx) - i\int_0^{kx} \frac{(pA)}{c(kp)}\,d\phi]\,, \quad (2)$$

where u is the bispinor, solution of the Dirac equation for a free electron with four-momentum p. Here q is the averaged four-momentum of the electron within the field:

$$q^\mu = (\frac{Q}{c}, \vec{q}) = p^\mu - \frac{\overline{A^2}}{2(kp)c^2}k^\mu, \quad (3)$$

where, $\overline{A^2}$ denotes the time-averaged square of the four-vector potential of the laser field. The square of this four-vector is Lorentz invariant.

The first-order Born transition amplitude for the Coulomb scattering of a Volkov electron reads:

$$T_{f/i} = \frac{iZ}{c}\int d^4x \overline{\psi_f}\frac{\gamma^0}{|\vec{x}|}\psi_i. \quad (4)$$

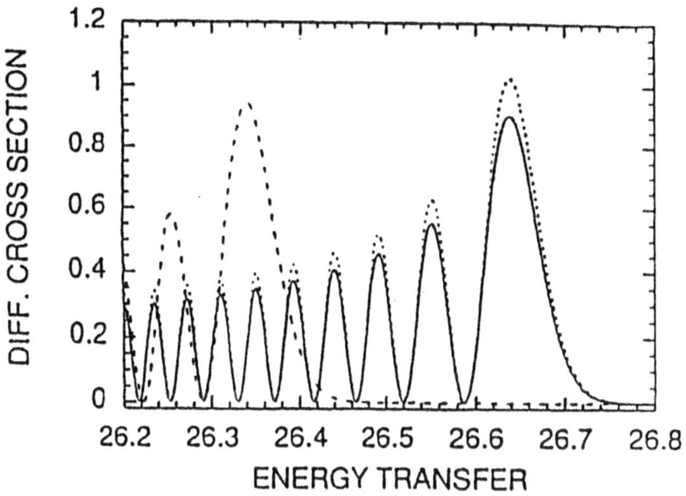

Figure 2. Envelope of the absorption edge part of the differential cross section $d\sigma/do$ scaled in 10^{-9} a.u. as a function of energy transfer scaled in units of $1000\,\omega$ for an electrical field strength of $E = 1$ a.u. or vector potential $A = 3186$ a.u.. The initial electron energy is $W = 100$ a.u.. The solid line denotes the result for electrons, the short dashed one the differential cross section for spinless particles and the long dashed one the result for the nonrelativistic limit.

Though cumbersome, the calculation can be carried out exactly, the final result being given as a complicated expression in terms of Bessel functions [13], see also [14]. We note that the final formula reduces, in the non relativistic limit, to a known expression first derived in the sixties by Bunkin and Fedorov, [15].

The influence of relativity on the differential cross section is clearly evidenced in the high intensity regime, for scattering events in the course of which large numbers of IR photons are exchanged. For instance, for incoming $2.7keV$ electrons, at $I = I_0 = 3.5\times10^{16}W/cm^2$ up to $\approx 2.6\times10^4$ photons from a Nd: Yag laser can be absorbed in a 90° scattering event. Under such circumstances, noticeable differences are already observed between the nonrelativistic and relativistic treatments, see Fig. 2. The observed differences can be ascribed to the mass-shift experienced by the electron embedded within the field, [10]:

$$m_* = m\sqrt{1 + \frac{\langle A^2\rangle}{c^4}}. \tag{5}$$

This is, of course, amplified at $I = 1.2\times10^{18}\ W/cm^2$, i.e. at an intensity well beyond the atomic unit I_0. Then, spin-dependent effects begin to show

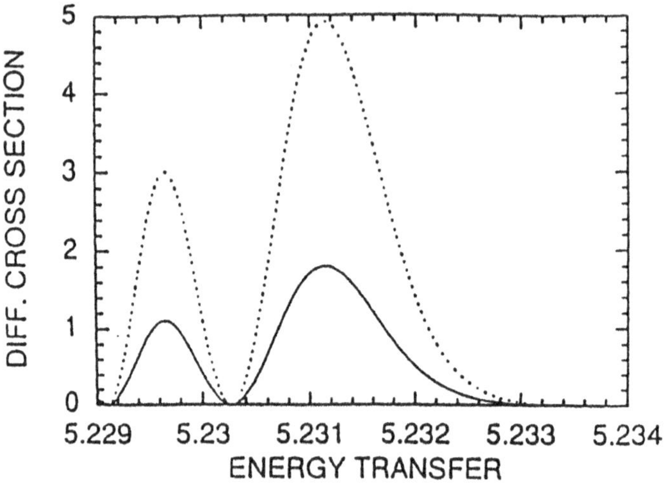

Figure 3. Envelope of the absorption edge part of the differential cross section $d\sigma/do$ scaled in 10^{-14} a.u. as a function of energy transfer scaled in units of $10^5\omega$ for an electrical field strength of $E = 5.89$ a.u. or vector potential $A = 18769$ a.u. . The solid line denotes the result for electrons and the dashed line sketches the result for spinless particles.

up, as a result from both the standard spin-orbit coupling and from the coupling with the magnetic component of the field, see Fig. 3.

In conclusion, it appears that relativity can affect significantly the dynamics of laser- assisted collisions at intensities achieved by several laser sources currently operated worldwide, [13].

4. Relativistic effects in photoionization spectra

A weakly bound electron submitted to an intense pulse of IR radiation can be ionized by a variety of mechanisms, from multiphoton absorption to tunnel or even "over the barrier" ionization, depending of the intensity regime, [16]. At higher intensities tunnel ionization and over the barrier processes are dominant and, as a consequence, the ejected electrons are initially relatively slow. This implies that the changes induced by relativity in the corresponding photoionization spectra are expected to be small.

The situation can differ significantly in the presence of a pulse of UV or X-ray radiation. Indeed, a weakly bound electron submitted to a UV pulse is readily ionized, via one-photon absorption, even at low or moderate peak intensities. However, if the peak intensity of the radiation field is increased, several model simulations show that, very counter-intuitively, the ionization yields can *decrease*, [17]. This situation can be pictured as resulting of a *dichotomy* of the electron wave function strongly driven away from the nucleus in the high frequency field. The more intense is the field, the farther from the nucleus is the electron, thus accounting for the "stabilization" of the dressed atom. We have verified that this phenomenon does survive in the relativistic regime.

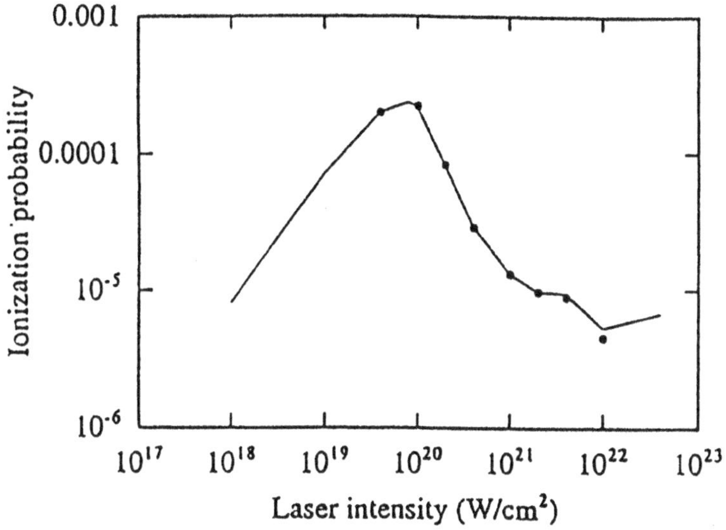

Figure 4. Ionization yield as a function of laser intensity for a radiation pulse with a linear turn-on of $100T_H$. The field frequency is $\omega_H = 10a.u.$. These yields are computed at $t = 1100T_H$. The line represents the results for the time-dependent Schrödinger equation treatment while the dots are the results of the Klein-Gordon equation treatment.

This is shown in the Figure 4, where ionization yields for a simplified model atom are reported as a function of the peak intensity of the field. Here the data are obtained from both time-dependent Schrödinger and Klein-Gordon treatments for a 1-dimensional "soft-Coulomb" potential,[18]:

$$V(x) = -\frac{1}{\sqrt{2+x^2}}. \tag{6}$$

with ground state binding energy $E_0 = -0.5\,a.u.$. For the sake of illustration, we have chosen the electric field envelope with linear turn-on of 100 high-frequency cycles $T_H = 2\pi/\omega_H$ durations. The dressing high frequency is $\omega_H = 10\,a.u.$, with field strength intensities around $F_{0H} \approx 10^3\,a.u.$, the corresponding excursion length being : $\alpha_0 = qF_{0H}/(m\omega^2) \approx 10\,a.u.$. These yields are computed at $t = 1100T_H$

Two interesting comments are in order: first, the stabilization phenomenon, originally demonstrated within the framework of the Floquet formalism, i.e. for constant amplitude fields, is also observed in numerical simulations for an explicitly time-dependent field. Regarding the role of relativity it should be understood that it is a qualified statement, as it has been established in a one-dimensional system which cannot account for spin and retardation effects. Up to know, it is not clear if such effects would affect the general trend observed here, [16]. Another remarkable observation is that the ionization yield is almost the same whether it is computed from a non-relativistic or from a fully relativistic treatment. This holds even at ultra-high intensities at which the dynamics of a free electron is certainly relativistic. Our results clearly indicate that, for a weakly bound electron, relativity does not affect quantitatively the dynamics of the ionization stage even in an ultra-strong field.

However, it is possible to probe such a relativistic dynamics with a second field with a lower frequency, so that multiphoton absorption, leading to ATI, can take place. In order to ionize such a stabilized atom with a significant probability, the second field must force the electron wave function to explore again the vicinity of the nucleus to be able to absorb energy (i.e. photons). This can be achieved by chosing parallel polarizations and field strength intensities and frequencies such that the characteristic excursion lengths α_0 for each field are comparable. We have chosen a "low" frequency $\omega_L = 0.2\, a.u.$ with a field strength intensity being around $F_{0L} \approx 0.3\, a.u.$. We note that the contribution of this additional field to the effective mass and to the relativistic parameter $\eta = \frac{qF_{0H}}{mc\omega_H}$ are negligible.

In the figure 5, we have compared the photoelectron energy spectrum lines as obtained from a non-relativistic (time-dependent Schrödinger) and from a relativistic (Klein-Gordon) treatment, [19]. The fields frequencies and intensities are $\omega_H = 10\, a.u.$, $I_H = 5 \times 10^{21}\, W/cm^2$ ($\eta = 0.28$), $\omega_L = 0.2\, a.u.$ and $I_L = 10^{15}\, W/cm^2$ The most salient feature is the notable difference observed in the locations of the ATI peaks. The non relativistic simulation displays regularly spaced peaks, separated from each other by $\omega_L = 0.2\, a.u.$, corresponding exactly to the frequency of the probe field. In contrast, the relativistic simulation reveals photoelectron peaks with spacings of $\omega' = 0.21\, a.u. = 1.05\omega_L$. We mention that a simulation (not shown), see [19], for a stronger high frequency field, $I_H = 4 \times 10^{22} W/cm^2$ ($\eta = 0.8$), exhibits ATI line spacing of $\omega' = 0.28\, a.u. = 1.24\omega_L$.

The observed spacing cannot be explained using a simple "Doppler shift" argument because the exchange takes place within the high frequency dressing field, i.e. while, not only the electron experiences a periodically accelerated motion but while it has also acquired an effective mass. As the ATI spectrum is recorded after the end of the pulse, one has to analyze how the absorbed energy is transferred into the kinetic energy of the detected "bare" electron. To this end, we model the process with the use of a "free" gaussian wavepacket, "dressed" by the high frequency relativistic field, the multiphoton absorption from the low frequency field being simulated by

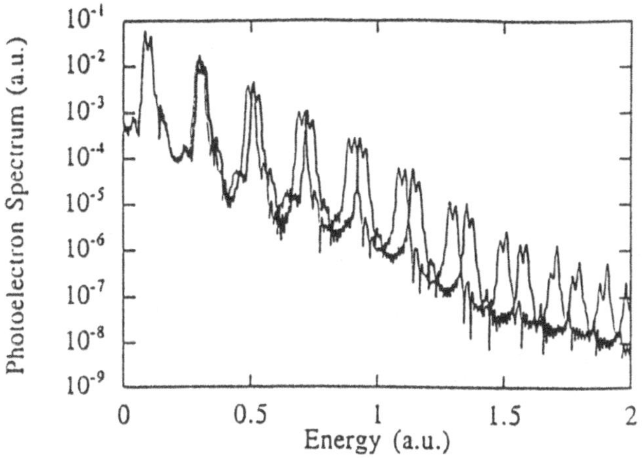

Figure 5. Two-color photoelectron spectrum for a radiation pulse containing two frequencies $\omega_H = 10\,a.u.$ and $\omega_L = 0.2\,a.u.$, with intensities $I_H = 5 \times 10^{21}\ W/cm^2$ and $I_L = 10^{15}\ W/cm^2$. The total duration of the high frequency pulse is $T = 28T_L$. The spectrum is obtained by solving the time-dependent Klein-Gordon equation (thick line) and for the sake of reference we have shown (thin line) the one obtained by solving the time-dependent Schrödinger equation.

a "kick" of magnitude p_0 in momentum, taking place at a random time during the pulse. The transferred energy will be, [19]:

$$\langle \Delta\mathcal{E}^t_{KG} \rangle \approx \frac{p_0^2}{2m} \langle \left[1 + \frac{q^2 A_H(t_0)^2}{m^2 c^4} \right]^{-3/2} \rangle, \tag{7}$$

while the detected energy is given by : $\langle \Delta\mathcal{E}^d_{KG} \rangle \approx \frac{p_0^2}{2m}$.

This implies that, as the transferred energy is a multiple of the low-frequency quantum, the detected electrons have kinetic energies as if they had absorbed photons with shifted frequencies :

$$\omega' \approx \frac{1}{\langle \left[1 + \frac{q^2 A_H(t_0)^2}{m^2 c^4} \right]^{-3/2} \rangle} \omega_L. \tag{8}$$

It is worth mentioning that there is no equivalent frequency shift within the framework of the Schrödinger theory.

Numerical values of the frequency shift computed from this formula are shown in Fig. 6, where it can be verified that they coincide with the one deduced from the spectral analysis of the final wave function.

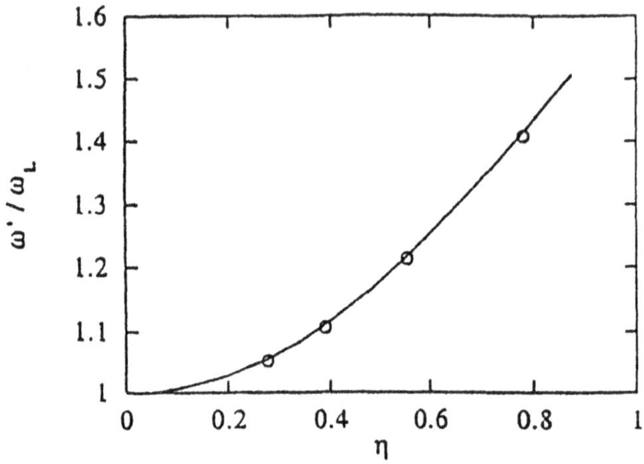

Figure 6. Frequency shift ω'/ω_L, Eq. (11), as a function of the relativistic parameter $\eta = (qF_{0H})/(mc\omega_H)$, with $\omega_H = 10\,a.u.$. The circles indicate the shift deduced from the full time-dependent calculation.

5. Conclusions

In the present paper, we have reported the results of very recent computations, either analytical or fully numerical, of the transition amplitudes and/or of the cross sections for different phenomena which are expected to be observed when atomic systems experience non-linear radiative processes in intense external fields. In each case, we have addressed the question of the relative importance of the effects of relativity and we have made explicit the conditions in which the signatures of relativistic effects and retardation could be evidenced.

We have considered in particular the case of multiphoton transitions, to be observed with the help of intense high frequency fields as produced by X-ray Lasers or Free-Electron Lasers (FEL). As a result of our analysis, we have shown that two-photon bound-bound transition amplitudes in high-Z hydrogenic systems are significantly affected by relativistic corrections, even for low values of the charge of the nucleus. For instance at $Z = 20$, the corrections amount to about 10%, a value much higher than what is observed for standard one-photon transitions in X-ray spectroscopy measurements for which the non-relativistic dipole (NRD) approximation agrees with the exact result to within 99% at comparable frequencies.

Relativity affects also significantly the scattering cross section of fast electrons in the Coulomb field of a nucleus, in the presence of an ultra-strong infrared laser. When the laser intensity becomes comparable or larger than the so-called atomic unit of intensity $I_0 = 3.5 \times 10^{16}\, W/cm^2$, notable corrections are observed in addition to the standard influence of the spin-orbit coupling which gives rise to the Mott scattering cross section. For a Dirac electron, the coupling of the spin with the magnetic component of the field contributes also to the changes observed in the energy spectra of the scattered electrons.

Eventually, we have shown also that in the presence of ultra-intense laser fields the dynamics and the amount of energy absorbed by an atom can be significantly modified by relativistic effects, when compared to the results of non relativistic simulations. Such effects are expected to be observable in some selected cases clearly identified.

Acknowledgements

The Laboratoire de Chimie Physique-Matière et Rayonnement is a Unité Mixte de Recherche, Associée au CNRS, UMR 7614, and is Laboratoire de Recherche Correspondant du CEA, LRC # DSM-98-16. Parts of the computations have been performed at the Centre de Calcul pour la Recherche (CCR, Jussieu, Paris) and at the Institut du Développement et des Ressources en Informatique Scientifique (IDRIS).

References

1. Grant, I.P., J. Phys. B: At. Mol. Opt. Phys. **7** (1974), 1458.
2. Desclaux J.P., Comp. Phys. Comm. **9** (1975), 31.
3. Bruneau, J., J. Phys. B: At. Mol. Opt. Phys. **16** (1983), 4135.
4. a recent reference is : Doll K., Pyykkö P. and Stoll H., J. Chem. Phys. **109** (1998), 2339.
5. Mourou, G.A., Barty, C.P.J., and Perry, M.D., Physics today **51** (1998), 22.
6. Recent references include : Zerne, R. *et al.* Phys. Rev. Lett. **79** (1997), 1006; Chang, Z. *et al.* Phys. Rev. Lett. **79** (1997), 2967; Spielmann, Ch. *et al.* Science **278** (1997), 661; Schnürer, M. *et al.* Phys. Rev. Lett. **80** (1998), 3236.
7. Bellini, M. *et al.* Phys. Rev. Lett. **81** (1998), 297.
8. Recent references can be found in: *X-Ray Lasers 1996*, Institute of Physics Conferences Series **151**, (1996).
9. Ron, A., Goldberg, I.B., Stein, J., Manson, S.T., Pratt, R.H., and Yin, R.Y., Phys. Rev. A **50** (1994), 1312.
10. Brown, L.S., and Kibble, T.W.B., Phys. Rev. **133** (1964), A705.
11. Szymanowski, C., Véniard, V., Taïeb, R., and Maquet, A., Phys. Rev. A **56** (1997), 700.
12. Maquet, A., Véniard, V., and Marian, T, J. Phys. B: At. Mol. Opt. Phys. **31** (1998), 3743.
13. Szymanowski, C., Véniard, V., Taïeb, R., Maquet, A., and Keitel, C.H., Phys. Rev. A **56** (1997), 3846.
14. Denisov, M.M., and Fedorov, M.V., Sov. Phys. JETP, **26** (1968), 779.
15. Bunkin, F.V., and Fedorov, M.V., Sov. Phys. JETP, **22** (1966), 974.
16. Protopapas, M., Keitel, C.H., Knight, P.L., Rep. Prog. Phys. **60** (1997), 389-485.
17. Gavrila, M., in *Atoms in Intense Laser Fields*, edited by M. Gavrila, Adv. Atom. Molec. Opt. Phys. Suppl. **1** (Acad. Press, San Diego, 1992), pp 435.

18. Q. Su and J. H. Eberly, Phys. Rev. A **44** (1991), 5997 and references therein.
19. Taïeb, R., Véniard, V., and Maquet, A., Phys. Rev. Lett. **81** (1998), 2882.

Part VII

Reactive Collisions and Chemical Reactions

SEMICLASSICAL CLOSE-COUPLING DESCRIPTION OF ELECTRON TRANSFER IN MULTICHARGED ION-ATOM COLLISIONS

J. CAILLAT AND A. DUBOIS
Laboratoire de Chimie Physique–Matière et Rayonnement
Université Pierre et Marie Curie
11, rue Pierre et Marie Curie - 75231 Paris Cedex 05, France

AND

J.P. HANSEN
Institute of Physics, University of Bergen
Allégaten 55, N-5007 Bergen, Norway

Abstract. Cross sections for electron transfer in collisions of atomic hydrogen with fully stripped carbon ions are studied for impact energies from 0.1 to 500 keV/u. A semi-classical close-coupling approach is used within the impact parameter approximation. To solve the time-dependent Schrödinger equation the electronic wave function is expanded on a two-center atomic state basis set. The projectile states are modified by translational factors to take into account the relative motion of the two centers. For the processes $C^{6+}+H(1s) \longrightarrow C^{5+}(nlm) + H^{+}$, we present shell-selective electron transfer cross sections, based on computations performed with an expansion spanning all states of $C^{5+}(n$=1-6) shells and the H(1s) state.

1. Introduction

Modeling and diagnostics of laboratory and astrophysical plasmas require accurate informations on the cross sections for the inelastic processes occurring during the collisions between multiply-charged ions and neutral atoms. In fusion plasmas, the knowledge of such cross sections for the collisions of atomic hydrogen with the fully stripped plasma impurity ions (commonly carbon and oxygen nuclei) is of great importance for impact energies from 0.1 to 1000 keV/u.

J. Maruani et al. (eds.), New Trends in Quantum Systems in Chemistry and Physics, Volume 2, 121–131.

Semi-classical models to describe the electronic processes occuring during heavy particle collisions [1]-[5] are appropriate for impact energies above the eV range, as the de Broglie wavelength associated with the internuclear motion is very small in comparison with the size of the scattering region. The relative motion of the nuclei may therefore be described by a classical approach, whereas a quantum treatment is required for the electrons.

Atomic units are used, except where stated otherwise.

2. Theory

2.1. THE IMPACT PARAMETER CLOSE-COUPLING METHOD

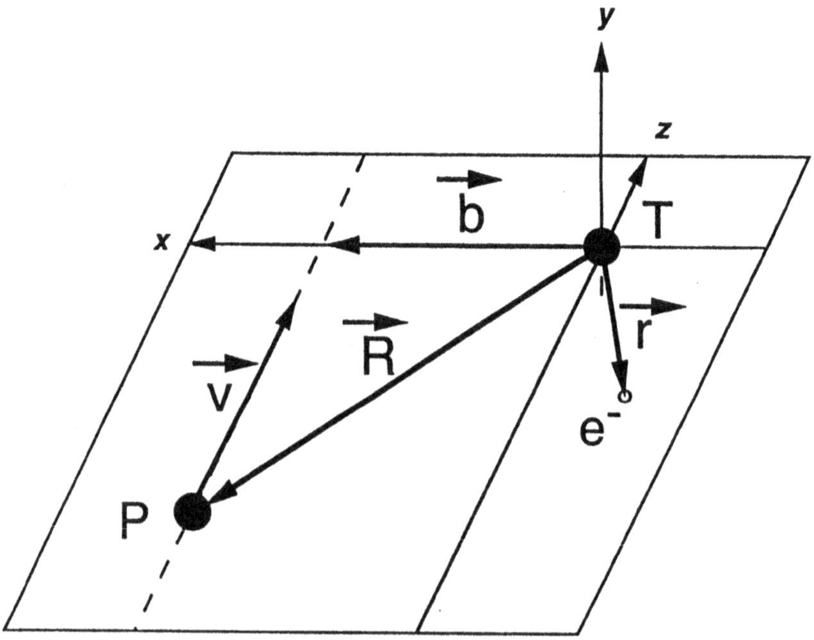

Figure 1. Collision system and coordinate frame: **T** and **P** represent respectively target and projectile centers; the impact velocity $\vec{v}$ is parallel to the z-axis and the impact parameter $\vec{b}$ to the x-axis. $\vec{r}$ and $\vec{R}$ are the respective positions of the electron and the projectile relatively to the target.

Within the semi-classical impact parameter method [2], the relative motion of the nuclei is described as a straightline constant velocity trajectory (figure 1)

$$\vec{R} = \vec{b} + \vec{v}t$$

$\vec{v}$ being the impact velocity and $\vec{b}$ the so-called *impact parameter*. For one-electron systems, the time-dependent Schrödinger equation is

$$[\hat{H}_{\text{el}} - i\frac{\partial}{\partial t}]\Psi(\vec{r},t) = 0 \tag{1}$$

with the electronic hamiltonian

$$\hat{H}_{\text{el}} = \hat{T} + \hat{V}_{Te}(r) + \hat{V}_{Pe}(|\vec{r} - \vec{R}(t)|) \tag{2}$$

Here $\hat{T}$ is the electron kinetic energy operator, $\hat{V}_{Te}$ and $\hat{V}_{Pe}$ are the potential operators for the interactions (coulombic or effective) between the electron and, respectively, the target and the projectile centers.

Asymptotically, the system consists of two isolated atomic systems. This suggests an expansion of the electronic scattering wave-function Ψ on a two-center atomic basis

$$\Psi(\vec{r},t) = \sum_{j=1}^{N_T} c_j^T(t)\phi_j^T(\vec{r})e^{-i\epsilon_j^T t} + \sum_{j=1}^{N_P} c_j^P(t)\phi_j^P(\vec{r}-\vec{R})e^{-i\epsilon_j^P t}F(\vec{r},t) \tag{3}$$

where the $\phi_j^{T(P)}$ are eigenfunctions of the isolated target (projectile) hamiltonian. The *electron translation factor* (ETF)

$$F(\vec{r},t) = e^{-i\frac{1}{2}v^2 t}e^{i\vec{v}\cdot\vec{r}} \tag{4}$$

takes into account the motion of the projectile center relatively to the target center [3, 4].

Inserting this expansion into the time-dependent Schrödinger equation leads to a set of coupled differential equations,

$$i\text{S}\dot{\text{C}} = \text{HC} \tag{5}$$

where C is the vector of the $c_j^X(t)$ coefficients, $\dot{\text{C}}$ is the vector of their first derivatives with respect to time, S is the overlap matrix and H the coupling matrix. It is necessary to evaluate one-center matrix elements

$$\begin{aligned} S_{ij}^{TT} &= S_{ij}^{PP} = \delta_{ij} \\ H_{ij}^{TT} &= \langle\phi_i^T|V_{Pe}|\phi_j^T\rangle \\ H_{ij}^{PP} &= \langle\phi_i^P|V_{Te}|\phi_j^P\rangle \end{aligned} \tag{6}$$

and two-center matrix elements

$$\begin{aligned} S_{ij}^{TP} &= \langle\phi_i^T|e^{i\vec{v}\cdot\vec{r}}|\phi_j^P\rangle e^{-i(\epsilon_j^P+\frac{1}{2}v^2-\epsilon_i^T)t} \\ S_{ij}^{PT} &= (S_{ji}^{TP})^* \\ H_{ij}^{TP} &= \langle\phi_i^T|V_{Te}e^{i\vec{v}\cdot\vec{r}}|\phi_j^P\rangle e^{-i(\epsilon_j^P+\frac{1}{2}v^2-\epsilon_i^T)t} \\ H_{ij}^{PT} &= \langle\phi_i^P|V_{Pe}e^{-i\vec{v}\cdot\vec{r}}|\phi_j^T\rangle e^{-i(\epsilon_j^T-\epsilon_i^P-\frac{1}{2}v^2)t} \end{aligned} \tag{7}$$

The solution of the coupled-channel equations (5) leads to the total cross sections, *i.e.* for a transition from the initial target state i ($\Psi(t \to -\infty) = \psi_i^T$) to any atomic state f when the collision is over

$$\sigma_{i\to f}(v) = 2\pi \int_0^\infty bdb\, P_{i\to f}^v(b) \tag{8}$$

$P_{i\to f}^v(b)$ is the transition probability from state i to state f, for a collision velocity v and an impact parameter b. Within the present framework

$$P_{i\to f}^v(b) = |c_f(t \to +\infty)|^2 \tag{9}$$

2.2. IMPLEMENTATION

The implementation of such a model mostly depends on the choice of the atomic orbitals. Linear combinations of Slater Type Orbitals are natural and moreover allow a good description of one-center matrix elements even at large internuclear distances. However, a complete analytical calculation of the two-center integrals cannot be performed due to the ETF, and time consuming numerical integrations [6, 7] are required (demanding typically 90% of the total CPU time).

We use an alternative to this method, that enables a fast and accurate evaluation of the two-center integrals. Analytical integration is possible when linear combinations of Gaussian Type Orbitals are used to describe the atomic states [1, 9, 10]. Imperfect behavior of such gaussian functions at large distances does not affect the results, since the two-center matrix-elements (7) have an exponential decay for increasing internuclear distances. For example, for integrals H_{ij}^{TP} and S_{ij}^{TP} expressed in cartesian coordinates, one has to evaluate expressions such as

$$I = \int_{-\infty}^{+\infty} dx_T \int_{-\infty}^{+\infty} dy_T \int_{-\infty}^{+\infty} dz_T\, r_T^{2n^T} r_P^{2n^P} e^{-\alpha^T r_T^2 - \alpha^P r_P^2} e^{i\vec{v}\cdot\vec{r}_P} x_T^{u^T} y_T^{v^T} z_T^{w^T} x_P^{u^P} y_P^{v^P} z_P^{w^P} \tag{10}$$

This triple integral can be worked out

$$\begin{aligned} I \ = \ & \lim_{\vec{a}\to\vec{0}} \lim_{\vec{c}\to\vec{0}} \left(-i\frac{\partial}{\partial a_x}\right)^{u^T} \left(-i\frac{\partial}{\partial a_y}\right)^{v^T} \left(-i\frac{\partial}{\partial a_z}\right)^{w^T} \\ & \left(-i\frac{\partial}{\partial c_x}\right)^{u^P} \left(-i\frac{\partial}{\partial c_y}\right)^{v^P} \left(-i\frac{\partial}{\partial c_z}\right)^{w^P} \left(-\frac{\partial}{\partial \alpha^T}\right)^{n^T} \left(-\frac{\partial}{\partial \alpha^P}\right)^{n^P} J \end{aligned} \tag{11}$$

with

$$J = \begin{cases} \left(\frac{\pi}{A}\right)^{\frac{3}{2}} e^{\frac{B^2}{4A}} e^{-\alpha^P R^2} e^{-i\vec{c}\cdot\vec{R}} & \text{for } \tilde{S}_{ij}^{TP} \\ 2\frac{\pi^{\frac{3}{2}}}{B\sqrt{A}} e^{\frac{B^2}{4A}} e^{-\alpha^P R^2} e^{-i\vec{c}\cdot\vec{R}} erf\left(\frac{B}{2\sqrt{A}}\right) & \text{for } \tilde{H}_{ij}^{TP} \end{cases} \quad (12)$$

$$\vec{B} = i(\vec{a}+\vec{c}+\vec{v}) + 2\alpha^P\vec{R}$$
$$B = (B_x^2 + B_y^2 + B_z^2)^{\frac{1}{2}}$$
$$A = \alpha^T + \alpha^P$$

where $erf()$ is the complex error function.

A FORTRAN code has been writen to perform the successive symbolic differenciations of the J functions (12) for any combination of atomic states. The I integrals (11) are then evaluated by numerical calculations. This code was inserted in the scattering program developed by the group [7, 8] and tested on the H-H$^+$ benchmark system: we obtained results in excellent agreement with those computed in a full Slater Type Orbital treatment, but with a typical factor 10 gain in CPU time.

3. Electron transfer in C^{6+}- H$(1s)$ collisions

Laboratory plasmas temperature is highly influenced by unavoidable presence of impurities (multicharged ions). To model this requires accurate knowledge of capture cross sections for collisions between atomic hydrogen and fully stripped ions – such as C^{6+}and 0^{8+} [11]. R. K. Janev *et al.* [12] have presented a set of recommanded cross sections for those systems at collision energies between 0.1 and 1000 keV/u, by fitting several theoretical calculations and a few experimental results. Nevertheless, due to a lack of data or to severe disagreements among data for most of the capture levels the fits do not cover the whole energy range.

In the following, we present total cross sections defined by

$$\sigma_{i\to n}(v) = \sum_{l=0}^{n-1} \sum_{m=-l}^{+l} \sigma_{i\to nlm}(v)$$

for the electron transfer to C^{5+}(n=3-6) shells. Our results obtained with the present method are to be compared with previous calculations and recommanded data.

The basis set (B_1) used in the present work is composed of all n=1-6 C^{5+}(nlm) states and the H$(1s)$ initial state. In order to check basis convergence, we have performed calculations with the basis augmented by the H$(2lm)$ states (B_2), for three impact energies, 0.3, 10 and 300 keV/u. The

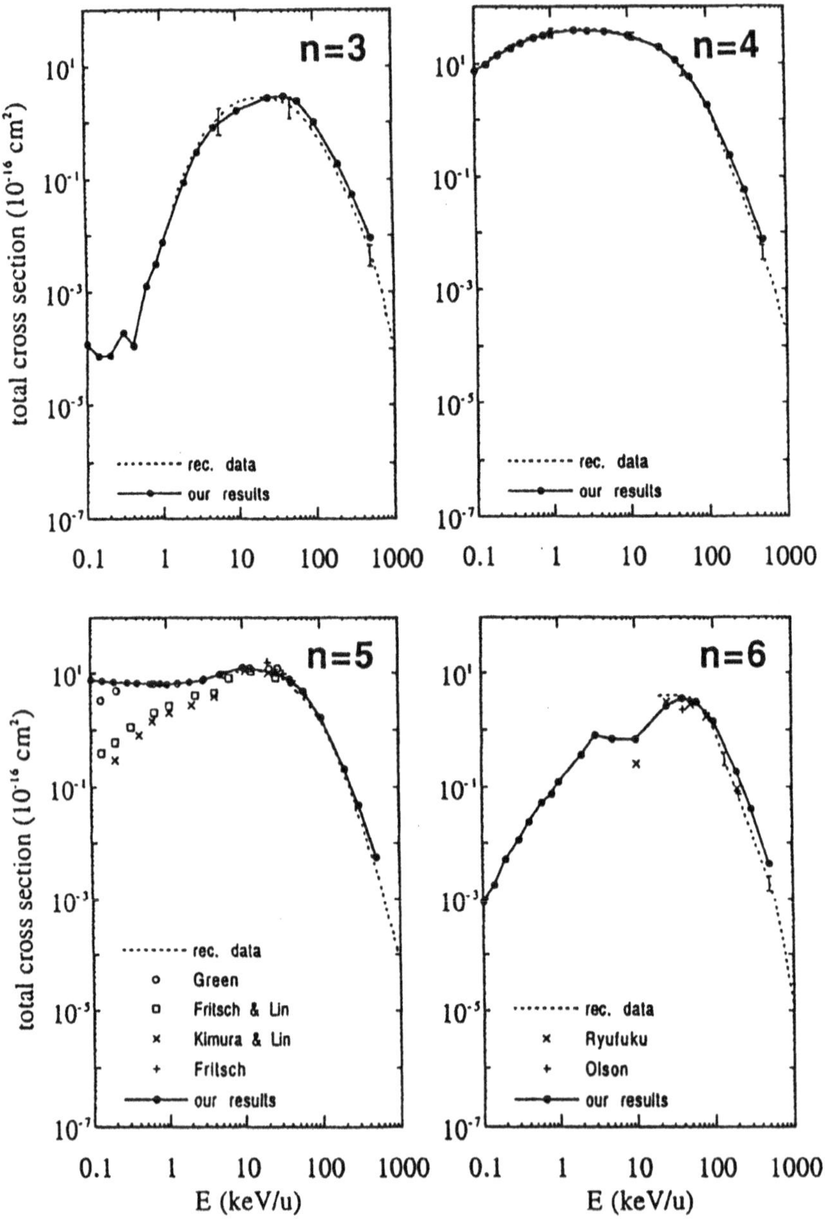

Figure 2. Total capture cross sections from H(1s) to C^{5+}(n=3-6) shells as a function of impact energy.

difference between results using the two bases are less than 5 percent. In agreement with former studies [13], this shows the weak couplings between excitation and capture processes, even though excitation becomes more important than capture in the high energy range [14]. The B_1 basis does not allow the description of ionization which dominates the other inelastic processes above typically 100 keV/u [14, 15]. However, the goal of the present study is the description of electron capture where data are not available – *i.e.* at low impact energies . The B_1 basis is then appropriate.

3.1. RESULTS

In figure 2 we present the results of our calculations performed at 19 collision energies from 0.1 to 500 keV/u, together with the recommanded fits of Janev *et al.* [12] and some of the available theoretical results.

The fits of Janev *et al.* [12] stem from a compilation of the results obtained with different theoretical approaches: (i) semi-classical close-coupling methods with a development of the wave function on atomic orbitals (Fritsch and Lin [16]), molecular orbitals (Green *et al.* [17]), or both (Kimura and Lin [18], (ii) pure classical model – *i.e.* the Classical Trajectory Monte Carlo method (Olson and Schultz [19]) – and (iii) perturbative quantum approach (Belkić *et al.* [20]). In order to get precise fits, theoretical results accuracy was estimated according to many criteria, most important being the domain of validity of each technique.

It appears that electron transfer to $C^{5+} n$=4 and 5 levels are dominant on the covered energy range. Total cross sections for both levels are of the same order: around 10^{-15} cm^2 below 10 keV/u, they decrease down to 10^{-19} cm^2 at 500 keV/u. Capture cross sections for $C^{5+} n$=3 and 6 levels present a typical bell-shape with a maximum of the order of 10^{-16} cm^2 at about 50 keV/u.

As expected, we note a systematic overestimation of our cross sections compared to the recommanded data beyond 100 keV/u where ionization is dominant. Nevertheless, in this energy range, capture processes are rare, and the corresponding cross sections steeply decay with increasing impact velocities. It appears at lower collision energies that our results are very satisfactory.

As shown in figure 2, recommanded cross sections for electron transfer to n=3 level are given only for impact energies above 5 keV/u: our results match the fit within the error bars. For capture to $C^{5+}(n=4)$ level, whose fit covers the whole energy range, we get a perfect agreement with Janev *et al.* recommanded data. Concerning capture to n=6 level, the fit has not been set below 20 keV/u due to a lack of theoretical results. However our results compare well with the recommanded data in the energy range from

20 to 100 keV/u.

A special attention should be given for n=5 capture level. The fit was not set below 60 keV/u because of severe discrepancies between the different theoretical data. Beyond 60 keV/u, our results match the recommanded data. For lower collision energies, they are in agreement with Green *et al.* results [17], and in strong disagreement with Fritsch and Lin [16] and Kimura and Lin [18] calculations.

3.2. INTERPRETATION

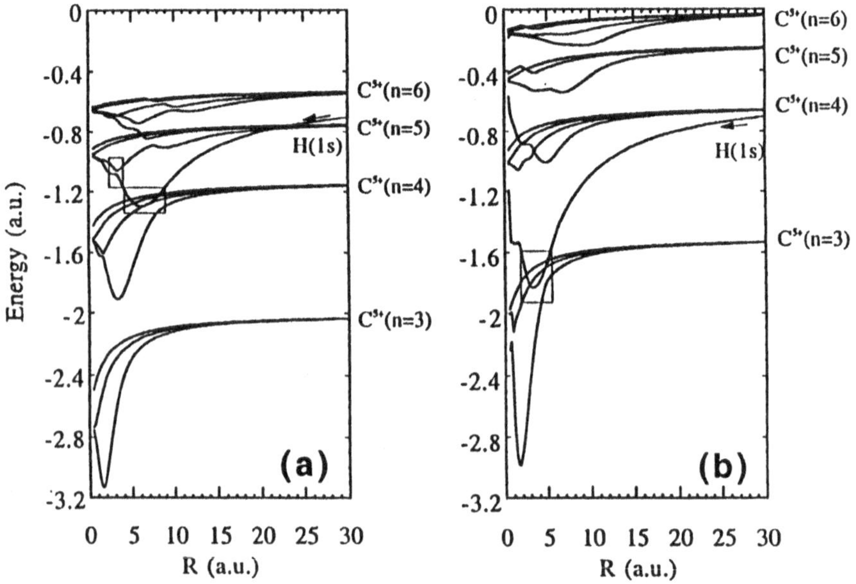

Figure 3. Energy curves obtained by diagonalisation of $\hat{H}_{el}$ on the two-center atomic basis, at (a) zero collision energy and (b) 25 keV/u. On each diagram, the curve correlated to the H(1s) state is indicated by an arrow. The important avoided crossings discussed in the text are framed.

The general behaviors of the cross sections presented in figure 2 can be analysed using the energy diagrams obtained by diagonalisation of the electronic hamiltonian $\hat{H}_{el}$ (2) on the ETF-augmented atomic basis set (3) at several impact velocities. We show in figure 3(a) and 3(b) the diagrams obtained for two velocities, respectively 0 and 1 a.u. Note that for the former one, the curves are equivalent to the adiabatic potential energy diagram of the CH^{6+} molecular ion. On the other hand, the main features of the graphs obtained at non-zero velocity stem from the translation of the curves correlated at infinity to the projectile states by the corresponding kinetic energy term $\frac{1}{2}v^2$ (*cf.* equation 4), *i.e.* 0.5 a.u. for figure 3(b). We make the assumption that the important transitions occur around avoided crossings between the H(1*s*) and the projectile capture states correlated curves, at small internuclear distances, typically $R < 15$ a.u.

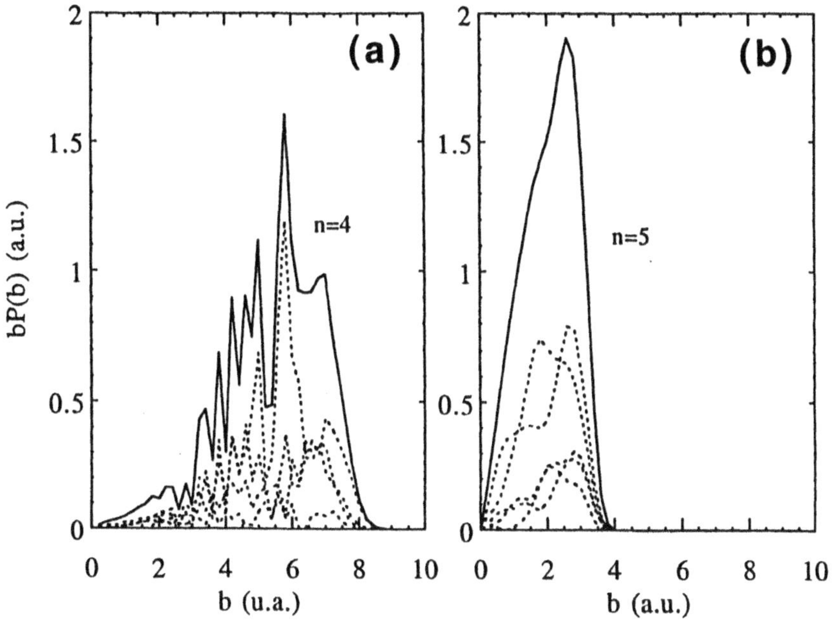

Figure 4. The reduced probability $bP^{v}_{i\to f}(b)$ of electron capture in (a) n=4 and (b) n=5 C^{5+} atomic levels from H(1*s*), at collision energy of 0.1 keV/u (v=0.06 a.u.). Full lines: total capture to n levels; dotted lines: nl sublevels capture details.

We can first explain the maximum localised around 25 keV/u for the n=3 capture cross section. There is indeed in figure 3(a) no direct avoided crossing between the curve correlated to H(1*s*) state and those correlated to $C^{5+}(n=3)$ states. The corresponding cross section is therefore small at low impact energies. However, due to the energy shift of $\frac{1}{2}v^2$, such avoided crossings (framed in figure 3(b)) appear at $R \simeq 3$ a.u. We can support this interpretation by noting that the transition probability $bP(b)$ to $C^{5+}(n=3)$ states presents a maximum at 25 keV/u for an impact parameter b=3 a.u. (graph not presented).

The evolution of the capture cross section to n=6 level cannot be interpreted in such a way, as the curves correlated to $C^{5+}(n=6)$ states never cross the ones correlated to the initial state at internuclear distances less than 15 a.u.

The quite flat behaviors of the cross sections for electron capture to the n=4 and 5 shells at low energies can also be understood in the view of figure 3. The many avoided crossings occuring around 6 a.u. between the H(1s) and C^{5+}(n=4) correlated curves (framed in figure 3(a)), are responsible for the capture to projectile n=4 states. This is consistent with the very oscillatory structure of the reduced probability presented in figure 4(a). Those avoided crossings are not present beyond 10 keV/u (v=0.63 a.u.), where the corresponding capture cross section starts to decay. Capture to C^{5+}(n=5) is a two-step mechanism through the avoided crossings framed around 3 a.u. in figure 3(a) which couple n=4 and n=5 C^{5+}states. This is also clearly illustrated by the fast decay of $bP(b)$ for this reaction (figure 4(b)): impact parameters greater than 4 a.u. cannot lead to the avoided crossings of interest.

Finally, one can explain the discrepancies observed at low impact energies between the different theoretical calculations of n=5 capture cross section. The lowest cross sections stem from semi-classical models using coulombic repulsion for the description of heavy particles relative motion (Fritsch and Lin [16], Kimura and Lin [18]). Such trajectories might increase the closest approach distances so that the avoided crossings leading to C^{5+}(n=5) levels are not reached - in contrary to the trajectory models such as in Green *et al.* work [17] and the present study [21]. This difference may explain the order of magnitude of disagreement between the proposed models.

4. Concluding remarks

We have presented results which extend and confirm Janev *et al.* recommanded data for the C^{6+}– H(1s) collision system. The problems raised by the choice of the internuclear trajectory in semi-classical methods were shortly discussed. A more detailed study regarding those questions is in progress [21].

References

1. Bransden, B. H. and McDowell, M. R. C. (1992) *Charge Exchange and the Theory of Ion-Atom Collisions*, Oxford Science Publications, Oxford.
2. McCarroll, R. (1982) Charge exchange and ionization in ion atom collisions , in F. A. Gianturco (ed.), *Atomic and molecular collision theory* NATO ASI Series B, Plenum Publishing Corporation, New York, pp. 165–244.
3. Hansen, J. P. and Dubois, A. (1996), Cross Sections for Electron Capture in H-B^{q+} and He-B^{q+} (q=1,3,5) collisions, *Phys. Scripta*, **T62**, pp. 55–58.
4. Hansen, J. P., Kocbach, L., Synnes, S. A., Wang, J. B. and Dubois, A. (1998) Origin of the strucures in the excitation cross sections in ion-atom collisions, *Phys. Rev.*

A, **57**, R4086–R4089.

5. Barat, M. (1976) The Quasi-Molecular Model in Heavy Particle Collisions, in F. J. Wuilleumier (ed.), *Photoionization and other probes of many-electron interactions*, Plenum Publishing Corporation, New York, pp. 229–253.
6. Shakeshaft R., (1975) A note on the exchange integrals in the impact-parameter treatment of heavy-particle collisions, *J. Phys. B*, **8**, L134–L136.
7. Hansen, J. P. and Dubois, A. (1991) Procedure for analytical and numerical calculation of Coulombic one- and two-center integrals, *Comp. Phys. Com.*, **67**, pp. 456–464.
8. Dubois, A., Nielsen, S.E. and Hansen, J.P. (1993) State selectivity in $H^+ - Na(3s/3p)$ collisions: differential cross sections, alignment and orientation effects for electron capture *J. Phys. B*, **26**, pp. 705–721.
9. Kocbach, L. and Liska, R., (1995) Matrix elements of travelling gaussians for applications in the theory of atomic collisions, private communication.
10. Errea, L. F., Mendez, L. and Riera A. (1979) Evaluation of molecular integrals in a basis of travelling orbitals, *J. Phys. B*, **12**, pp. 69–82.
11. Fonck, R. J., Darrow, D. S. and Jaehnig, K. P. (1984) Determination of plasma-ion velocity distribution via charge-exchange recombination spectroscopy, *Phys. Rev. A*, **29**, pp. 3288–3309.
12. Janev, R. K., Phaneuf, R. A., Tawara H. and Shirai, T. (1993) Recommanded cross sections for state-selective electron capture in collisions of $C^{6+}$and 0^{8+}ions with atomic hydrogen, *At. Data and Nucl. Data Tables*, **55**, pp. 201–232.
13. Toshima, N. (1994) Unexpectedly large contribution of C^{5+}deep bound states in $C^{6+} + H$ collisions, *J. Phys. B*, **27**, L49–L54.
14. Toshima, N. (1994) Ionization and charge transfer of atomic hydrogen in collision with multiply charged ions, *Phys. Rev. A*, **50**, pp. 3940–3947.
15. Rivarola, R. D., Fainstein, P. D. and Ponce, V. H. (1989) Electron Ionization by Impact of Carbon and Oxygen Ions on Hydrogen and Helium Atoms, *Phys. Scripta*, **T28**, pp. 101–105.
16. Fritsch, W. and Lin, C. D. (1984) Atomic-orbital-expansion studies of electron transfer in bare-nucleus Z (Z=2,4-8) –hydrogen-atom collisions, *Phys. Rev. A*, **29**, pp. 3039–3051.
17. Green, T. A., Shipsey, E. J. and Browne, J. C. (1982) modified method of perturbed stationary states. IV. Electron-capture cross sections for the reaction $C^{6+} + H(1s) \longrightarrow C^{5+}(nl) + H^+$, *Phys. Rev. A*, **25**, pp. 1364–1373.
18. Kimura, M. and Lin, C. D. (1985) Unified treatment of slow atom-atom and ion-atom collisions. II. Applications to $H^+ + H$ and $C^{6+} + H$ collisions, *Phys. Rev. A*, **32**, pp. 1357–1362.
19. Olson, R. E. and Schultz, D. R. (1989) n, l Distributions for Electron-Capture from $H(1s)$ by $C^{6+}$and 0^{8+}, *Phys. Scripta*, **T28**, pp. 71–76.
20. Belkić, D., Gayet, R. and Salin, A. (1992) Cross sections for electron capture from atomic hydrogen by fully stripped ions, *At. Data and Nucl. Data Tables*, **51**, pp. 59–150.
21. Caillat, J., Dubois, A. and Hansen, J. P. (1999) *J. Phys. B*, in preparation.

SINGLE AND DOUBLE ELECTRON CAPTURE IN BORON COLLISION SYSTEMS

M.C. BACCHUS-MONTABONEL[1] AND P. HONVAULT[2]

1) Laboratoire de Spectrométrie Ionique et Moléculaire, CNRS et Université Lyon I, 43 bd du 11 Novembre 1918, 69622 Villeurbanne Cedex, France
2) Laboratoire PALMS, Université Rennes I, Campus de Beaulieu, 35042 Rennes Cedex, France

Abstract. Although single electron capture remains generally the main process in most ion-atom charge-transfer reactions, double and eventually multiple electron capture may be important in the understanding of such processes. An illustration of these mechanisms is presented on the example of the B^{2+} + H and B^{4+} + He collisions. A complete ab-initio molecular treatment of the potential energy curves and coupling matrix elements followed by a semiclassical collision dynamics has been performed for these systems. An adiabatic representation providing an unambiguous description of the molecular states has been used throughout this work. The results compare well to experimental data and improve markedly previous theoretical work.

1. Introduction

An accurate understanding of the electron capture processes by multiply charged ions from atomic hydrogen and helium is of considerable interest in controlled thermonuclear fusion research. In particular, systems involving boron ions have attracted increasing attention as boron is known to be more and more used as an impurity for cooling fusion reactors. Single electron capture remains the main process in most ion-atom charge-transfer reactions, as for example in the B^{2+}+ H collision system, nevertheless double and even multiple electron capture may be also important in the understanding of such reactions [1-4]. As far as a charge-transfer mechanism is concerned, single and double electron capture at low collision energies are dominated by specific curve crossing transitions and may thus be described by the same approach. But the treatment of double electron capture reactions is considerably complicated by the number of channels generally involved in the process, as well as the possible interaction with Rydberg states or with the continuum. In some cases, however, double capture channels may be populated directly and a complete treatment may be performed. This is the case for the

J. Maruani et al. (eds.), New Trends in Quantum Systems in Chemistry and Physics, Volume 2, 133–142.

B^{4+}+ He reaction. This collisional system has been investigated theoretically within the framework of the semiclassical close-coupling formalism using different model potential approaches [2,3] which lead to a discrepancy of about a factor 5 for the double capture cross section values. We have thus performed an alternative study of this system by means of a full molecular expansion method, focusing our attention on the double electron capture process.

2. Single electron capture: the B^{2+} + H collision

Relatively little attention has been paid on this system. From an experimental point of view, it has been investigated on a quite large collision energy range, about 0.8-40 keV, by McCullough et al. [5] and Crandall et al. [6], and at higher energies by Goffe et al. [7]. But only the $B^{2+}(1s^22s)^2S + H(1s) \rightarrow B^+(1s^22s^2)^1S + H^+$ ground state reaction has been considered theoretically by Crothers and Todd [8] using the phase integral interpretation of the two-state exponential model. As pointed out by McCullough et al. [5], other transitions should be taken into account in order to explain the experimental measurements.

A complete treatment has thus been undertaken, taking into account all the possible radial and rotational transitions occuring from $[B^{2+}(1s^22s)^2S + H(1s)]^{1,3}\Sigma^+$ entry channels. The adiabatic states involved in such a process are shown below (Fig. 1):

$B^{2+}(1s^22s)^2S + H(1s)$	$^{1,3}\Sigma^+$	(4)
$B^+(1s^22s2p)^1P + H^+$	$^1\Sigma^+, ^1\Pi$	(5)
$B^+(1s^22s2p)^3P + H^+$	$^3\Sigma^+, ^3\Pi$	(6)
$B^+(1s^22s^2)^1S + H^+$	$^1\Sigma^+$	(7)

The potential energy curves of the Σ and Π states have been determined by means of MCSCF + CI calculations using the CIPSI algoritm [9] with a threshold η=0.005 for the perturbation contribution to the wavefunctions. The basis of atomic functions used in the calculation is a (6s3p1d) basis of contracted gaussian functions for boron [10] and a (5s3p) basis for hydrogen [11] allowing a good description of atomic levels (Table 1).

The radial coupling matrix elements between all pairs of states of same symmetry have been calculated by means of the finite difference technique:

$$g_{KL}(R) = \langle \psi_K | \partial/\partial R | \psi_L \rangle = \lim_{\Delta \to 0} 1/\Delta \langle \psi_K(R) | \psi_L(R) \rangle,$$

with the parameter Δ = 0.0012 a.u. as previously tested and using the boron nucleus as origin of electronic coordinates. The rotational coupling matrix elements $\langle \psi_K | iL_y | \psi_L \rangle$ between Σ and Π states have been calculated directly from the quadrupole moment tensor as well as the translation factor corrections [12].

TABLE 1. Comparison of calculated atomic levels (eV) with experimental data

Atomic levels	Theory	Experiment [13]
$B^{2+}(1s^2 2p)^2P$	31.146	31.135
$B^{2+}(1s^2 2s)^2S$	25.099	25.137
$B^{+}(1s^2 2p^2)^1D$	12.953	12.691
$B^{+}(1s^2 2p^2)^3P$	12.482	12.265
$B^{+}(1s^2 2s2p)^1P$	9.173	9.100
$B^{+}(1s^2 2s2p)^3P$	4.651	4.631
$B^{+}(1s^2 2s^2)^1S$	0.0	0.0

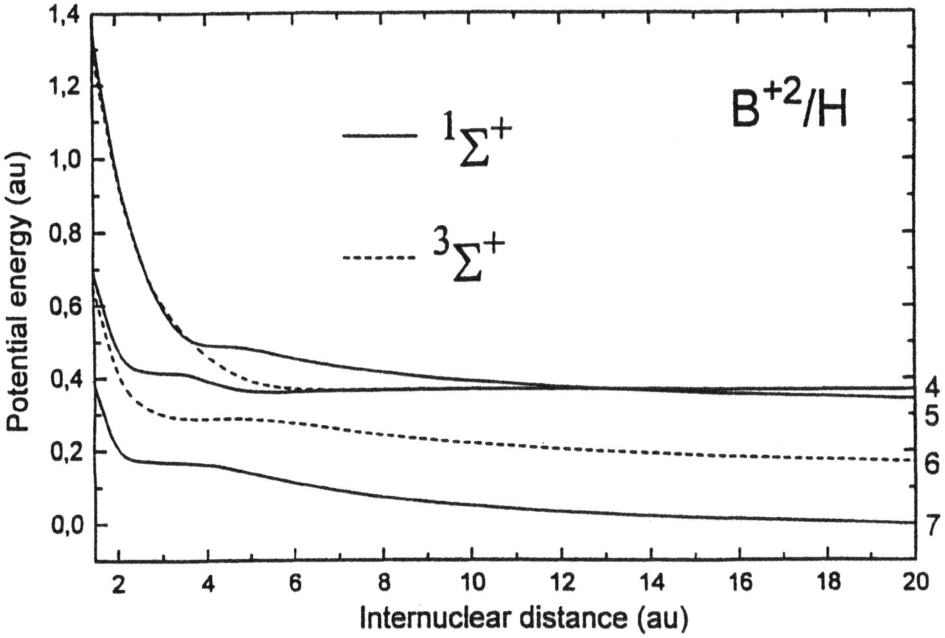

Figure 1. Adiabatic potential energy curves for the $^1\Sigma^+$ and $^3\Sigma^+$ states of $(BH)^{2+}$. The different channels are labeled as in the main text.

The process is mainly driven by the $^{1,3}\Sigma^+$ adiabatic states. Their potential energy curves are presented in Figure 1. The $^1\Sigma$ states show clearly two avoided crossings between the entry channel and the $[B^+(2s2p)^1P + H^+]$ level, a very diabatic one at about R = 13 a.u. and a wider one at about R = 3.7 a.u. which interferes with the avoided crossing between the entry channel and the $[B^+(2s^2)^1S$ +

$H^+]$ $^1\Sigma^+$ state. The $^3\Sigma^+$ states exhibit only an avoided crosing at about R = 5.4 a.u. between the entry channel and the $[B^+(2s2p)^3P + H^+]$ state.

The collision dynamics has been performed in the 0.1-190 keV incident-ion energy range by means of a semiclassical formalism using the EIKONXS program [14], available for these energies and accounting for translation effects by means of the common translation factors as proposed by Errea et al. [15]. Radial and rotational couplings induce three transitions between $^1\Sigma^+$ states (Fig. 2):

$$B^{2+}(1s^22s)^2S + H \rightarrow B^+(1s^22s^2)^1S + H^+ \qquad {}^1\Sigma^+ - {}^1\Sigma^+ \tag{1}$$
$$B^{2+}(1s^22s)^2S + H \rightarrow B^+(1s^22s2p)^1P + H^+ \qquad {}^1\Sigma^+ - {}^1\Sigma^+ \tag{2}$$
$$B^{2+}(1s^22s)^2S + H \rightarrow B^+(1s^22s2p)^1P + H^+ \qquad {}^1\Sigma^+ - {}^1\Pi \tag{3}$$

The partial cross sections are presented in Figure 2. The main process corresponds obviously to reaction (1) which was the only one considered by Crothers and Todd [8], but the effect of rotational coupling is clearly not negligible and induces a strong weight of the $[B^+(2s2p)^1P + H^+]$ state. Such an importance for a reaction occuring by rotational coupling only is quite noteworthy. On the contrary the process (2) induced by radial coupling is quite negligible, except at higher energies (E > 20 keV).

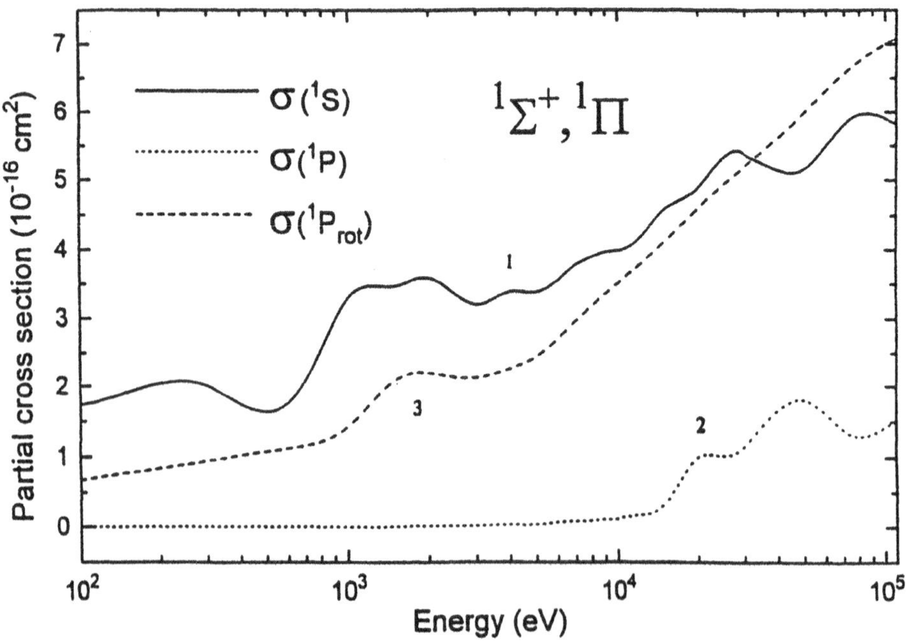

Figure 2. Partial charge-transfer cross sections for the $^1\Sigma^+$ and $^1\Pi$ states of BH^{2+} as a function of incident-ion energy. The notations are the same as in the main text.

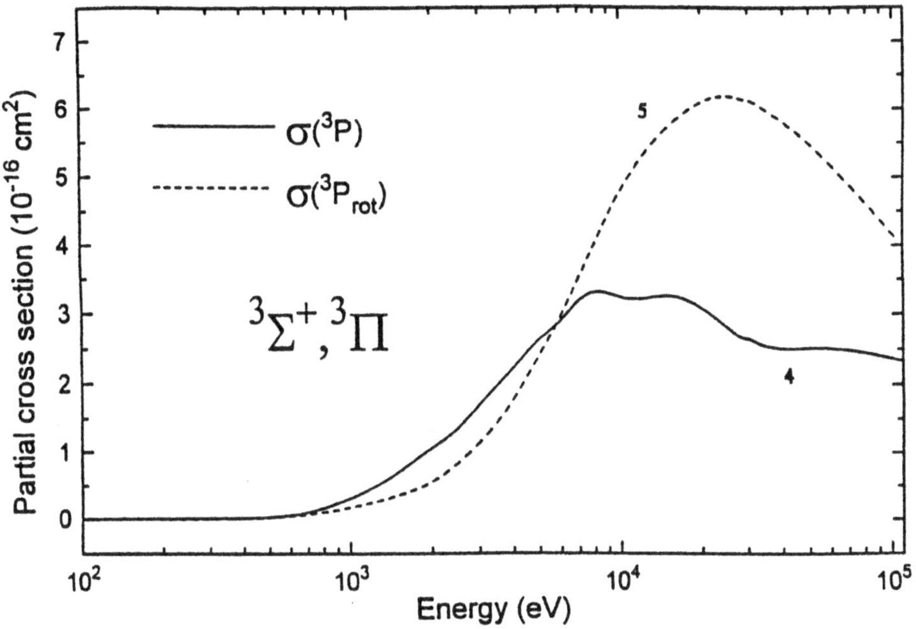

Figure 3. Partial charge-transfer cross sections for the $^3\Sigma^+$ and $^3\Pi$ states of BH^{2+} as a function of incident-ion energy. The notations are the same as in the main text.

For triplet states, two transitions may be considered (Fig. 3):

$$B^{2+}(1s^22s)^2S + H \rightarrow B^+(1s^22s2p)^3P + H^+ \qquad {}^3\Sigma^+ - {}^3\Sigma^+ \qquad (4)$$

$$B^{2+}(1s^22s)^2S + H \rightarrow B^+(1s^22s2p)^3P + H^+ \qquad {}^3\Sigma^+ - {}^3\Pi \qquad (5)$$

which also lead to quite significant partial cross sections for keV and higher energies and exhibit an important rotational effect. This analysis show clearly that an approach considering only the ground state reaction (process 1) could not at all account for the collision mechanism. This may be visualized on the total cross sections which can be compared directly to experimental measurements. The total cross sections calculated with respect to the probabilities 1/4 and 3/4 for the electron to be in a singlet or triplet state are displayed in Figure 4. They show a complete shift of our values compared to Crothers and Todd calculations corresponding to the influence of triplet states and account of rotational effects. The shape of the variation of the total cross section is however globally smoothed in our calculation and the maximum remains underestimated compared to experimental measurements. This could be related to a possible influence of higher levels correlated to the excited B^{2+}(2p) entry channel which are relatively close in energy [1] and might be populated by means of rotational coupling.

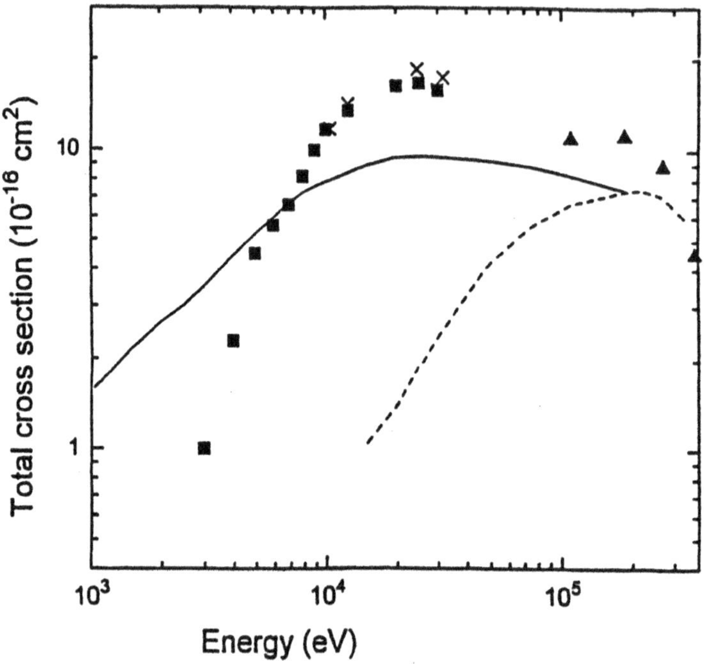

Figure 4. Comparison of our total charge-transfer cross section for B^{2+}/H with previous theoretical calculations and experimental measurements: ------ σ_{tot} (this work); - - - Crothers and Todd [8]; Y McCullough et al. [5]; × Crandall et al. [6]; Δ Goffe et al. [7].

3. Double electron capture: the B^{4+} + He collision

The main features as described above have been used for the determination of the potential energy curves of the B^{4+} + He collisional system, the radial and rotational couplings, as well as the translation factor corrections. Nevertheless, a fair description of the $B^{2+}(1s2\ell2\ell')$ levels involved in the double capture process being necessary, a larger basis of (6s4p2d) gaussian functions, previously optimized on B^{3+}(1s2s) [16] has been taken for boron. Helium has been described by means of a (4s1p) contracted basis set [10].

All the $^2\Sigma$ and $^2\Pi$ potential energy curves correlated to the entry channel $[B^{4+}(1s) + He(1s^2)]\ ^2\Sigma$, the $[B^{3+}(1s2\ell) + He^{+}(1s)]$ and the $[B^{2+}(1s2\ell2\ell') + He^{2+}]$ configurations have been calculated [4]. In Figure 5 we present results obtained for the $^2\Sigma$ states, which exhibit strong avoided crossings involving double exit channels at respectively R = 4.2, 5.65, 6.6, 7.78 and 12.1 a.u., corresponding to very narrow energy separations – respectively 8.12 10^{-4} a.u., 2.47 10^{-3}a.u., 4.82 10^{-4} a.u., 5.94 10^{-4} a.u. and 3.3 10^{-5} a.u.. The asymptotic values show a good agreement with experimental [13] and theoretical data taking into account relativistic correction terms [17], especially for exit channels (Table 2).

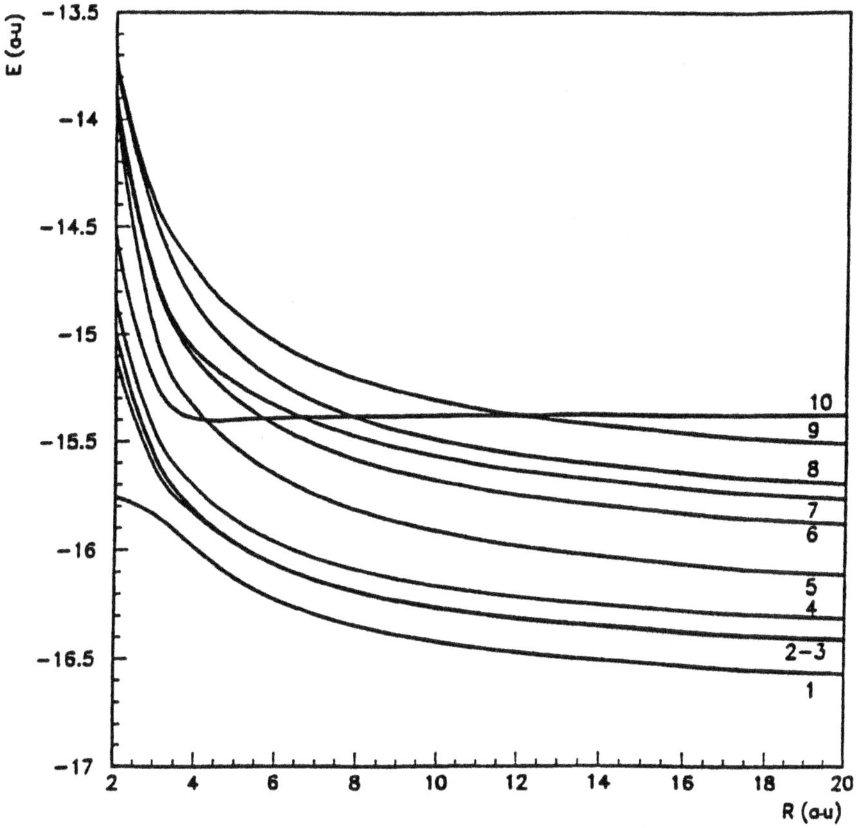

Figure 5. Adiabatic potential energy curves for the $^2\Sigma$ states of BHe^{4+}: 1-2, Σ states dissociating to $\{ B^{3+}(1s2s)^{1,3}S + He^{+}(1s) \}$; 3-4, Σ states dissociating to $\{ B^{3+}(1s2p)^{1,3}P + He^{+}(1s) \}$; 5, Σ state dissociating to $\{ B^{2+}(1s2s^2)^2S + He^{2+} \}$; 6-7, Σ states dissociating to $\{ B^{2+}(1s2s2p)^2P + He^{2+} \}$; 8, Σ state dissociating to $\{ B^{2+}(1s2p^2)^2D + He^{2+} \}$; 9, Σ state dissociating to $\{ B^{2+}(1s2p^2)^2S + He^{2+} \}$; 10, Σ state dissociating to $\{ B^{4+}(1s) + He(1s^2) \}$.

Using these data, we have performed a collisional treatment in the semi-classical framework [14] in the 0.1-0.7 a.u. velocity range (2.7-130 keV E_{lab} energy range) including radial and rotational couplings as well as electron translation effects. The cross sections of double electron capture are presented in Figure 6 (solid curve) and compared directly to the previous calculations of Fritsch and Lin [2] (dotted curve) and Hansen and Taulbjerg [3] (broken curve) in the same energy range. As far as absolute values are concerned, our work is seen to be in better agreement with the partial cross sections obtained by Fritsch and Lin [2], especially for the capture on the B^{2+}(1s2s2p) level for which the accordance is quite good. Nevertheless, the variations of the partial, and accordingly total cross sections calculated by Hansen and Taulbjerg [3] appear to be in good agreement with ours, although their values remain underestimated at all velocities: in particular, the complete collapse of the total cross section at lower energies does not seem realistic. Our approach, which takes into account the three electron interactions, gets rid of the frozen core approximation description of the $B^{2+}(1s2\ell2\ell')$ levels, the correlation between the 1s and 2ℓ orbitals being involved in the configuration interaction calculation. It seems well adapted to describe the $B^{2+}(1s2\ell2\ell')$ levels.

TABLE 2. Comparison of asymptotic energy values with experimental [13] (single-electron capture levels) and theoretical data [17] (double-electron capture levels) (in a.u.)

Level	This work	Experiment [13]
B^{4+} + He	1.345	1.331
$B^{3+}(1s2p)^1P + He^+$	0.258	0.257
$B^{3+}(1s2p)^3P + He^+$	0.161	0.161
$B^{3+}(1s2s)^1S + He^+$	0.157	0.155
$B^{3+}(1s2s)^3S + He^+$	0.0	0.0
Level	**This work**	**Calculations [17]**
B^{4+} + He	0.938	0.971
$B^{2+}(1s2p^2)^2S + He^{2+}$	0.605	0.600
$B^{2+}(1s2p^2)^2D + He^{2+}$	0.420	0.423
$B^{2+}(1s2s2p)^2P + He^{2+}$	0.349	0.340
	0.234	0.236
$B^{2+}(1s2s^2)^2S + He^{2+}$	0.0	0.0

4. Concluding remarks

This work provides accurate potential energy curves as well as coupling matrix elements for the B^{2+}/H and B^{4+}/He systems. From the molecular point of view, it appears important to involve all levels correlated to the entry channels in the collision dynamics and, in particular, to take into account rotational effects, which might be quite important. The results concerning the double electron capture process in the (B^{4+} + He) collision point out the limitations of the potential approach model, especially to account for open shell levels, for which more elaborate calculations are necessary.

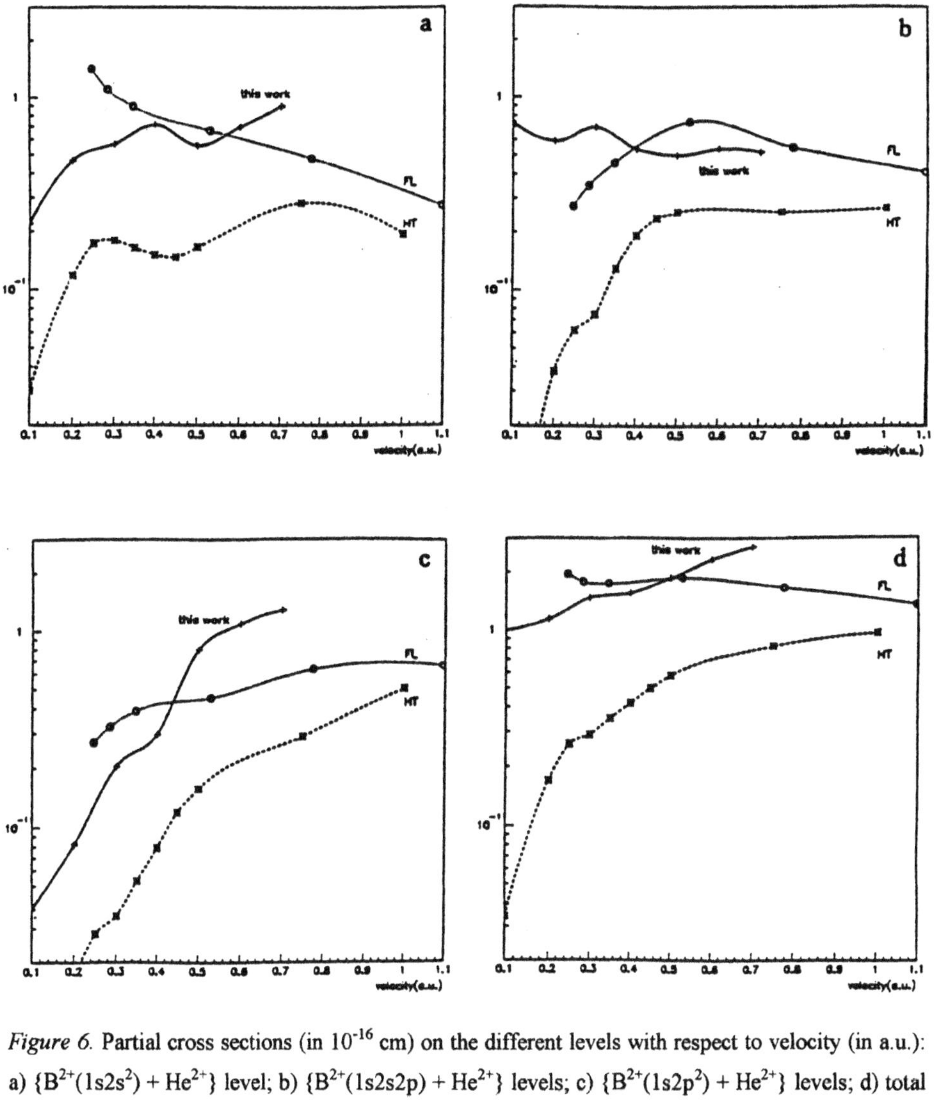

Figure 6. Partial cross sections (in 10^{-16} cm) on the different levels with respect to velocity (in a.u.): a) $\{B^{2+}(1s2s^2) + He^{2+}\}$ level; b) $\{B^{2+}(1s2s2p) + He^{2+}\}$ levels; c) $\{B^{2+}(1s2p^2) + He^{2+}\}$ levels; d) total double electron capture cross section, ------ this work; …… Fritsch and Lin [2]; - - - Hansen and Taulbjerg [3].

References

1. Honvault, P., and Bacchus-Montabonel, M.C. (1997) *Mol. Phys.* **91**, 223.
2. Fritsch, W., and Lin, C.D. (1992) *Phys. Rev. A* **45**, 6411.
3. Hansen, J.P., and Taulbjerg, K. (1993) *Phys. Rev. A* **47**, 2987.

4. Bacchus-Montabonel, M.C. (1996) *Phys. Rev. A* **53**, 3667.
5. McCullough, R.W., Nutt, W.L., and Gilbody, H.B. (1979) *J. Phys. B* **12**, 4159.
6. Crandall, D.H., Phaneuf, R.A., and Meyer, F.W. (1979) *Phys. Rev. A* **19**, 504.
7. Goffe, T.V., Shah, M.B. and Gilbody, H.B. (1979) *J. Phys. B* **12**, 3763.
8. Crothers, D.S.F., and Todd, N.R. (1980) *J. Phys. B* **13**, 547.
9. Huron, B., Malrieu, J.P. and Rancurel, P. (1973) *J. Chem. Phys.* **58**, 5745.
10. Fraija, F., Bacchus-Montabonel, M.C., and Gargaud, M. (1994) *Z. Phys. D* **29**, 179.
11. Chambaud, G., Millié, Ph., Ridard, J., and Lévy, B. (1979) *J. Phys. B* **12**, 221.
Husinaga S. (1965) *J. Chem. Phys.* **42**, 1293.
12. Fraija, F., Allouche, A.R., and Bacchus-Montabonel, M.C. (1994) *Phys. Rev. A* **49**, 272.
13. Bashkin, S., and Stoner, J.O. (1975) *Atomic Energy Levels and Grotrian Diagrams* (North-Holland, Amsterdam).
14. Allan, R.J., Courbin, C., Salas P., and Wahnon, P. (1990) *J. Phys. B* **23**, L461.
15. Errea, L.F., Mendez, L., and Riera, A. (1982) J. *Phys. B* **15**, 101.
16. Bacchus-Montabonel, M.C., and Fraija, F. (1994) *Phys. Rev. A* **49**, 5108.
17. Davis, B.F., and Chung, K.T. (1985) *Phys. Rev. A* **31**, 3017; (1984) **29**, 1878.
Chung, K.T., and Bruch, R. (1983) *Phys. Rev. A* **28**, 1418.

THEORETICAL STUDY OF THE INTERACTION OF CARBON DIOXIDE WITH Sc, Ti, Ni, AND Cu ATOMS

F. MELE[1], N. RUSSO[1], M. TOSCANO[1] AND F. ILLAS[2]

1) Dipartimento di Chimica, Universita della Calabria, I-87030 Arcavacata di Rende (CS), Italy
2) Department de Quimica Fisica i Centre de Recerca en Quimica Teorica, Facultat de Quimica, Universitad de Barcelona, C/ Marti i Franques 1, E-08028 Barcelona, Spain

Abstract. Density functional theory (DFT) using both gradient-corrected (PWP) and hybrid (B3LYP) functionals has been used to investigate the geometrical structures, harmonic vibrational frequencies and binding energies of the $ScCO_2$, $TiCO_2$, $NiCO_2$, and $CuCO_2$ systems. Eight possible coordination modes and the species produced by the insertion of the metal into the C-O bond have been considered. Results show that the interactions of copper and nickel atoms with CO_2 are endothermic processes, while scandium and titanium are able to form stable complexes. In addition, we found that the Ti atom inserts spontaneously into the C-O bond while a barrier of 6.4 kcal/mol is required for Sc insertion.

1. Introduction

Carbon dioxide is the output of different industrial processes and, in many cases, is discharged into atmosphere heavily perturbing the natural environmental equilibria of the Earth. Part of the atmospheric CO_2 is removed by green plants that are able to use the solar energy to transform it into carbohydrates through a photosynthesis cycle. A large amount of CO_2 is retained in the atmosphere, so, it can be considered the most abundant source of carbon. In 1988 it was estimated that the total amount of CO_2 in the atmosphere and in the oceans represents 10^{14} tons of carbon while carbonates represents 10^{16} tons of carbon [1,2]. The possibility to activate and transform carbon dioxide in other useful products is of enormous importance from the industrial, environmental, geochemical and oceanographic points of view. In spite of the large amount of the available CO_2, only few processes, using carbon dioxide as raw material, have been developed in the synthetic chemical industry [3,4].

J. Maruani et al. (eds.), New Trends in Quantum Systems in Chemistry and Physics, Volume 2, 143–165.

The difficulty to transform CO_2 into other organic compounds lies in its high thermodynamic stability. Typical activation energies for the dissociation and recombination of CO_2 are of 535 and 13 kJ/mol, respectively [5]. The activation can occur by photochemical or electrochemical processes, by catalytic fixation or by metal-ligand insertion mechanisms. As documented in different reviews, organometallic compounds, metallo-enzyme sites and well defined metallic surfaces are able to activate carbon dioxide [6-16].

The coordination of CO_2 on a metallic centre seems to be the most important steep in its reduction [6]. Due to the importance of the metal-CO_2 interaction as an elementary and fundamental step in the metal-catalysed conversion of carbon dioxide, many experimental and theoretical studies have been devoted to the investigation of the energy, coordination modes and spectroscopic constants of this process. From an experimental point of view, $AuCO_2$ [17], $AgCO_2$ [18], and first-row transition-CO_2 species [19-20] have been studied using different spectroscopies. From these studies it emerged that some first-row transition metal atoms (i.e. Fe, Co, Ni, Cu) form M-CO_2 complexes, while other elements (i.e. Ti, V, Cr, Mo, W) insert spontaneously into CO bond, yielding OMCO systems. Theoretical investigations exist for Ni [20], Sc [21,22], Cr [23], Cu [24], Pd [25] and Ti [27] interacting with a single CO_2 molecule to form stable complexes. Other related studies concern the alkali-metal-carbon dioxide systems [28-33].

However, these studies do not allow a homogeneous comparison between the reactivity of CO_2 towards different transition metal atoms. Therefore, we have considered of interest to investigate simultaneously several M-CO_2 (M= Sc, Ti, Ni, Cu) species at the same level of theory. In particular we have examined eight different coordination modes, including some situations never previously investigated. This is the case of the metal insertion into the CO bond, which has been considered only for Sc and Ti.

Calculations were carried out using the gradient-corrected functionals of Perdew and Wang (for exchange) [34] and of Perdew (for correlation) (PWP) [35], not yet used previously in the investigation of these systems, and the Becke (exchange) [36] and Lee, Yang and Parr (correlation) (B3LYP) [37] hybrid functional, used until now only for Sc-CO_2. The isotropic hyperfine coupling constants for the radical species Sc-CO_2 and OScCO are reported for the first time.

For all these systems, the geometries, harmonic vibrational frequencies and bonding analysis for the stable species, relative to the potential energy surface, are reported.

2. Method

A modified version of the deMon DFT code [38] in which molecular orbitals are given as linear combinations of Gaussian-type atomic functions was used for the PWP computations. In order to localise the extreme points on the potential energy surfaces the Broyden-Fletcher-Goldfarb-Shanno minimisation algorithm [39] was used. Saddle points were searched by the Abashkin-Russo algorithm [40]. For C, O (7111/411/1*) triple zeta type orbitals in conjunction with (4,4;4,4) auxiliary basis set were employed [41]. For scandium and titanium all electron (63321/5211*/41+) and (63321/531*/41+) orbital and (5,5;5,5) and (5,5;5,5) auxiliary bases [41] were used, respectively. For Ni [42] and Cu [43] atoms the Huzinaga-type model core potentials (MCP) that allow to treat explicitly the $3p^6 3d^{10} 4s^2$ and $3p^6 3d^{10} 4s^1$ valence electrons, respectively, were chosen. The corresponding orbital and auxiliary basis sets are (311/31/311) and (3,4;3,4) for Ni [42] and (311/31/311) and (5,5;5,5) for Cu [43]. In all the computations the FINE grid option was used, namely 194 angular grid points and 64 radial grid points per atom.

Calculations using the hybrid B3LYP exchange-correlation functional were performed using the Gaussian 94 package [44]. Internal 6-31G* basis sets were used for C and O and 6-311G for Sc, Ti, and Ni. For copper the LANL2DZ [45] pseudopotential was used.

3. Results and Discussion

For all examined systems we have considered eight coordination modes, which are schematically shown in Figure 1.

These are: η^1 C or side-on C (a), η^2 O,O (b), η^1 C or side-on C linear (c), η^2 C,O (d), η^1 O or end-on linear (e), η^1 O or end-on (f), η^1 O or end-on cis (g), and η^1 O or end-on trans (h), respectively. For simplicity, the results for each system are reported and discussed separately.

3.1. $ScCO_2$ COMPLEX

The possible formation of a $ScCO_2$ complex has been the subject of previous theoretical studies [21,22]. In a first one, four coordination modes corresponding to **a**, **b**, **d** and **f** in Figure 1 and two different spin states, doublets and triplets, were considered at the HF level. The **b** and **d** forms in the doublet states have also been studied at the DFT level, employing the Becke-Perdew exchange-correl-ation potential [21]. For the more stable species, **b** and **d**, the binding energies were rigorously determined by performing single-point MCPF, QCISD(T) and MCPF(ANO) computations. The second work [22] examines the η^2 O,O coordi-nation mode in detail, using different DFT potentials (BP, BLYP, BHLYP and B3LYP) and also the CCSD(T) method.

We have optimised the structure of the lowest doublets and quartets of all the possible coordination modes (**a-h**). PWP and B3LYP functionals give only two minima, corresponding to the η^2O,O and η^2C,O coordination modes, in both the doublet and quartet potential energy surfaces (PES). In addition a transition state (TS), corresponding to the **g** structure, results from the B3LYP computation on doublet states.

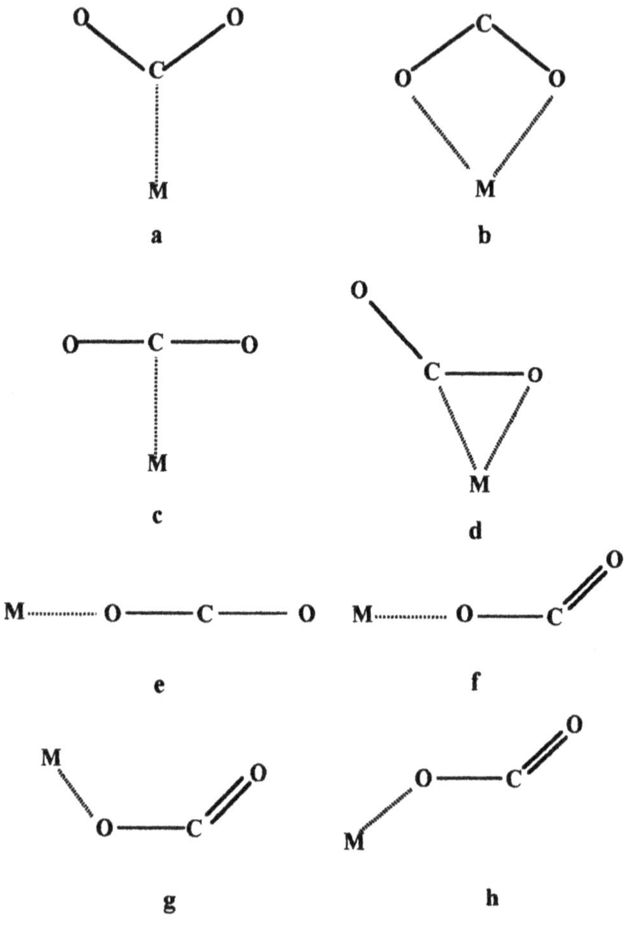

Figure 1. Possible metal-CO_2 coordination modes used as starting points in the geometry optimisation.

Geometrical parameters and relative energies of minima are reported in Table 1 together with other theoretical data [21,22]. The geometries obtained are close to those previously reported and we would like to underline the agreement of our PWP and B3LYP results with those yielded by the CCSD(T) computations.

TABLE 1. PWP, (B3LYP), [BP], {CCSD(T)}, and [[MCPF]] geometrical parameters and relative energies (ΔE) for $ScCO_2$ minima. Distances are in Å, angles in degrees, and relative energies in kcal /mol. BP and CCSD(T) results are from refs. 21 and 22, respectively

	η^2C,O		η^2O,O	
	^{2}A'	^{4}A'	$^2A_1$	$^4A_1$
R_{Sc-C}	2.155 (2.118) [2.116]	2.419 (2.281)	1.961 (2.324) [2.310]	2.515 (2.423)
R_{Sc-O}	1.924 (1.951) [1.875]	2.052 (2.094)	1.961 (1.935) [1.992] {1.945}	(2.163)
R_{O-C}	1.407 (1.385) [1.432]	1.292 (1.271)	1.363 (1.344) [1.365] {1.380}	1.275 (1.259)
$R_{C=O}$	1.208 (1.254) [1.199]	1.209 (1.203)	1.363 (1.344) [1.365] {1.380}	1.275 (1.259)
OCO	111.0 (133.5) [126.4]	141.5 (139.6)	110.2 (112.7) [112.4] {110.3}	126.0 (139.6)
ΔE	0.0 (5.0) [[2.4]]	26.9 (25.5) /	3.1 (0.0) [[0.0]]	7.5 (28.4) [[16.8]]

a) from ref. 21; b) from ref. 22

The inspection of the relative energies reveals that at the PWP level the more stable species results to be the ^{2}A' η^2C,O form followed by the $^2A_1$ η^2O,O one at 3.1 kcal/mol. On the contrary, B3LYP stabilises the $^2A_1$ η^2O,O on the ^{2}A' η^2C,O isomer by 5.0 kcal/mol. Taking into account the energy correction for zero point vibrational energy further reduces these values and, if we consider that the BSSE corrections for the two species differ by 0.4 kcal/mol (lower for $^2A_1$ η^2O,O) [21] the difference between the two coordination modes is very small. Previous BP computations favoured η^2C,O by 1.5 kcal/mol and the MCPF, QCISD/(T) and MCPF(ANO) levels of theory favour η^2O,O by 2.4, 2.6 and 1.2 kcal/mol, respectively. From the present computations the fact that the $^2A_1$ η^2O,O and ^{2}A' η^2 C,O structures are very close in energy is further underlined. Both forms can be present, and particular physicochemical conditions may stabilise one over the other.

The binding energies of the $^2A_1$ η^2 O,O complex with respect to the Sc (^2D)+ CO_2 ($^1\Sigma_g^+$) dissociation limit are reported in Table 2. Comparisons with many high-level theoretical results using different basis sets and methodological approximations are possible. It is obvious in Table 2 that, opposite to common belief, B3LYP results for D^0 may be strongly dependent on the basis set. Thus, B3LYP /6-31G* gives a D^0 that is lower by 9.1 kcal/mol than that computed by using the D95+* basis set. On the other hand, the B3LYP values obtained employing the TZVP and aug-cc-pVTZ basis sets appear to be rather similar. A minor effect of the basis set on the D^0 values can be underlined in the case of BP/TZVP and BP/D95+* computations. The results obtained with different functionals and same basis sets show that both PWP/TZVP and BP/TZVP gradient-corrected potentials give D^0 values (33.1 kcal/mol and 38.5 kcal/mol) strongly overestimated with respect to the B3LYP/TZVP one (26.5 kcal/mol). Furthermore, comparison between BP/D95+*, B3LYP/D95+* and CCSD(T)/D95+* again shows that the gradient-corrected potential gives the highest D^0.

TABLE 2. Binding energies (kcal /mol) for the $ScCO_2$ complex from different levels of theory

Method	D^0
PWP/TZVP	33.1
BP/TZVP	38.5
B3LYP/TZVP	26.5
B3LYP/6-31G*	20.3
BP/D95+*	40.9
B3LYP/D95+*	29.4[a]
B3LYP/aug-cc-pVTZ	25.9[a]
CCSD(T)/D95+*	19.9[b]
MCPF/aug-cc-pVTZ	24.2[b]

a) from ref. 22; b) from ref. 21

On the basis of these results and in the absence of experimental data the choice of the best basis set is not an easy task. Therefore, as a compromise between reliability and reasonable computational cost we have employed mainly the 6-31G* and TZVP basis sets in the hybrid and gradient-corrected computations, respectively.

Let us now discuss the binding nature in the two $^2A'$ η^2 C,O and $^2A_1$ η^2 O,O minima. The Mulliken net charges obtained by B3LYP and PWP functionals are reported in Figure 2. Both exchange-correlation functionals indicate there is a considerable charge transfer from Sc to the CO_2 oxygen atoms. This transfer is more pronounced in the η^2 O,O coordination mode. From the electrostatic point of view this form is more stabilised and the ionic character appears to be larger in the B3LYP computations. Although the charge values derived from Mulliken analysis are not always credible, they are largely used in literature. In our case, because the amount of charge transfer is noticeable, the possible lack of precision in the reproduction of absolute values should not change the essential of the physical phenomena. Although the formation of $ScCO_2$ complex can be essentially explained by the electron transfer from the metal to the antibonding orbital of free CO_2, there is also a covalent contribution due to the bonding between the metal d_{xz} orbital and CO_2 in both the coordination modes. The covalent character of this orbital is clearly seen from the representation given in Fig. 3 for both η^2 O,O and η^2 C,O forms.

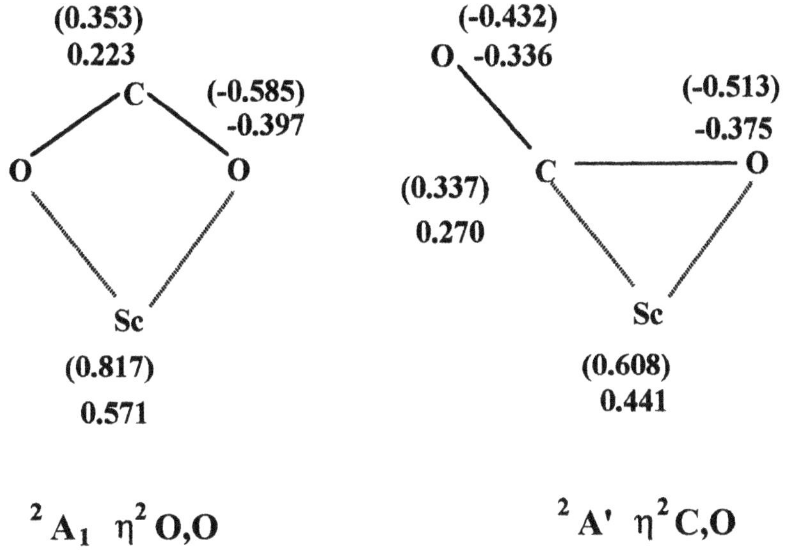

Figure 2. PWP (B3LYP) Mulliken net charges for the two lowest energy minima of the $ScCO_2$ complex.

Mulliken analysis shows there is some charge back-donation from the occupied orbitals of the ligand to the metal. On the basis of these data the preferred coordination mode arises from a delicate balance between these contributions. Two methods used attribute different weights to the ionic and covalent factors and thus result in a different absolute minimum.

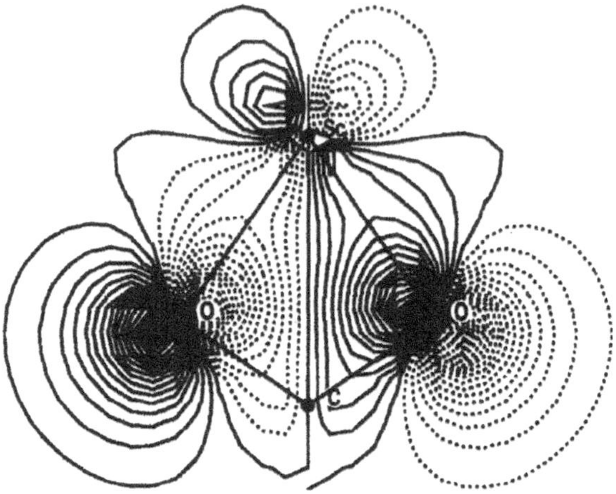

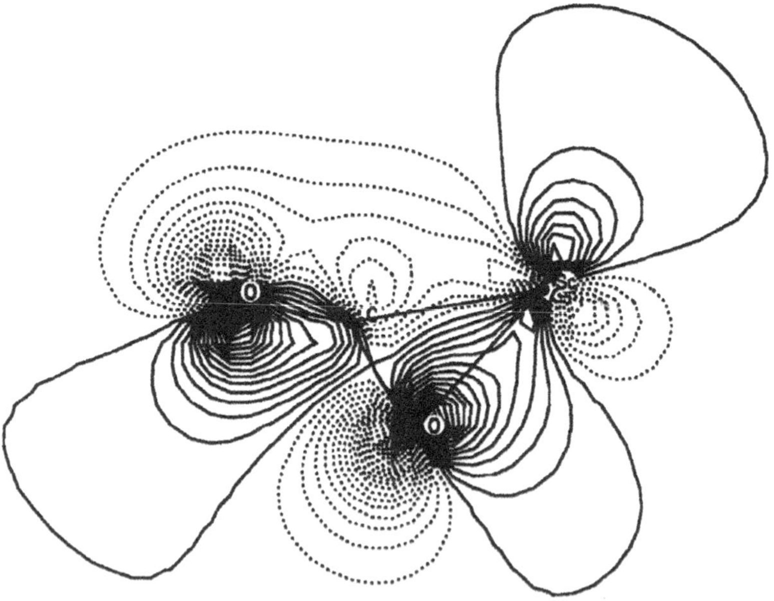

Figure 3. B3LYP $4b_2$ and $9a_1$ molecular orbitals of η^2 O,O and η^2 C,O $ScCO_2$ isomers, respectively. Dark shading represents positive coefficients (cutoff=0.425, step=0.025).

Next, we comment the vibrational frequencies obtained from the two DFT approaches. Results are reported in Table 3. In the η^2 C,O form, these frequencies are quite different. In particular, the one involving the direct interaction with the metal is sensibly smaller, because of the CO-bond lengthening due to metal-ligand interaction.

TABLE 3. PWP, (B3LYP), [BP] and {CCSD(T)} harmonic vibrational frequencies (ω in cm^{-1}) for $ScCO_2$ minima. BP and CCSD(T) results from refs. 21 and 22, respectively

	η^2C,O		η^2O,O	
	$^2A'$	$^4A'$	$^2A_1$	$^4A_1$
ω_1	292 (333)	(202)	293 (319)	(363)
	[302]		[315] {313}	
ω_2	363 (429)	(363)	478 (418)	(398)
	[349]		[360] {432}	
ω_3	460 (511)	(403)	497 (473)	(616)
	[368]		[487] {502}	
ω_4	674 (739)	(717)	827 (767)	(761)
	[615]		[745] {820}	
ω_5	901 (1213)	(1196)	929 (1061)	(1360)
	[832]		[927] {1057}	
ω_6	1740 (1724)	(1901)	978 (1065)	(1910)
	[1754]		[983] {1146}	

3.2. $TiCO_2$ COMPLEX

For the interaction of Ti with carbon dioxide we have considered the titanium in the $s^2d^2(^3F)$ and s^1d^3 (^5F) electronic configurations: the electronic state is given in parentheses although it must be pointed out that the present calculations for the isolated atom do not use spherical symmetry and, therefore, do not only provide an approximation to the atom states. Because of the presence in the literature of an accurate study on this interaction [26], performed employing the BP gradient-corrected functional, our study is limited to the investigation using the B3LYP functional. Three minima (**b**, **d** and **e**) and one transition state (**a**) were found on the PES for the triplet, while three minima (**b**, **d** and **g**) and two TS (**a** and **h**) were present on the quintet PES. Geometrical parameters and energetic data for the minima are reported in Table 4.

The absolute minimum (η^2 C,O) is followed by the η^2 O,O at 3.5 kcal/mol in the triplet PES. The η^2 C,O and η^2 O,O minima on the quintet PES lie at 14.0 and 18.8 kcal/mol, respectively. Our data disagree with those resulting from BP computations [26]. In fact, B3LYP points towards the η^2 C,O isomer as the absolute minimum, while BP does not found this isomer as a stable species in the triplet PES. The binding energy of this species is 7.8 kcal/mol with respect to the ground-state fragments Ti (^{3}F) and CO_2 ($^1\Sigma_g^+$) at infinite separation. The D^0 of the most stable species in the quintet PES are 19.5 (η^2 C,O) and 14.7 (η^2 O,O) kcal/mol. If one considers that the calculated value for the $s^2d^2 \rightarrow s^1d^3$ excitation energy of titanium is 25.7 kcal/mol, the minima structures on the quintet states are practically unbound with respect to the fragments in their ground state. Previous HF [27] and BP [26] calculations give, for the most stable isomers, different binding energies. In particular, BP estimates the binding energy for η^2 O,O coordination mode to be 30 kcal/mol once corrected for the basis set superposition error.

TABLE 4. B3LYP geometrical parameters and relative energies (ΔE in kcal/mol) for $TiCO_2$ minima. Distances are in Å and angles in degrees

	η^2C,O		η^2C,O		η^1O	
Parameter	triplet	quintet	triplet	quintet	triplet	quintet
R_{Ti-C}	2.057	2.221	2.365	2.408	/	2.942
R_{Ti-O}	1.985	2.145	1.943	2.138	2.216	1.955
R_{O-C}	1.279	1.249	1.378	1.256	1.185	/
$R_{C=O}$	1.206	1.200	/	/	1.185	/
OCO	133.2	146.1	110.5	62.4	180.	134.1
ΔE	0.0	14.0	3.5	18.8	26.5	19.2

We suggest that the discrepancy between B3LYP (11.3 kcal/mol) and previous theoretical binding energy predictions can be explained by considering that HF and BP ionisation potentials (IPs) of titanium are far from the experimental value: the HF value is underestimated by about 46.12 kcal/mol and the BP one is overestimated by about 23.06 kcal/mol. With the B3LYP potential, the obtained value is very close to the experimental one (153.58 versus an experimental value of 157.27 kcal/mol). The error in the IP evaluation can significantly affect the amount of charge transfer from the metal to the ligand, with serious consequences both on the preferred coordination mode and on the binding energy value.

Mulliken population analysis shows that the bonding in the two η^2 C,O and η^2 O,O minima have considerable ionic character (see Figure 4). As in the case of the $ScCO_2$ complex, the amount of charge transfer is also larger in the η^2 O,O form and, consequently, the OCO valence angle in this isomer is smaller than in the η^2 C,O one (110.5° versus 133.2°). Finally, one notices that for the quintet state minima the charge transfer is less pronounced than in the corresponding triplet ones.

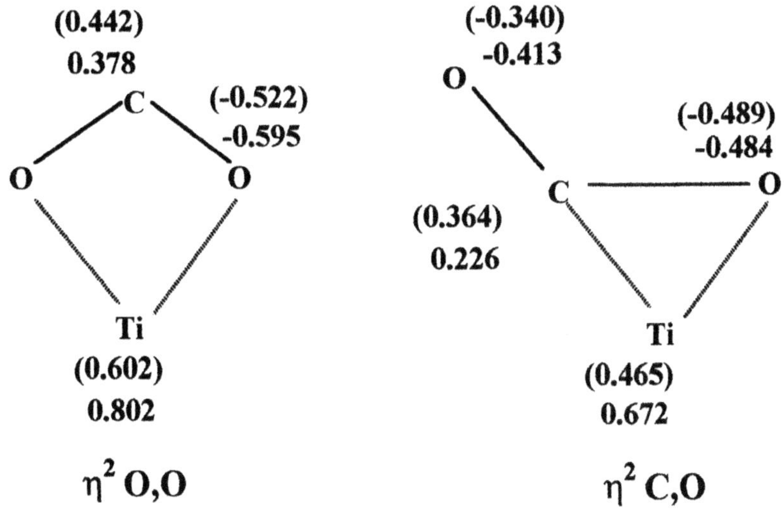

Figure 4. Mulliken net charges for η^2C,O and η^2O,O minima in the triplet (quintet) states.

In Table 5 we report the B3LYP vibrational frequencies for the two most stable isomers in triplet and quintet states. As expected, the vibrational frequencies that are more affected by the coordination modes are the two CO stretching modes. The splitting of these two frequencies is larger in the η^2 O,O isomer because both oxygen atoms are directly involved in the metal-ligand bond formation. In this case the agreement between the previous BP and present B3LYP frequency values is rather good.

3.3. $NiCO_2$ COMPLEX

Both singlet and triplet spin states were considered for the eight coordination modes depicted in Figure 1. Results from B3LYP and PWP potentials show the singlet PES has only two critical points corresponding to the **a** and **d** isomers. Vibrational analysis indicates that the former is a transition state and that the second is a minimum. In addition, hybrid functional predicts **e** as a second order transition state with two negative eigenvalues. All other starting geometries collapse to **a** and **d** or yield the Ni and CO_2 dissociation products.

In the triplet PES, B3LYP and PWP level calculations agree in predicting that structure **b** is a minimum while **a** and **d** are two transition states. Comparison with a previous DFT study [20] performed employing the BP exchange-correlation functional and considering all electron and double zeta basis set, reveals some disagreement. These earlier computations found a further TS (**b**) on the singlet PES and a different minimum (**d** instead of **b**) on the triplet one. Present results together with the corresponding BP data [20] are reported in Table 6. As a general trend, one notes that different functionals give very similar geometries, showing the two computation approaches are unaffected by the different descriptions of the Ni atom (PWP data were obtained using a Ni^{+16} model core potential while BP ones used an all-electron basis set).

Although all methods give the $^1A_1$ η^2 C,O as the most stable structure, the relative energy of the $^3A_1$ η^2O,O form is quite different. The PWP and B3LYP binding energies, taking into account the zero point corrections and the BSSE value proposed by Galan et al. [20], are 4.9 and -15.6 kcal/mol, respectively. These are significantly smaller than those previously obtained using the BP functional (17.5 kcal/mol). Different hypotheses can be suggested to justify these discrepancies, but most likely the weakly bound systems are not well described by the gradient-corrected functionals [46].

TABLE 5. B3LYP (BP) harmonic vibrational frequencies (ω in cm^{-1}) for the minima of the $TiCO_2$ complex. BP results from ref. 26

	η^2C,O	η^2O,O	η^2C,O	η^2O,O
	triplet		quintet	
ω_1	327	277 (235)	200 (200)	275 (257)
ω_2	411	290 (390)	346 (364)	375 (306)
ω_3	501	397 (507)	400 (398)	394 (342)
ω_4	709	676 (699)	689 (658)	783 (711)
ω_5	1118	889 (937)	1201 (1113)	1385 (1281)
ω_6	1565	947 (1008)	1968 (1008)	1610 (1547)

An additional argument that may be used to explain the differences mentioned above is that the computation of the ground state energy of the Ni atom (as well as of other transition metals) presents, at the DFT level, some problems due to the inadequate description of the atomic states [47-49]. The three values of the binding energy are quite different but all indicate that the $NiCO_2$ system is really a weakly bound system. This fact is confirmed by the experimental difficulty to generate a sensible amount of 1:1 $NiCO_2$ complex [19]. In diluted argon matrices (a condition that is closer to the in-vacuo computations) there is no evidence for the formation of the complex. A similar situation occurs when nickel atoms are condensed in a clean CO_2 matrix [19]. Only in N_2 matrices do infrared spectra [20] reveal nickel-carbon dioxide complexes: as it has been shown, the coordination of nitrogen on nickel increases its capability to coordinate to a CO_2 molecule.

In order to check whether $NiCO_2$ formation can occur without energy barriers we have explored the energetic process following the Ni--OCO reaction coordinate, fixing the Ni-O distance and optimising all other geometrical parameters. Results show that the complex formation occurs without energy barrier as in the case of a previous study [20] in which CO_2 geometry was kept fixed during the path. The analysis of the valence orbitals of the η^2C,O ground state reveals that only the 6a' and 7a'orbitals (see Figure 5) are responsible for the metal-ligand bonding.

In the former molecular orbital the 2p orbital of CO_2 is predominant (82 and 75 % at PWP and B3LYP levels, respectively) while the second is formed mainly by the $3\pi^*$ orbital of CO_2 and the 3d orbitals of Ni (54 and 58% for PWP and B3-LYP, respectively). Further insight can be obtained from Mulliken net charges and bond order analysis. Again, both computational methods indicate a substantial amount of charge transfer from nickel to CO_2. In particular, the net charges on Ni are 0.30 and 0.48 at PWP and B3LYP levels, respectively, and those on carbon are 0.38 (PWP) and 0.36 (B3LYP). The oxygen near the metal has a net charge of -0.35 or -0.48 at PWP or B3LYP level, respectively. These values indicate a complex with a dative π bond, in agreement with a previous analysis [20].

Fourier Transform Infrared (FTI) experiments in different matrices individuate a series of infrared frequencies assigned using the computed BP harmonic force field [20]. The following frequencies have been correlated to the side-on complex: ω(C=O)=1815, ω(C-O)=1131, ω(OCO)=744 and ω(NiO)=410 cm^{-1}. The corresponding PWP (B3LYP) values are: ω(C=O)=1919 (1977), ω(C-O)= 1081(1134), ω(OCO)=699(724) and ω(NiO)=417(502) cm^{-1}. Considering that the experiments are carried out on matrices, the agreement is reasonably good.

TABLE 6. Optimised geometries (bond lengths in Å and valence angles in degrees), harmonic vibrational frequencies (ω in cm^{-1}) and relative energies (ΔE in kcal/mol) for the minima of the $NiCO_2$ complex obtained at PWP, (B3LYP) and [BP] levels. BP results from ref. 20

	η^2C,O	η^2O,O
Parameter	$^1A_1$	$^3A_1$
$R_{Ni\text{-}C}$	1.808 (1.801) [1.820]	2.310 (2.340)
$R_{Ni\text{-}O}$	1.893 (1.815) [1.862]	2.061 (2.086)
$R_{O\text{-}C}$	1.278 (1.279) [1.288]	1.251 (1.224)
$R_{C=O}$	1.201 (1.193 [1.201]	1.251 (1.224)
OCO	149.6 (144.5) [144.6]	129.9 (135.4)
ω_1	1919 (1977) [1901]	1664 (1679)
ω_2	1081 (1134) [1078]	1203 (1291)
ω_3	699 (727) [702]	601 (718)
ω_4	499 (577) [532]	429 (448)
ω_5	417 (502) [523]	320 (250)
ω_6	235 (316) [284]	207 (162)
ΔE	0.0 (0.0) [0.0]	33.1 (16.8)

Following the previous experience [24] we have considered for this system the lowest doublet spin state. Full geometry optimisations of the eight coordination modes, carried out at the PWP and B3LYP levels, give only a minimum in the PES, that corresponds almost to the **g** coordination in Figure 1. In the two structures the geometrical parameters obtained with the two functionals differ considerably. The Cu-C distance is 2.887 Å with B3LYP and 2.104 Å with PWP. A difference also affects the CuOC valence angle: the PWP and B3-LYP values are 142.5° and 119.9°, respectively. The OCO angle measures about 180° in both cases. The origin of these large differences in geometrical parameters lies in the weak bond character of the metal-ligand interaction in this complex (vide infra).The calculated binding energies (D^0) with respect to the Cu (^2S) + CO_2 ($1\Sigma_g^+$) dissociation limits are both endothermic being -0.62 and -2.40 kcal/mol at B3LYP and PWP levels, respectively. If for this system the basis set superposition eror is about -0.5 kcal/mol, the endothermicity is further increased precisely by that amount. Since no significant charge transfer is observed (net charge on Cu is 0.04) and due to the very long Cu-O distance, one can conclude that only van der Waals interactions are possible between Cu and CO_2.

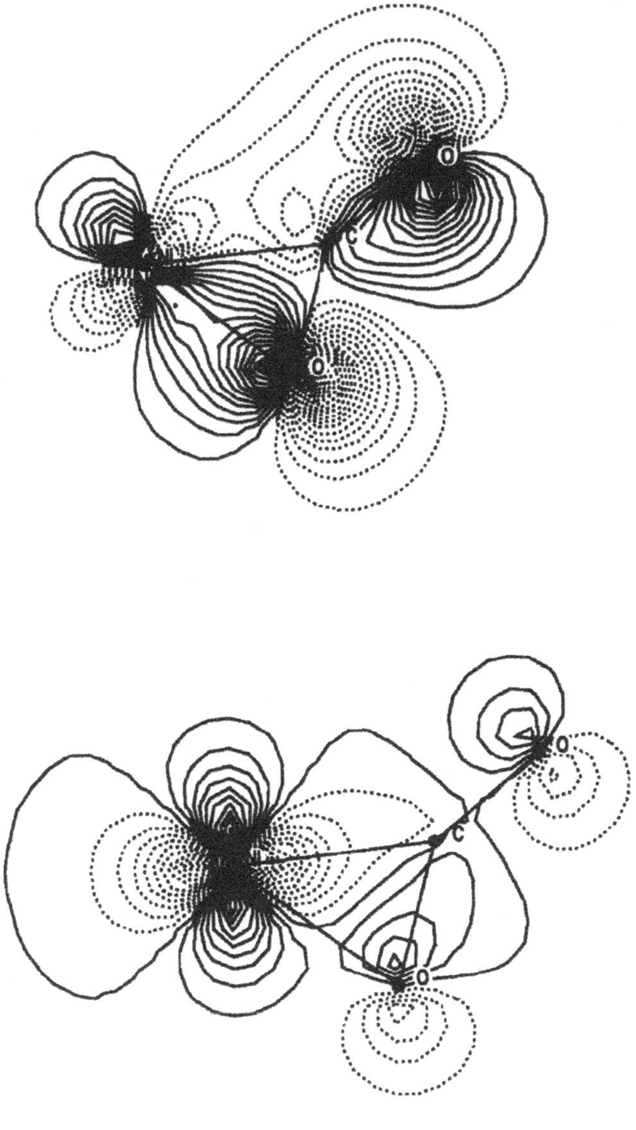

Figure 5. B3LYP 6a' (up) and 7a' (down) molecular orbitals of $^1A'$ η^2C,O of the $NiCO_2$ complex. Dark shadings represent positive coefficients (cutoff=0.425, step=0.025).

3.4. $CuCO_2$ COMPLEX

A further confirmation derives from the vibrational analysis, which gives frequencies similar to those of the free carbon dioxide with two low additional vibrational frequencies arising from interaction with the copper atom. These results are in agreement with IR spectra on the $CuCO_2$ system [19], which indicate the formation of a weakly bound complex. Unfortunately, this complex is thermolable and isotopic measurements conclude that the results are matrix dependent. The observed frequencies were assigned to C=O, C-O stretching modes (1716, 1215 cm^{-1}) and to OCO bending mode (718 cm^{-1}). None of these is present in our simulated spectra. In fact we obtain, at B3LYP (PWP) level, the frequencies: 28 (27), 52 (53), 636 (600), 640 (625), 1366 (1316), and 2425 (2344) cm^{-1}. The presence of a strong matrix effect in the experimental conditions may be the reason for this difference.

A previous configuration interaction study [24] on this complex suggested the presence of three minima corresponding to η^2 O,O, η^2 O, and η^1 C coordination modes with very low stabilisation energy. Because this work was performed without energy minimisation, gradient procedure and vibrational analysis, it is not possible to argue if they are all real minima on the PES.

3.5. OScCO AND OTiCO INSERTION COMPLEXES

Up to now, we have considered the interaction of M (Sc, Ti, Ni or Cu) with CO_2 to give a MCO_2 complex. In the case of scandium and titanium atoms, the interaction with carbon dioxide lead to stable complexes. The point addressed now focusses on the possible insertion of these two atoms in a CO bond of CO_2 to give oxmetal-carbonyl (OMCO) products. Indeed, both functionals indicate the presence of stable OMCO species for Sc and Ti atoms. The results are reported in Figure 6 and Tables 7 and 8.

For scandium, B3LYP and PWP agree in predicting the ground state of the OScCO system to be $^2A''$ followed by $^4A''$ at 53.9 and 57.6 kcal/mol at B3LYP and PWP levels, respectively. As shown in Figure 6, the two methods give comparable geometrical structures, PWP bond lengths being longer and the OScO valence angle smaller than the corresponding B3LYP ones. This discrepancy is essentially due to the different basis sets used in the computations. The binding energies with respect to Sc (^2D) + CO_2 ($^1\Sigma_g^+$) and ScO ($^2\Sigma^+$) + CO ($^1\Sigma^+$) dissociation limits are both positive (see Table 7). In particular the endothermicity of the former reaction is higher at both levels of theory. We note that PWP strongly overestimates the binding energies and this effect is probably due to the poor description of the Sc atomic energy. The same error source accounts for the difference in the stability between the insertion and molecular complexes obtained by the two methods.

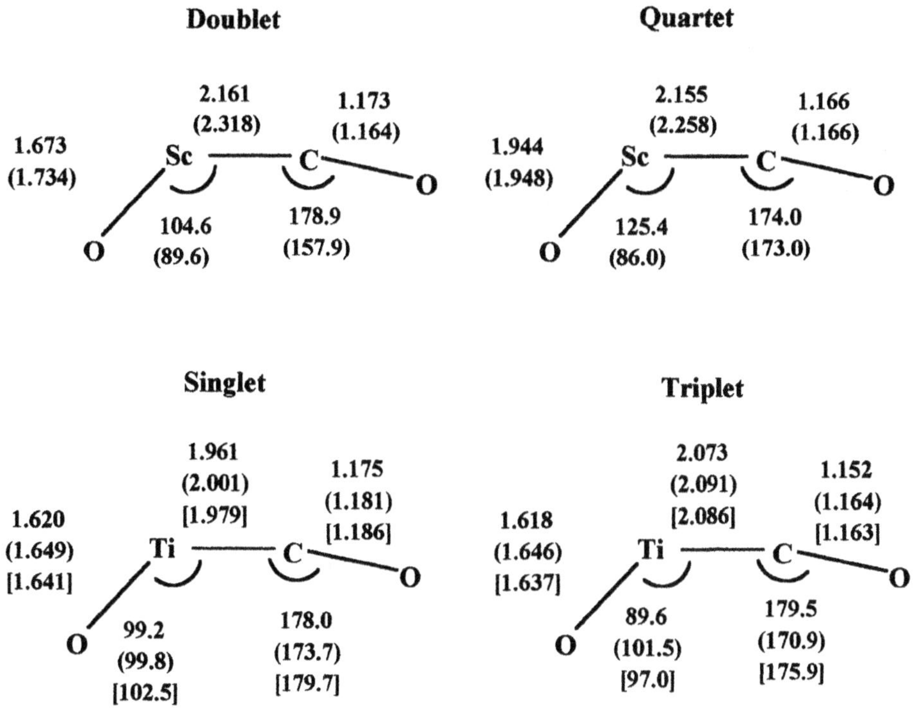

Figure 6. B3LYP, (PWP) and [BP} geometrical parameters for the OScCO and OTiCO systems. Distances are in Å and angles in degrees.

TABLE 7. B3LYP (PWP) harmonic vibrational frequencies (cm^{-1}) for the ground ($^2A''$) and first excited ($^4A''$) states of the OScCO complex. Relative energy (ΔE) and binding energies (D_1 and D_1) are in kcal/mol.

frequency	$^2A''$	$^4A''$
ω_1	92 (87)	113 (107)
ω_2	177 (161)	234 (329)
ω_3	295 (289)	280 (267)
ω_4	424 (410)	368 (538)
ω_5	957 (980)	630 (601)
ω_6	1986 (1970)	1955 (1940)
ΔE	0.0 (0.0)	53.9 (57.6)
D_1*	18.4 (41.7)	-35.5 (-15.9)
D2**	7.3 (11.5)	-46.6 (46.1)

*relative to the OScCO ($^2A'$) => Sc(^2D) + CO_2 ($^1\Sigma_g^+$) process

**relative to the OScCO ($^2A''$) => ScO ($^2\Sigma^+$) + CO ($^1\Sigma^+$) process

We have explored in more detail the insertion mechanism, at the B3LYP level. The resulting PES is depicted in Figure 7.

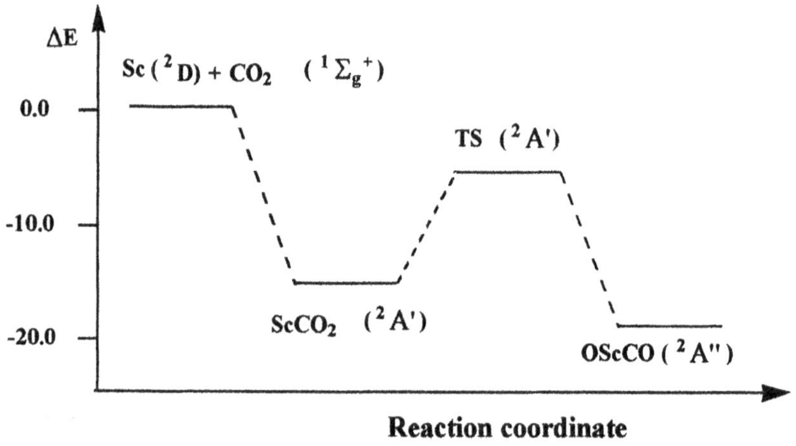

Figure 7. Potential energy surfaces for the insertion mechanism of Sc into the C-O bond of carbon dioxide. Relative energy (ΔE) is in kcal/mol.

As expected, a transition state (TS) connects the $ScCO_2$ (^{2}A') molecular species and the OScCO (^{2}A") insertion product. The geometrical parameters of the TS are reported in Figure 8. This TS is characterised by an imaginary frequency of -336 cm^{-1} and a C-O distance of 1.726Å, which means that the C-O bond is already broken. The energy barrier is found to be 6.4 kcal/mol. Accordingly, the Sc insertion seems to be possible with a relatively low energetic cost, giving a stable molecular species only in its ground doublet state.

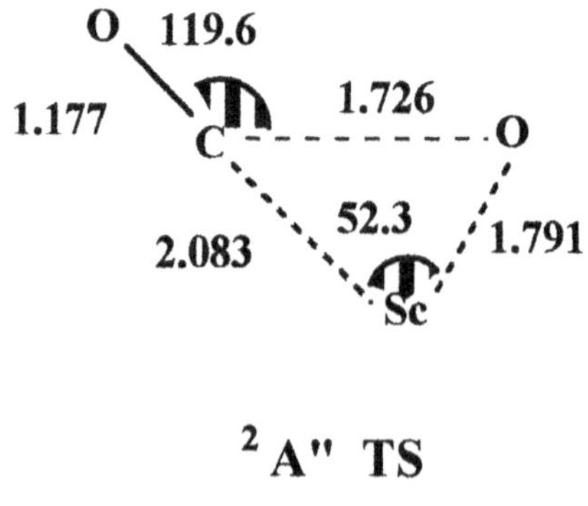

Figure 8. Geometrical structure of the transition state for Sc insertion into a C-O bond of carbon dioxide. Distances are in Å and angles in degrees.

There is experimental evidence [50] that Ti, as well as transition-metal atoms from the left-end side of the periodic table (V and Cr), inserts spontaneously into the CO bond, giving the OTiCO system. Some IR frequencies [50] have been measured in CO_2 matrices, and been assigned to a possible OTiCO (CO_2) complex. Papai, Mascetti and Fournier have recently studied the Ti + CO_2 reaction, using DFT with the BP exchange-correlation functional [26]. Their conclusion was that Ti insertion into the CO bond of CO_2 takes place without any activation barrier.

We examined the insertion of Ti in CO_2, considering different spin multiplicities (singlet, triplet, quintet). Because the quintet state lies at very high energy with respect to the singlet and triplet (more than 50 kcal/mol at both B3LYP and PWP levels), we report only the results for the first two electronic states. Structures are shown in Figure 5 and harmonic vibrational frequencies and energetic parameters are reported in Table 8.

For this element the agreement between the geometrical parameters obtained at both PWP and B3LYP levels is better than in the corresponding scandium system. Our structures are also very close to the BP counterparts. The $^3A''$ state is the most stable, but the $^1A'$ excited state lies only at 1.2 (PWP) or 9.6 (B3LYP) kcal/mol from it. A spacing of 2.2 kcal/mol is found using the BP exchange-correlation functional [26].

Using the CO and TiO ground-state optimised geometries, OTiCO ($^3A''$) $\rightarrow$ TiO ($^3\Delta$) + CO ($^1\Sigma^+$), the OTiCO binding energy is found to be 18.1 (B3LYP) or 28.8 (PWP) kcal/mol. Comparison with the corresponding BP binding energy (23.0 kcal/mol) indicates an agreement between the two gradient-corrected calculations, but these results appear to be overestimated with respect to the value obtained with the hybrid B3LYP method. Considering the Ti (^3F) + CO_2 ($^1\Sigma_g^+$) $\rightarrow$ OTiCO ($^3A''$) reaction, the differences between B3LYP and PWP energies (see Table 8) are again evidenced. These are probably due to the poor description of the Ti atomic ground state at the PWP level. Here we underline that both hybrid and gradient-corrected functionals give the same qualitative results from the thermodynamic point of view.

The B3LYP and PWP electronic structures of the OTiCO insertion product appear to be very similar to the BP one [26]. The net charge distribution analysis shows a charge transfer between Ti and O in the Ti-O bond, and inspection of the molecular orbitals reveals that only two molecular orbitals (15a' and 5a") include contributions from all atoms while others have pure TiO or CO character. The composition of 15a' and 5a" orbitals shows that the Ti-C bond arises from CO $\rightarrow$ Ti σ-donation (15a') and simultaneously Ti $\rightarrow$ CO π-back donation from the HOMO (5a").

The calculated harmonic vibrational frequencies can be compared with two experimental IR bands at 1867 and 953 cm^{-1}. The three computed frequencies are similar and indicate that the two experimental bands can be associated to the TiO and especially the CO stretching modes of the $^1A'$ excited state.

TABLE 8. B3LYP (PWP) [BP] harmonic vibrational frequencies (cm^{-1}) for the ground ($^3A''$) and first excited ($^1A'$) states of the OTiCO complex. Relative energy (ΔE) and binding energies (D_1 and D_1) are in kcal/mol. BP results from ref. 29

frequency	$^1A'$	$^3A''$
ω_1	140 (253) [190][a]	127 (242) [230]
ω_2	296 (332) [326]	236 (272) [241]
ω_3	363 (393) [365]	364 (349) [339]
ω_4	517 (486) [495]	408 (429) [398]
ω_5	1041 (1053) [967]	1027 (1053) [991]
ω_6	1985 (1895) [1881]	2061 (1976) [1986]
ΔE	9.6 (1.2)	0.0 (0.0)
D_1*	19.7 (66.2)	29.3 (67.4)
D_2**	8.5 (27.6)	18.1 (28.8)

* relative to the $TiCO_2$ ($^3A'$) → $Ti(^3F)$ + CO_2 ($^1\Sigma_g^+$) process
**relative to the OTiCO ($^3A''$) → TiO ($^3\Delta$) + CO ($^1\Sigma^+$) process

Considering the low separation energy between the $^3A''$ and $^1A'$ states, the hypothesis that, in the reaction of laser-ablated Ti, TiO and TiO_2 with CO, the excited $^1A'$ state can be generated, becomes plausible. This is consistent with the B3LYP results for the PES of the insertion mechanism of Ti into the C-O bond of carbon dioxide, which shows that no barrier is necessary for this process, in agreement with the experimental indication [49].

3.6. HYPERFINE COUPLING CONSTANTS

Isotropic hyperfine coupling constants (A_{iso}) were computed at the B3LYP all-electron level, using the optimised geometry for the $ScCO_2$ and OScCO lowest isomers. The following expression has been considered for the calculation:

$$A_{iso} = \frac{8\pi}{3}\frac{g_e}{g_0} g_N \beta_N \sum (P^{\alpha}_{\mu,\nu} - P^{\beta}_{\mu,\nu}) \langle \varphi_\mu | \delta(r_{kN}) | \varphi_\nu \rangle$$

where P^α and P^β are the computed density matrices, β_N is the nuclear magneton, g_e, g_0 and g_N are the free-electron, electron-in-radical and nuclear magnetogyric ratios, respectively, and δ is the Dirac operator. Here it is assumed that $g_0 = g_e$.

Since no experimental data exist on these systems, our values can be only a guideline for possible EPR measurements. A summary of the results is reported in table 9.

TABLE 9. Isotropic hyperfine coupling constants (A_{iso}) for $ScCO_2$ and OScCO lowest isomers

Atom	η^2C,O	η^2O,O	OScCO
Sc	920.03	981.91	-125.72
C	12.02	32.47	18.09
O	-6.30*	0.87	4.85*
O	-2.12	0.87	-27.28

*) values referred to the oxygen atom directly bound to the metal

Previous studies [51,52] have shown the reliability of B3LYP calculations in A_{iso} predictions, even for transition metal-containing systems. Therefore the identification of the most stable form could experimentally be done, considering the different values of A_{iso} for the Sc atom in the molecular complexes and in the insertion product.

4. Concluding remarks

From the systematic work described in the previous sections several features can be outlined. First, hybrid B3LYP and gradient-corrected PWP functionals predict the same qualitative behaviour in the M (Sc, Ti, Ni and Cu) + CO_2 reactions, but quantitative differences are often found. With respect to the reactivity of the different transition-metal atoms studied here, it has been shown that Cu and Ni give weakly bound complexes with CO_2 while Ti and Sc are able to form MCO_2 stable complexes and OMCO insertion products. In the first case, the insertion occurs without any energy barrier while in the second, a small barrier of 6 kcal/mol is found.

Acknowledgements

This work was performed with the financial help of the Italian MURST, Universita' della Calabria, and Spanish Ministerio de Educacion y Ciencia under CICyT project PB98-1216-CO2-01. N. R. and F. M. acknowledge support for their stay in Barcelona through TMR contract ERB FMGE CT95 0062 held by the *Centre de Supercomputacio de Catalunya* (CESCA) *and Centre Europeu de Parallelisme de Barcelona* (CEPBA).

References

1. W. Keim (ed.) *Catalysis in C_1 Chemistry* (Reidel, Dordrecht, 1983).
2. M. E. Vol'pin and I. S. Kolomnikov, *Pure Appl. Chem.*, **33** (1973) 567.
3. S. Inoue and N. Yamazaki (eds.) *Organic and Bioinorganic Chemistry of Carbon Dioxide* (Kodansha, Tokyo, 1982).

4. R. P. A. Sneeden, *J. Mol. Catal.*, **17** (1982) 349.
5. A. W. Yau and H. O. Pritchard, *Can. J. Chem.*, **57** (1979) 1731.
6. P. Braunstein, D. Matt and D. Nobel, *Chem. Rev.*, **88** (1988) 747.
7. A. Behr (eds.) *Carbon Dioxide Activation by Metal Complexes* (VCH, Berlin, 1988).
8. M. Aresta and J. V. Schloss (eds.) *Enzymatic and Model Carboxylation and Reactions for Carbon Dioxide Utilization* (Kluwer, Dordrecht, 1990).
9. H.-J. Freund and M. W. Roberts, *Surf. Sci. Rep.*, **25** (1996) 1.
10. E. W. Plummer, C. T. Chen, W. K. Ford, W Eberhardt, R. P. Messmer and H.-J. Freund, *Surf. Sci.*, **158** (1985) 58.
11. D. A. Palmer and R. Van Eldik, *Chem. Rev.*, **83** (1983) 651.
12. P. G. Jessop, T. Ikariya and R. Noyori, *Chem. Rev.*, **95** (1995) 259.
13. S. Sakaki and A. Dedieu, *Inorg. Chem.*, **26** (1987) 3278.
14. H.-J. Freund and R. P. Messmer, *Surf. Sci.*, **172** (1986) 1.
15. G. H. Jeung, *Chem. Phys. Lett.*, **232** (1995) 319.
16. D. R. Salahub and N. Russo (eds.) *Metal Ligand Interaction. From Atoms to Clusters to Surfaces* (Kluwer, Dordrecht, 1992).
17. H. Huber, D. McIntosh and G. A. Ozin, *Inorg. Chem.*, **16** (1977) 975.
18. G. A. Ozin, H. Huber and D. McIntosh, *Inorg. Chem.*, **17** (1978) 1472.
19. J. Mascetti and M. Tranquille, *J. Phys. Chem.*, **92** (1988) 2177.
20. F. Galan, M. Fouassier, M. Tranquille and J. Mascetti, *J. Phys. Chem.*, **101** (1997) 2626.
21. M. Sodupe, V. Branchadell and A. Oliva, *J. Phys. Chem.*, **99** (1995) 8567.
22. L. Rodriguez-Santiago, M. Sodupe and V. Branchadell. *J. Chem. Phys.*, **105** (1996) 9966.
23. G. H. Jeung, *Mol. Phys.*, **65** (1988) 669.
24. R. Caballol, E. Sanchez Marcos and J. Barthelat, *J. Phys. Chem.*, **91** (1987) 1328.
25. S. Sirois, M. Castro and D. R. Salahub, *Int. J. Quantum Chem.*, **S28** (1994) 645.
26. I. Papai, J. Mascetti and R. Fournier, *J. Phys. Chem.*, **101** (1997) 4465.
27. G. H. Jeung, *Mol. Phys.*, **67** (1989) 747.
28. Y. Yoshioka and K. Jordan, *Chem. Phys. Lett.*, **84** (1981) 370.
29. K. Jordan, *J. Phys. Chem.*, **88** (1984) 2459.
30 J. Bentley and I. J. Carmichael, *J. Phys. Chem.*, **89** (1985) 4040.
31. F. Ramondo, N. Sanna, L. Bencivenni and F. Grandinetti, *Chem. Phys. Lett.*, **180** (1991) 369.
32. F. Moscardo and E. San-Fabian, *Theor. Chim. Acta*, **70** (1986) 297.
33. F. Mele, N. Russo and F. Illas, *Chem. Phys. Lett.*, **295** (1998) 409.
34. J. P. Perdew and Y. Wang, *Phys. Rev. B*, **33** (1986) 8800.
35. J. P. Perdew, *Phys. Rev. B*, **33** (1986) 8822.
36. A. D. Becke, *J. Chem. Phys.*, **98** (1993) 5648.
37. C. Lee, W. Yang and R. G. Parr, *Phys. Rev. B*, **37** (1988) 785.
38. A. St. Amant, PhD Thesis, Université de Montreal, 1992.
39. C. C. Broyden, *J. Inst. Math. Appl.*, **6** (1970) 76; R. Fletcher, *Comput. J.*, **13** (1970) 317; D Goldfarb, *Math. Comput.*, **24** (1970) 23; D. F. Shanno, *Math. Comput.*, **24** (1970) 647.
40. Y. Abaskin and N. Russo, *J. Chem. Phys.*, **100** (1994) 4477.
41. N. Godbout, D. R. Salahub, J. Andzelm and E. Wimmer, *Can. J. Chem.* **70**, (1992) 560.
42. A. Piluso, A. Goursot and D. R. Salahub, *Inorg. Chem.*, **29** (1990) 1545.
43. V. Musolino, N. Russo and M. Toscano, *Phys. Lett.*, **165A** (1992) 377.
44. M. J. Frisch, G. W. Trucks, H. B. Schlegel, P. M. W. Gill, B. G. Johnson, M. A. Robb, J. R. Cheeseman, T. A. Keith, G. A. Petersson, J. A. Montgomery, K. Raghavachari, M. A. Al-Laham, V. G. Zakrzewski, J. V. Ortiz, J. B. Foresman, J. Cioslowski, B. B. Stefanov, A. Nanayakkara, M. Challacombe, C. Y. Peng, P. Y. Ayala, W. Chen, M. W. Wong, J. L. Andreas, E. S. Reploge, R. Gomperts, R. L. Martin, D. J. Fox, J. S. Binkley, D. J. Defrees, J. Baker, J. P. Stewart, M. Head-Gordon, C. Gonzales, J. A. Pople (1995) Gaussian, Inc., Pittsburg, PA.
45. P. J. Hay and W. R. Wadt, *J. Chem. Phys.*, **82** (1985) 270; 284; 299.
46. A. Martinez, A. M. Köster and D. R. Salahub, *J. Phys. Chem.* A, **101** (1997) 1532.

47. N. Russo and D. R. Salahub (eds.) *Metal Ligand Interaction: Structure and Reactivity* (Kluwer, Dordrecht, 1995).
48. T. Ziegler, *Chem. Rev.*, **91** (1990) 651.
49. D. P. Chong (ed.) *Recent Advances in Density Functional Methods*. Parts I and II (World Scientific, Singapore, 1995 and 1997).
50. G. V. Chertihin and A. Andrews, *J. Am. Chem. Soc.*, **117** (1995) 1595.
51. V. Barone, R. Fournier, F. Mele, N. Russo and C. Adamo, *Chem. Phys. Lett.*, **237** (1995) 189.
52. F. Mele, N. Russo, M. Toscano and J. Rubio, *J. Mol. Struct. THEOCHEM*, in press.

Part VIII
Condensed Matter

RECURRENT VARIATIONAL APPROACH APPLIED TO THE ELECTRONIC STRUCTURE OF CONJUGATED POLYMERS

STÉPHANE PLEUTIN
Max-Planck-Institut für Physik Komplexer Systeme, Dresden, Germany

ERIC JECKELMANN
Fachbereich Physik, Philipps-Universität Marburg, Marburg, Germany

MIGUEL A. MARTÍN-DELGADO
Departamento de Física Teórica I, Universídad Complutense, Madrid, Spain

AND

GERMAN SIERRA
Instituto de Matemáticas y Física Fundamental, C.S.I.C., Madrid, Spain

Abstract. In this note, first the Recurrent Variational Approach (RVA) is introduced by using as example a non–trivial spin–model, the spin–1/2 antiferromagnetic two–leg–ladder. Then, a first application of this scheme to the electronic structure of conjugated polymers is proposed. An analogy between the two–leg ladder and the dimerized chain described by the Extended Peierls–Hubbard Hamiltonian is underlined and comparisons are made with DMRG calculations. The proposed ansatz is useful to get some analytical insight and it is a good starting point for further improvements using the RVA.

1. Introduction

Conjugated molecules have attracted the attention of chemists and physicists since the early days of quantum mechanics. A remarkable example is given by Hückel who developed in the thirties the first independent–electron theory applicable to polyenes [1]. In the sixties, J.A. Pople and S.H. Walm-

J. Maruani et al. (eds.), New Trends in Quantum Systems in Chemistry and Physics, Volume 2, 169–187.

sley [2], introduced by using this simple theory in their pioneering work what would become twenty years later the solitonic excitations. Following this line, Su, Schrieffer, Heeger and, independently, Rice, proposed a model specifically devoted to the polyacetylene – known as the SSH model – where the role of the electron–phonon interaction is especially emphasized [3, 4]; here, the polyacetylene is thought of as a Peierls insulator and the solitons, polarons and bipolarons as the major relevant excitations in this system. A very nice review of these early studies is included in the book of L. Salem [5] and a review about the SSH model can be found in the next reference [6].

As a counterpart to this one–electron theory, the major role of electron correlation was soon recognized. In the fifties, the celebrated Pariser–Parr–Pople (PPP) Hamiltonian was suggested as a possible candidate for the studies of conjugated oligomers [7, 8]. Later, Ovchinnikov et al. suggested to consider conjugated polymers as Mott–Hubbard instead of Peierls insulators [9]. Then, one has to face a N–body problem with a strong Coulomb interaction with a non–zero range; it was subject to many numerical works such as exact diagonalizations of very small oligomers [10], selected CI calculations [11] and, quite recently, Density Matrix Renormalization Group studies [12, 13, 14, 15]. The PPP Hamiltonian was also a preferential model to support the development of new methods before applying them to more realistic models; a very good example of such work is given by the Coupled–Cluster method first implemented in quantum chemistry for the PPP Hamiltonian [16].

Today, it is well recognized that both the electron–phonon and electron–electron interactions are equally important [17]. Moreover, interchain interaction and disorder are also of importance. Therefore, the problems one has to face to fully understand conjugated polymers stay quite challenging as it is the case for most of the recent organic compounds including high Tc superconductors. Despite a large amount of work, from the early thirties to now, the main questions remain awaiting an answer! For instance, a comprehensive picture for the ground state of these systems is still needed; it is the goal of this work to propose a relatively simple and, we believe, promising, way to understand the ground state of conjugated polymers.

More generally, one dimensional correlated electron systems appear over the years as a growing and important field in condensed matter theory. One main reason of that comes from the fact that these systems serve as a theoretical laboratory to explore new methods of solution, analytically or numerically. A second main reason – we would say more physical – comes from the fact that over the years, more and more experimental realisations of these systems appear. New concepts emerge from such studies as the ones included in the phenomenology of Luttinger–liquids [18].

Among the technical methods proper to the one dimensional geometry, one may cite the Bethe ansatz [19], the bosonization techniques [18], and, more recently, the Density–Matrix Renormalization Group (DMRG) method [20, 21] and a closely related scheme which is directly considered in this note, the Recurrent Variational Approach (RVA) [22, 21]. The two first methods are analytical and the third one is numerical; the RVA method is in between.

The RVA, presented here, is a variational method invented recently in the context of spin–ladders [22]. It is a method very closely related to the DMRG scheme with the main differences coming from the fact that only one state is retained as the best candidate for the ground state in the RVA on the contrary to the DMRG which considered much more states [21]. Most often, the results obtained with it are less accurate than the DMRG ones, but it is much easier to get a physical insight into the problem; the analysis of the results is simpler and the RVA could provide a "physical" picture of some phenomena which, we believe, are of utmost importance in the field of conjugated polymers where the most important ingredient are still not fully recognized [23].

This note is organized as follows. In section II, the RVA is presented using the example of the two–leg spin ladders. In section III, the RVA method is applied to the study of the dimerized chain described by the Extended Peierls–Hubbard model. A nice similarity between the two–leg ladder and this system is pointed out and comparisons with DMRG calculations are made. We conclude in section IV and give some reasonable perspectives.

2. RVA method applied to two–leg spin ladders

A spin ladder is an array of coupled spin chains. The horizontal chains are called the legs, the vertical ones, rungs. In the case of spin one–half antiferromagnet spin–ladders, these systems show a remarkable behaviour in function of the number of leg: there is a gap in the excitation spectrum of even–leg ladders and, on the contrary, no gap in the excitation spectrum of odd–leg ladders. In terms of correlation lengths, this means that there is short (long) –range spin correlation in even (odd) –leg ladder (see [24] for a review).

The even–leg ladders show a spin-liquid ground state described by a Resonance Valence Bond (RVB) state introduced by P.W. Anderson m[25]; such systems may be efficiently described by using a direct–space method as the DMRG and RVA ones. In this note, we will consider only the simplest even–leg ladder, the two–leg ladder, and we will describe it with the simplest RVB state, the Dimer–RVB state where the elementary singlets are between two nearest–neighbour sites. This problem is easily solved by using the RVA method [22]. Moreover, with the same scheme, it is possible straightforwardly to go beyond the simple Dimer–RVB state by considering more extended elementary singlets [22], and to study more complex systems as the four–leg ladders for instance [26]. In principle, this method should be relevant for every one–dimensional system with exponentially decreasing correlation lengths as conjugated polymers.

In this note we consider the simplest case given by the two–leg ladder (see figure (1)). The corresponding Hamiltonian is the following

$$H = J \sum_{k=1}^{N-1} (\mathbf{S}_k \cdot \mathbf{S}_{\overline{k+1}} + \mathbf{S}_{k+1} \cdot \mathbf{S}_{\overline{k}}) + J' \sum_{k=1}^{N} \mathbf{S}_k \cdot \mathbf{S}_{\overline{k}} \tag{1}$$

where J (J') are the exchange integral for the leg (rung) with $J > 0$ ($J' > 0$). N denotes the number of rungs of the ladder and open periodic boundary conditions have been assumed.

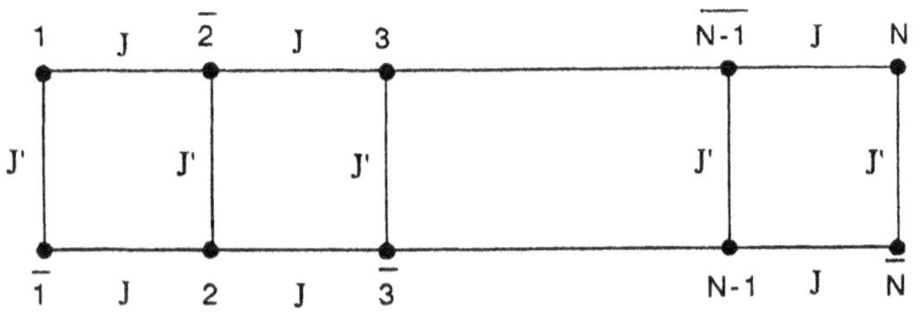

Figure 1. The two–leg ladder with an AF–Heisenberg model described by two coupling constants: J for the links in the legs, and J' for the vertical rungs.

In the limit where J' is much larger than J, the ground state of (1) is simply given by the product of singlets localized along the rungs (see fig. (3)). Starting from this limit, one may include fluctuations around it to treat the general case described by the Hamiltonian (1). A minimal way to do it, is to consider instead of singlet on rungs, pairs of nearest neighbour singlets on the leg; these local fluctuations are called resonon in the rest of the paper. Therefore, the resonance mechanism – in the sense of Pauling [27] – we will consider in the following are represented in figure (2).

With the two selected local configurations, the singlet localized on a rung and the resonon, we have to generate the corresponding Hilbert space; the configurations are characterized by the number of resonon, M, and their positions, $\{x_i\}$, as it is shown in the next examples. The Hilbert space is split according to the number of resonon; we have then to consider

- the zero–resonon sector with one unique state $|0\rangle$ (Fig. 3),
- the one–resonon sector with the states $\{|x_1\rangle\}$ (Fig. 4),
- the two–resonon sector with the states $\{|x_1, x_2\rangle\}$ (Fig. 5),

etc...

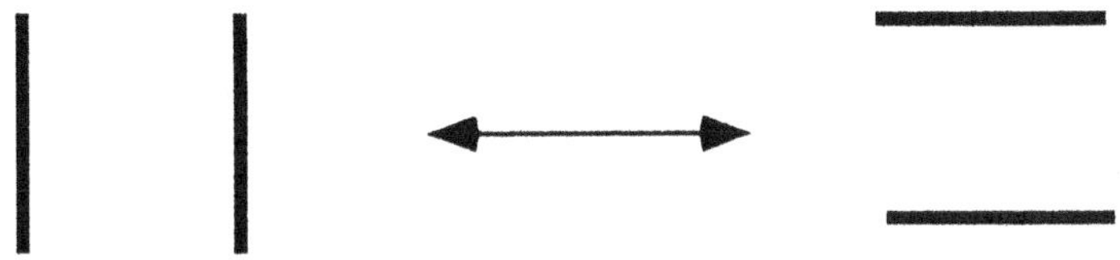

Figure 2. The basic bond resonance mechanism between horizontal and vertical bonds in an elementary plaquette of four sites.

Figure 3. The zero–resonon state $|0\rangle$ in a two–leg ladder. It is made of all vertical rung singlets. It is the reference state in the strong coupling limit $J' \gg J$.

- the $\frac{N}{2}$–resonon sector with one unique state $|x_1, x_2, \ldots, x_M\rangle$ (Fig. 6).

With this classification of states we may write the following generalized Dimer–RVB state for a two–leg spin ladder,

$$|\psi(u)\rangle_{RVB} = \sum_{M=0}^{[N/2]} u^M \cdot \sum_{\{x_1,x_2,\ldots,x_M\}} |x_1, x_2, \ldots, x_M\rangle \tag{2}$$

where the sum is taken over all sectors with M resonons and u is the amplitude for a resonon. Here u is a variational parameter to be determined upon minimization of the ground–state energy. As a result, it is a function of the ratio of couplings, namely, $u = u(J/J')$.

Since we kept only two very localized building blocks, it is possible to generate the state (2) in a recursive manner which we present now.

2.1. RECURRENCE RELATIONS FOR THE WAVE FUNCTION

Denoting by $|N\rangle \equiv |\psi(u)\rangle_{RVB}$ a generic RVB state (2) for a two–leg ladder of N rungs, there are only two possibilities or *movements* to create generic RVB states of higher length, namely,

- addition of one vertical rung to create the state $|N + 1\rangle$,
- addition of one pair of horizontal bonds (resonon) to create $|N + 2\rangle$.

From these arguments we can establish that the Dimer–RVB states (2) satisfy a *recursion relation* (RR) given by,

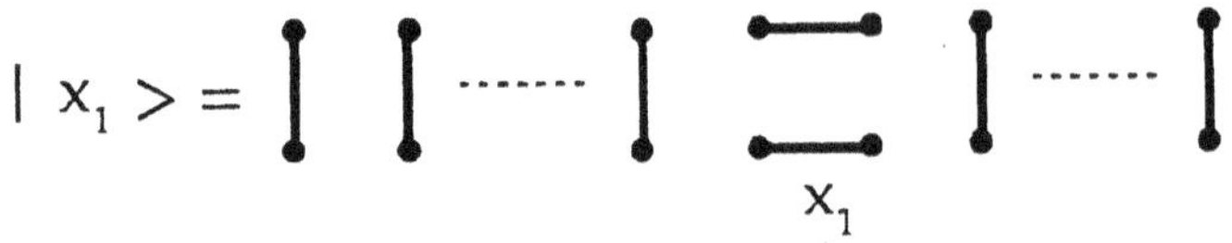

Figure 4. A generic one–resonon state $|x_1\rangle$ in a two–leg ladder.

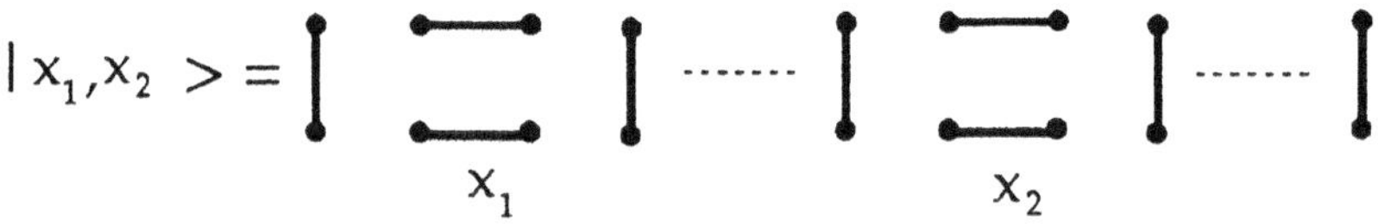

Figure 5. A generic two–resonon state $|x_1, x_2\rangle$ in a two–leg ladder.

$$|N+2\rangle = |N+1\rangle \otimes |\phi_1\rangle_{N+2} + u\,|N\rangle \otimes |\phi_2\rangle_{N+1,N+2} \tag{3}$$

where the state denoted by $|\phi_1\rangle_{N+2}$ is a vertical rung at position $N+2$ and $|\phi_2\rangle_{N+1,N+2}$ is made up of a pair of horizontal bonds located between the rungs at $(N+1, N+2)$, i.e.,

$$|\phi_2\rangle_{N+1,N+2} = (N+1, \overline{N+2})(N+2, \overline{N+1}) \tag{4}$$

where $(N, \overline{M})$ denotes a singlet between the sites N and $\overline{M}$.

Using (3) one can generate recursively the Dimer RVB state $|N\rangle$ from previous states with lower length and estimate variationally the ground–state energy as,

$$e_N^{(g.s.)} \equiv \frac{\langle N|H_{AFM}^{two\ leg}|N\rangle}{(2N)\langle N|N\rangle} = e(N, J'/J; u) \tag{5}$$

The value of u is fixed by minimization of (5).

2.2. RECURRENCE RELATION FOR THE NORM

To compute the norms of the states (2) let us define

$$Z_N = \langle N|N\rangle \tag{6}$$

The two selected local configurations are not orthogonal, it is then convenient to define an auxiliary function Y_N as follows,

$$Y_N = ({}_N\langle\phi_1| \otimes \langle N-1|)|N\rangle \tag{7}$$

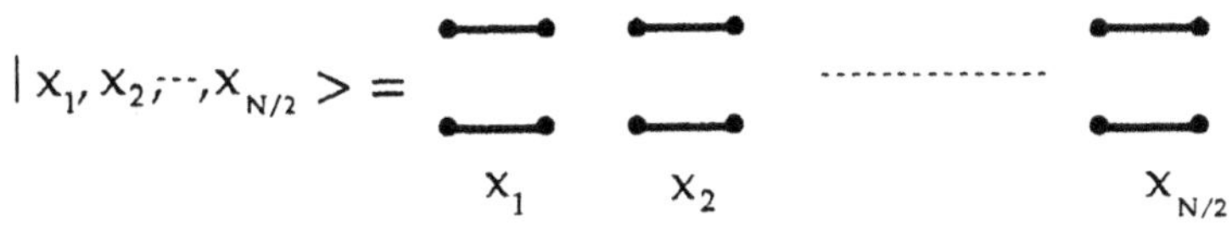

Figure 6. The all–resonon state $|x_1, x_2, \ldots, x_{\frac{N}{2}}\rangle$ in a two–leg ladder.

The second order RR's for the states leads (3) to a closed set of RR's for the overlaps Z_N, Y_N , namely,

$$\begin{aligned} Z_{N+2} &= Z_{N+1} + uY_{N+1} + u^2 Z_N \\ Y_{N+2} &= Z_{N+1} + \tfrac{u}{2} Y_{N+1} \end{aligned} \tag{8}$$

where we have made use of the result,

$$_{N+1}\langle\phi_1|\phi_2\rangle_{N,N+1} = \frac{1}{2}|\phi_1\rangle_N \tag{9}$$

The RR's (8) together with the initial conditions

$$Z_0 = Z_1 = 1 \qquad Y_0 = 0,\ Y_1 = 1 \tag{10}$$

determine Z_N and Y_N for arbitrary values of N as functions of u.

Note that instead of introducing the function Y_N, it is also possible to make first linear combination of our building blocks, making used of (9) [28], in order to guaranty the orthogonality between the selected local configurations. The overlap would then be given by only one recurrent relation as it will be the case in the next section.

2.3. RECURRENCE RELATION FOR THE EXPECTATION VALUE OF THE HAMILTONIAN

The expectation values of the Hamiltonian can be determined following the same steps outlined above. To this end, we introduce the quantities

$$E_N = \langle N|H_N|N\rangle \qquad D_N = ({}_N\langle\phi_1| \otimes \langle N-1|)H_N|N\rangle \tag{11}$$

To obtain the RR's satisfied by E_N, D_N, one splits the Hamiltonian H_N of a two–leg ladder of length N (1) into two pieces:

$$H_N = H_{N-1} + H_{N-1,N} \tag{12}$$

where H_{N-1} is the Hamiltonian of length $N-1$ and $H_{N-1,N}$ is the rest of the whole H_N Hamiltonian, $H_{N-1,N} \equiv H_N - H_{N-1}$, which is made of one vertical rung and two horizontal links. With this splitting and using (3) and (6), we obtain

$$\begin{aligned} E_{N+2} = & \; E_{N+1} + J'\epsilon_0 Z_{N+1} + u(D_{N+1} + (2J+J')\epsilon_0 Y_{N+1}) \\ & + u^2(E_N + 2J\epsilon_0 Z_N) \\ D_{N+2} = & \; E_{N+1} + J'\epsilon_0 Z_{N+1} + \tfrac{u}{2}(D_{N+1} + (2J+J')\epsilon_0 Y_{N+1}) \end{aligned} \tag{13}$$

where $\epsilon_0 = -3/4$ is the lowest eigenvalue of the operator $\mathbf{S}_1 \cdot \mathbf{S}_2$.

The initial conditions for E_N and D_N are

$$E_0 = 0,\ E_1 = J'\epsilon_0 \quad D_0 = 0,\ D_1 = J'\epsilon_0. \tag{14}$$

The RR's for the energies involve the norms of the states and depend both on u and the coupling constants J', J.

By working with orthogonal building blocks [28], it is also possible to go back to only one recurrent equation for the expectation value as for the norm; this will be the case in section 3.

2.4. RESULTS FOR THE VARIATIONAL GROUND–STATE ENERGY

The ground–state energy for a ladder of length N is estimated. In the thermodynamic limit $N \to \infty$ one can find a closed expression for the density energy per site [22],

$$e_\infty = \lim_{N\to\infty} \frac{1}{2N}\frac{E_N}{Z_N} = \frac{R(\alpha)}{2\alpha Q'(\alpha)P(\alpha)} \tag{15}$$

in terms of three polynomials P, Q, R evaluated at the biggest root α of the cubic polynomial $Q(y)$. Finally one looks for the absolute minimum of e_∞ by varying the parameter u. In Table 1 we show the ground–state energies per site for different values of the coupling constant ratio J/J', varying through strong, intermediate and weak coupling regimes.

In the strong coupling regime $J/J' < 1$ the RVA states give a slightly better ground–state energy than the mean field result. This latter state produces rather unphysical results for $J/J' > 1$, which does not occur in our case. The RVA results compare also well with the exact results for $J' \sim J$.

The RR's can also be used to compute the spin correlator $\langle \mathbf{S}_i \cdot \mathbf{S}_j \rangle$ which has an exponential decay behaviour $\exp(-|i-j|/\xi)$, with ξ the spin correlation length which satisfies the equation

$$u^3\mathcal{L}^3 - (2+u)u^2\mathcal{L}^2 - (2+4u)u^2\mathcal{L} + 4u^3 = 0 \tag{16}$$

where $\mathcal{L} = e^{1/\xi}$. For $J = J'$ one finds $\xi = 0.737$, which can be compared with its exact value given by 3.2. The latter results imply that at the isotropic

TABLE 1. The values $-e_\infty^{\mathrm{MF}}/J'$ are mean field (MF) values taken from [29], while $-e_\infty^{\mathrm{Lan}}/J'$ are Lanczos values taken from [30]

J/J'	u	$-e_\infty^{\mathrm{RVA}}/J'$	$-e_\infty^{\mathrm{MF}}/J'$	$-e_\infty^{\mathrm{Lan}}/J'$
0	0	0.375	0.375000	
0.2	0.128521	0.383114	0.382548	
0.4	0.323211	0.40835	0.405430	
0.6	0.578928	0.44853	0.442424	
0.8	0.87441	0.499295	0.489552	
1	1.18798	0.556958	0.542848	0.578
1.25	1.58519	0.63518	0.614473	0.6687
1.66	2.21853	0.772172	0.738360	0.8333
2.5	3.39153	1.06915	1.002856	1.18
5	5.9777	1.99285		2.265

point the bonds extended over three rungs are quite important. The simplest improvement of the Dimer–RVB state is to add a bond of length $\sqrt{5}$, analogue to a knight move, shown in figure (7); this first improvement leads to a third order RR. The ground state energy per site of this state is given by -0.5713 while the spin correlation length is 0.959. Hence there is an improvement in both quantities but still one needs longer bonds.

Figure 7. Local configurations extended over three rungs (the knight move).

The RVA method has already been applied to several model systems: to the Heisenberg two–leg ladder as we have seen above [22] and to two generalizations of it, the t-J [32] and the Hubbard [33] two–leg ladders where the t-J and the Hubbard Hamiltonian, respectively, are defined on the lattice represented in the figure (1); it has been also applied to other ladders, the four–leg ladder [26] and the diagonal ladders [34], and to the spin–1 chain [35]. In this latter case, very accurate calculations have been made by considering several local configurations (instead of our two retained here); the results are compared with Density Matrix Renormalization Group (DMRG) calculations [31]. We see that, in any case, the comparison are very good

TABLE 2. Here m is total the number of multiplets, N_A is the number of independent variational parameters, d_J is the number of multiplets with spin J, $e_\infty^{\mathrm{MP,DMRG}}$ is the g.s. energy density of the matrix–product state (DMRG), $1-P_m$ is the probability of the states truncated out in the DMRG and ξ^{MP} is the spin correlation length of the MP state. The exact results are given by $e_\infty = 1.4014845$ and $\xi = 6.03$ [31]

m	N_A	$d_{1/2}$	$d_{3/2}$	$d_{5/2}$	$-e_\infty^{\mathrm{MP}}$	$-e_\infty^{\mathrm{DMRG}}$	$1-P_m$	ξ^{MP}
1	0	1	0	0	1.333333	1.333333	1.6×10^{-2}	0.910
2	2	1	1	0	1.399659	1.369077	1.4×10^{-3}	2.600
3	4	2	1	0	1.401093	1.392515	1.3×10^{-5}	3.338
4	7	2	2	0	1.401380	1.401380	1.6×10^{-5}	3.937
5	10	2	2	1	1.401443	1.401436	7.6×10^{-6}	4.085
6	13	2	3	1	1.401474	1.401468	1.3×10^{-6}	4.453

and even sometimes the RVA method gives better results than the DMRG (see Table 2). A more complete review about RVA can be found in the following reference [21].

3. RVA method applied to conjugated polymers

The low–energy properties of conjugated polymers involve π–electrons. Usually, they are described by the use of effective models as the well–known Pariser–Parr–Pople Hamiltonian [7, 8]. This model includes both electron–phonon interaction (semi–classically) and a long range Coulomb interaction. The characteristic values for each kind of energetic contribution, kinetic, electron–phonon and electron–electron terms are approximatively of the same order of magnitude [17]; in this regime, the study of the PPP Hamiltonian – or some other short versions of it – are really not obvious.

Compared to the previous model used in section (2), the spin 1/2 antiferromagnetic Heisenberg model for a two–leg ladder, the PPP – or related – model is much more complicated: the long–range part of the Coulomb interaction is very difficult to treat and the dynamical variable are still not only spin but electron which involves spin and charge degree of freedom. However, due to the alternance between monomers, which can be large, and the link (the single bond) between the monomers, we will underline in this work a remarkable similarity between conjugated polymers and two–leg ladder. A similar wave function than (2) are proposed here for conjugated polymers - the rung of the ladder is played by the monomer and the resonon excitation of the two–leg ladder by a set of intermonomer fluctuations [36, 37]. We believe this wave function a natural and good starting point for further refinements.

We choose as model system the simplest conjugated polymers – the trans–polyacetylene. This compound shows a dimerized structure with an alternance between double bond (1.35Å) and single bond (1.45Å); the monomer is then simply a double bond (see figure (8)). The π electrons are assumed to be effectively describe not by the full PPP Hamiltonian but by the Extended Peierls–Hubbard model (EPH) [17, 14], a short version of it, for simplicity

$$\begin{aligned} H = -\textstyle\sum_{n,\sigma}(t-(-1)^n\frac{\Delta}{2})(c^{\dagger}_{n+1,\sigma}c_{n,\sigma}+c^{\dagger}_{n,\sigma}c_{n+1,\sigma})+ \\ U\textstyle\sum_n n_{n\uparrow}n_{n\downarrow}+V\sum_n n_n n_{n+1} \end{aligned} \tag{17}$$

The operator $c^{\dagger}_{n,\sigma}$ ($c_{n,\sigma}$) creates (annihilates) an electron of spin σ at site n, $n_{n\sigma}=c^{\dagger}_{n,\sigma}c_{n,\sigma}$ and $n_n=n_{n\uparrow}+n_{n\downarrow}$. t is the nearest–neighbour hopping term without dimerization, Δ is the usual dimerization order parameter, U is the on-site Hubbard repulsion and V is the nearest–neighbour charge density – charge density interaction. We will note in the following for convenience, $t_d=t+\frac{\Delta}{2}$ and $t_s=t-\frac{\Delta}{2}$, the hopping term for the double and the single bond respectively.

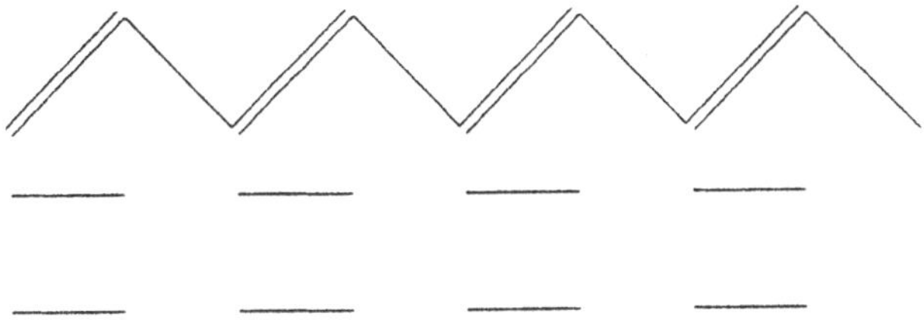

Figure 8. **Model for the trans–polyacetylene and the system of bonding – antibonding levels associated with each double–bond.**

Because of the dimerization, the one electron density is more pronounced on the double bonds than on the single bonds; this is more and more apparent when the dimerization increases. More generally, this is the case for every conjugated polymers where the one–electron density is peaked on the monomer region. In this sense, conjugated polymers are not strictly one dimensional systems but rather intermediate between quasi–zero and quasi–one dimensional systems [36, 37, 38, 39]. It seems then convenient to perform a unitary transformation which favours the orbitals localized on the monomers, in our case, on the double bonds. The new operators are then given by

$$B^{(\dagger)}_{n,\sigma}=\frac{1}{\sqrt{2}}(c^{(\dagger)}_{2n,\sigma}+c^{(\dagger)}_{2n+1,\sigma}) \quad , \quad A^{(\dagger)}_{n,\sigma}=\frac{1}{\sqrt{2}}(c^{(\dagger)}_{2n,\sigma}-c^{(\dagger)}_{2n+1,\sigma}) \tag{18}$$

These operators create or annihilate π electrons in the corresponding bonding or antibonding orbitals of the double bond n (see figure (8)).

As for the case of the two–leg ladder, we select the Local Configurations (LC) which are the much important for the ground state wave function. This selection of the relevant LC is made by combining energetic and symmetry considerations. For that purpose, we use the electron–hole symmetry operator, $\hat{J}$, to classify the LC. On the Fock space of a single site n, the action of this operator are summarized as follows [40]

$$\hat{J}_n \mid 0 >=\mid\uparrow\downarrow>, \quad \hat{J}_n \mid \sigma >= (-1)^n \mid \sigma >, \quad \hat{J}_n \mid\uparrow\downarrow>= -\mid 0 > \tag{19}$$

The electron–hole symmetry operators for several site is just given by the direct product of such single site operator $\hat{J} = \prod_n \hat{J}_n$.

The electronic configurations can be split into two classes of electron–hole symmetry denoted by (+) and (-). The ground state is in the (+) sector, $\hat{J} \mid GS >=\mid GS >$; therefore we select the lowest LC in energy which belong to this class of symmetry. We kept only four of such LC, represented in figure (9) which were shown as the most important ones in finite cluster diagonalizations [10]. We present briefly now these LC; their associated creation operator and their energy. They are classified in two categories following their extension: the LC localized on one double bond are called Molecular Configurations (MC) – to underline the analogy with molecular crystals; the LC extended over two nearest–neighbour double–bonds are named Nearest–Neighbour–Intermonomer–Fluctuations (NNIF). In such classification, the singlet on the rungs of the previous section enters in the first class and the resonon in the second one.

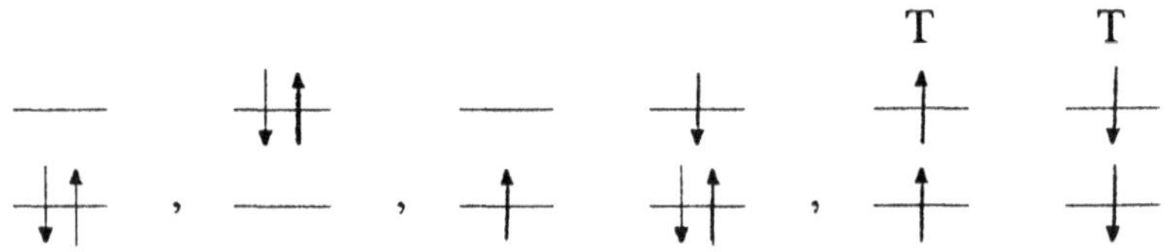

Figure 9. The set of local configurations considered in this work for the dimerized chain; a) the F-LC b) the D-LC c) the Ct_1-LC and d) the TT-LC made by combining two localized triplet symbolized by the T on the figure.

3.1. MOLECULAR CONFIGURATIONS (MC)

The MC are the LC completely localized on the monomers. They give an appropriated description of the system in the strong dimerized limit. There is two kinds of MC in the (+) electron–hole symmetry sector [36, 37].

- In the first, the monomer n is in its ground state; it is described by the following creation operator

$$F_n^\dagger = B_{n,\uparrow}^\dagger B_{n,\downarrow}^\dagger \tag{20}$$

For a reasonable choice of parameters, this is the lowest LC in energy so that we choose as reference state

$$\mid 0 >= \prod_n F_n^\dagger \mid Vacuum > \tag{21}$$

where $\mid Vacuum >$ is the state without any π electron. With respect to this reference state, the local operator $F_n^\dagger$ is simply given by the Unity I_n. We have named this LC, F-LC. In the following all the creation operators and the energies are defined with respect to $\mid 0 >$.
- In the second, the monomer n is doubly excited; this LC is associated with the following local operator

$$D_n^\dagger = A_{n,\uparrow}^\dagger A_{n,\downarrow}^\dagger B_{n,\uparrow} B_{n,\downarrow} \tag{22}$$

with energy given by $\epsilon_d = 4t_d$. This LC introduces electronic correlation inside a particular double bond. We have named this LC D-LC.

Notes that the LC where a monoexcitation take place in a double bond, let's call it S-LC, lower in energy than the D-LC, is in the (–) electron–hole symmetry sector; therefore, it will be a local constituent for excited states [36].

These two LC are the analogue of the singlet along the rung for the two–leg ladder.

3.2. NEAREST–NEIGHBOUR–INTERMONOMER–FLUCTUATIONS (NNIF)

With the two MC, the electrons are coupled by pairs on the double bonds and the resulting picture is the one of a molecular crystal in its ground state [36, 41, 39]; the analogue for the two–leg ladder is given by the state $\mid 0 >$ represented in figure (3) with only singlets along the rungs. This crude picture is obviously not adapted for conjugated polymers where π electrons are delocalized on the whole chain. As in the case of the two–leg ladder, fluctuations must be included in our treatment, we talk in this case of intermonomer fluctuations. This will be done in a somehow 'minimal' way by considering the two following LC, extending over two double bonds only [36, 37].

- In the first, one electron is transfered from one monomer to the nearest–neighbour one; this operation is associated with the local operator written as

$$Ct_n^\dagger = \tfrac{1}{2}(A^\dagger_{n+1,\uparrow}B_{n,\uparrow} + A^\dagger_{n+1,\downarrow}B_{n,\downarrow} - A^\dagger_{n,\uparrow}B_{n+1,\uparrow} - A^\dagger_{n,\downarrow}B_{n+1,\downarrow}) \tag{23}$$

with energy $\epsilon_{ct} = 2t_d + \frac{3}{4}V$. This LC introduces some nearest–neighbour intermonomer charge fluctuation reproducing in a minimal way the conjugaison phenomenom. We named this LC, Ct_1-LC.

- In the second, some intermonomer spin fluctuation is introduced by combining two n.n. triplets into a singlet; the corresponding operator is

$$\begin{aligned} TT_n^\dagger = \tfrac{1}{\sqrt{3}}(&A^\dagger_{n,\uparrow}B_{n,\downarrow}A^\dagger_{n+1,\downarrow}B_{n+1,\uparrow} + A^\dagger_{n,\downarrow}B_{n,\uparrow}A^\dagger_{n+1,\uparrow}B_{n+1,\downarrow} + \\ &\tfrac{1}{2}(A^\dagger_{n,\uparrow}B_{n,\uparrow}A^\dagger_{n+1,\uparrow}B_{n+1,\uparrow} + A^\dagger_{n,\uparrow}B_{n,\uparrow}A^\dagger_{n+1,\downarrow}B_{n+1,\downarrow} + \\ &A^\dagger_{n,\downarrow}B_{n,\downarrow}A^\dagger_{n+1,\uparrow}B_{n+1,\uparrow} + A^\dagger_{n,\downarrow}B_{n,\downarrow}A^\dagger_{n+1,\downarrow}B_{n+1,\downarrow})) \end{aligned} \tag{24}$$

with energy given by $\epsilon_{tt} = 4t_d - (U - V)$; this LC is named TT-LC.

There is more LC extended over two n.n. double–bonds within the (+) electron–hole symmetry class – and, of course, there is much more LC more extended – but we believe the two selected ones are the more important [37]. For instance, we saw that the S-LC is in the (−) class of symmetry, however, with the simple product form of the electron-hole symmetry operator, a LC with two S-LC – or more generally with an even number of S-LC – is in the (+) class of symmetry and should be considered here [37]. However, this kind of LC are sufficiently high in energy to be reasonably neglected.

These two LC are the analogue of the resonon configuration for the two–leg ladder.

3.3. RECURRENCE RELATION FOR THE GROUND STATE WAVE FUNCTION

With our choice of four LC, all possible electronic configurations are then build up. They are characterized by the number of D, Ct_1 and TT-LCs, n_d, n_{ct} and n_{tt} respectively, and by the positions of these different local configurations. The positions of the D, Ct_1 and TT-LCs are labelled by the coordinates $z(k)$ $(k = 1, .., n_d)$, $y(j)$ $(j = 1, .., n_{ct})$ and $x(i)$ $(i = 1, .., n_{tt})$ respectively; implicitly, the necessary non–overlapping condition between LC is supposed to be fulfilled all along the paper. The electronic configurations are then expressed as

$$\begin{aligned} &\mid x(1), ..., x(n_{tt}), y(1), ..., y(n_{ct}), z(1), ..., z(n_d) >= \\ &\prod_{i=1}^{n_{tt}} \prod_{j=1}^{n_{ct}} \prod_{k=1}^{n_d} TT^\dagger_{x(i)} Ct^\dagger_{y(j)} D^\dagger_{z(k)} \mid 0 > \end{aligned} \tag{25}$$

We look for a ground state of the following form

$$\mid GS >= \sum_{\{n_d,n_{ct},n_{tt}\}} u_f^{N-n_d-2(n_{ct}+n_{tt})} u_d^{n_d} u_{IF}^{n_{ct}+n_{tt}} u_{ct}^{n_{ct}} u_{tt}^{n_{tt}}$$
$$\prod_{k=1}^{n_d} TT^\dagger_{x(i)} Ct^\dagger_{y(j)} D^\dagger_{z(k)} \mid 0 > \quad (26)$$

where u_f, u_d, u_{IF}, u_{ct} and u_{tt} are parameters to be determined and where the summation runs over all the possible configurations.

This state is a kind of generalization of the Dimer–RVB state of the two–leg ladder. The singlet on the rung is replaced by a two component MC, which contain F-LC and D-LC, and the resonon is replaced by a two component NNIF, which contains Ct_1-LC and TT-LC. By analogy with the previous case, we can write a recurrent relation followed by the wave function (26); it is given by

$$|N >= u_f|N-1> \otimes |F_N> + u_d|N-1> \otimes |D_N> +$$
$$u_{IF}(u_{ct}|N-2> \otimes |Ct_{N-1,N}> + u_{tt}|N-2> \otimes |TT_{N-1,N}>) \quad (27)$$

where $|F_N>$ and $|D_N>$ are the MC created by $F^\dagger_N$ and $D^\dagger_N$ respectively and $|Ct_{N-1,N}>$ and $|TT_{N-1,N}>$ are the NNIF created by the $Ct^\dagger_{N-1}$ and $TT^\dagger_{N-1}$ respectively. There are two constraints due to normalization to be fulfilled by the coefficients, explicitely $u_f^2 + u_d^2 = 1$ and $u_{ct}^2 + u_{tt}^2 = 1$. Among the five parameters, three remain free and have to be determined variationally by minimalization of the total energy per unit cell.

3.4. RECURRENCE RELATIONS FOR THE ENERGY

The energy per unit cell for a polymer is given again by

$$\epsilon_\infty = \lim_{N\to\infty} \epsilon_N = \lim_{N\to\infty} \frac{1}{N}\frac{E_N}{Z_N} \quad (28)$$

where

$$E_N =< N|H|N > \quad and \quad Z_N =< N|N > \quad (29)$$

These two quantities, the expectation value and the norm of the wave function, are solutions of a set of coupled recurrent relations, as for the two–leg ladder case. In this case, they are simpler because the selected LC, MC and NNIF, are orthogonal; we don't have then to consider quantities as D_N and Y_N and, finally, we get only two coupled recurrent relations instead of four in the previous section

TABLE 3. Energies per site from DMRG and RVA calculations, $e_\infty^{\rm DMRG}$ and $e_\infty^{\rm RVA}$ respectively, for $U = 3t$ and $V = 1.2t$

Δ	$e_\infty^{\rm DMRG}/t$	$e_\infty^{\rm RVA}/t$
0.1	0.330670	0.396849
0.3	0.284416	0.327386
0.5	0.224344	0.254182
1.5	−0.175762	−0.164041

$$Z_N = Z_{N-1} + u_{IF}^2 Z_{N-2} \tag{30}$$

$$E_N = E_{N-1} + u_{IF}^2 E_{N-2} + \epsilon_m(u_f) Z_{N-1} + W_2(u_f, u_{IF}, u_{ct}) Z_{N-2} \tag{31}$$

where $\epsilon_m(u_f)$ is the correlated ground state energy of a double bond given by

$$\epsilon_m(u_f) = -4u_d^2 t_d + u_f u_d (U - V) \tag{32}$$

and $W(u_f, u_{IF}, u_{ct})$ expressed the energy of the NNIF-LC and their coupling with the MC

$$W_2(u_f, u_{IF}, u_{ct}) = u_{IF}^2 u_{ct}^2 \epsilon_{ct} + u_{IF}^2 u_{tt}^2 \epsilon_{tt} + u_{IF}^2 u_{tt} u_{ct} \sqrt{3} t_s + 2u_f^2 u_{IF} u_{ct} t_s \tag{33}$$

With the two recurrent relations, we associated the following natural initial conditions $Z_0 = Z_1 = 1$ and $E_0 = 0$, $E_1 = \epsilon_m(u_F)$.

The set of equations (30) and (31) together with the definition (32), (33) and the initial conditions aforementioned, can be solved analyticaly and are also really natural to implement on a computer.

In the following, comparisons are made with Density Matrix Renormalization Group results [20, 14, 21]. A finite system DMRG algorithm is used to calculate the ground state energy for several chain lengths up to four hundred sites and the energy per site is then extrapolated to the infinite lattice limit. The numerical error in the energy per site of the infinite lattice is estimated to be of the order of 10^{-4} or smaller. We compare the RVA results with these extrapolated values.

A reasonable choice of parameters for the polyacetylene is given by $U = 3t$ and $V = 1.2t$ [13, 15]. In table 3, we show the energy per site obtained from DMRG and RVA calculations for different values of the dimerization parameter Δ. For $\Delta = 2$, the chain is fully dimerized; the results given by the ansatz (26) is then very accurate per construction. Next, the accuracy of the RVA results decreases when the dimerization decreases; this appear clearly in the table 3. For values relevant for conjugated polymers, Δ larger than 0.25, the energy approaches 90% of the DMRG energy; we think these results satisfactory according to the relative simplicity of our proposed trial wave function (26): only four local configurations are retained! This wave function is, in a sense, an extrapolation of the wave function analysed in small cluster calculations [10], to an infinite lattice. Of course some improvements are suitable and may include more extended LC; by the way, works are currently in progress in that direction by using again the Recurrent Variational Approach [42]. Nevertheless, we believe the wave function (26) sufficient to get a first explanation of some physical phenomena as the very strong linear absorption due to excitonic states observed in polydiacetylenes [36].

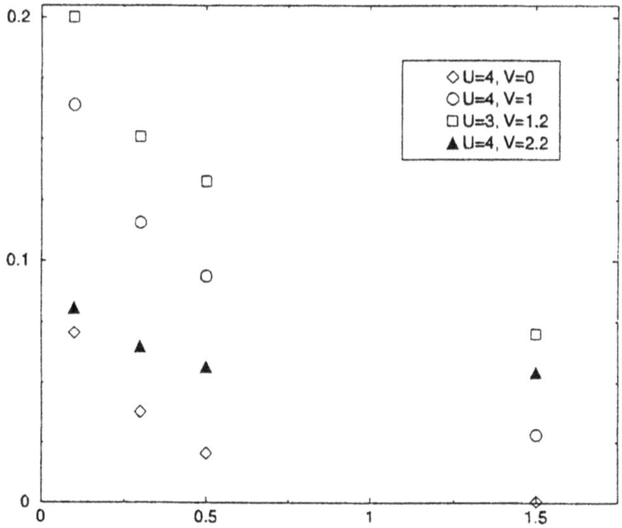

Figure 10. Relative errors of the RVA results compared to the DMRG ones. The unfilled symbols are for calculations where $U > 2V$; these choices give a BOW ground state. On the contrary, the filled triangles correspond to calculations where $U < 2V$; then a CDW ground state is stabilized.

Last, it is well known that the ground state of the EPH model shows a Bond Order Wave if $U > 2V$ and a Charge Density Wave in the other case [43]. In figure (10), we show the relative error of the RVA results compared to the DMRG ones for several choices of Coulombic parameters in function of the dimerization parameter Δ, in the two different regimes. Once again, the results show clearly that our ansatz is better when Δ increases. Moreover, its behaviour seems different in the two sides of the transition, this is natural since our ansatz is not appropriated to describe properly a Charge Density Wave.

4. Conclusions and Perspectives

To conclude, we have presented in this note the Recurrent Variational Approach and, more specifically, its first application to the electronic structure of conjugated polymers. The ansatz built here (26), shows some nice similarities with the trial state (3) made for the two–leg ladder.

By comparisons with Density Matrix Renormalization Group calculations, we think this trial wave function encouraging and a good starting point for further improvements [42]. Moreover, we believe such kind of simple wave functions useful to get some analytical insight into physical phenomena [36, 37].

Two natural extensions of this work are currently in progress: first, it is possible, in some extents, to improve the simple ansatz (26) by including more local configurations [42]; second, it is also possible to apply the method to other conjugated polymers of current interests as the poly–paraphenylene and the poly–paraphenylenevinylene.

References

1. E. Hückel, Z. Phys. **70**, 204 (1931).
2. J.A. Pople and S.H. Walmsley, Mol. Phys. **5**, 15 (1962).
3. W.P. Su, J.R. Schrieffer and A.J. Heeger, Phys. Rev. Lett. **42**, 1698 (1979).
4. M.J. Rice, Phys. Lett. **71A**, 152 (1979).
5. L. Salem, *Molecular Orbital Theory of Conjugated Systems* (Benjamin, London 1966).
6. A.J. Heeger, S. Kivelson, J.R. Schrieffer and W.P. Su, Rev. Mod. Phys. **60**, 781 (1988).
7. R. Pariser and R.G. Parr, J. Chem. Phys. **21**, 767 (1953).
8. J.A. Pople, Trans. Faraday Soc. **49**, 1375 (1953).
9. A.A. Ovchinnikov, I.I. Ukrainskii and G.V. Kventsel, Sov. Phys.-Usp. **15**, 575 (1973).
10. M. Chandross, Y. Shimoi and S. Mazumdar, Phys. Rev. **B 59**, 4822 (1999).
11. P. Tavan and K. Schulten, Phys. Rev. **B36**, 4337 (1987).
12. G.L. Bendazzoli, S. Evangelisti, G. Fano, F. Ortolani and L. Ziosi, J. Chem. Phys. **110**, 1277 (1999).
13. M. Boman and R.J. Bursill, Phys. Rev. **B57**, 15167 (1998).
14. E. Jeckelmann, Phys. Rev. **B57**, 11838 (1998).
15. Z. Shuai, J.L. Bredas, A. Saxena and A.R. Bishop, J. Chem. Phys. **109**, 2549 (1998).
16. J. Cizek and J. Paldus, J. Intern. J. Quantum Chem. **5**, 359 (1971).
17. D. Baeriswyl, D.K. Campbell and S. Mazumdar, in *Conjugated Conducting Polymers*, edited by H. Kiess (Springer-Verlag, Heidelberg, 1992), pp7-133.
18. F.D.M. Haldane, J. Phys. **C 14**, 2585 (1981).
19. H. Bethe, Z Physik **71**, 205 (1931).
20. S.R. White, Phys. Rev. **B48**, 10345 (1993).
21. *Density-Matrix Renormalization - A New Numerical Method in Physics*, Lecture Notes in Physics, eds I. Peschel, X. Wang, M. Kaulke, and K. Hallberg, Springer-

Verlag, 1999.
22. G. Sierra and M.A. Martín-Delgado, Phys. Rev. B **56**, 8774 (1997).
23. *Primary photoexcitations in Conjugated Polymers: Molecular Exciton versus Semiconductor Band Model*, edited by N.S. Sariciftci (World Scientific Publishing, Singapore, 1997).
24. E. Dagotto and T.M. Rice, Science **271**, 619 (1996).
25. P.W. Anderson Science **235**, 1196, 1987.
26. M. Roncaglia, G. Sierra and M.A. Martín-Delgado, Phys. Rev. **B60**, 12134 (1999).
27. L. Pauling, *The nature of the chemical bond*, Cornell University press (1960).
28. R. Eder, Phys. Rev. **B59**, 13810 (1999).
29. S. Gopalan, T.M. Rice and M. Sigrist, Phys. Rev. **B49**, 8901 (1994).
30. T. Barnes, E. Dagotto, J. Riera and E. S. Swanson, Phys. Rev. B **47**, 3196 (1993).
31. S.R. White and D.A. Huse Phys. Rev. B **48**, 3844 (1993).
32. G. Sierra, M.A. Martín-Delgado, J. Dukelsky, S. R. White and D. J. Scalapino, Phys. Rev.B **57**, 11666 (1998).
33. E.H. Kim, G. Sierra and D. Duffy, Phys. Rev. **B60**, 5169 (1999).
34. G. Sierra, M.A. Martín-Delgado, S.R. White, D.J. Scalapino, J. Dukelsky, Phys. Rev. **B59**, 7976 (1999).
35. J. Dukelsky, M.A. Martín-Delgado, T. Nishino and G. Sierra, Europhysics Lett. **43**, 457 (1998).
36. S. Pleutin and J.L. Fave, J. of Phys. : Condens. Matter **10**, 3941 (1998).
37. S. Pleutin and J.L. Fave, J. of Phys. : Condens. Matter **11**, 9629 (1999).
38. Z.G. Soos, M.H. Hennessy and D. Mukhopadhyay, in *Primary photoexcitations in Conjugated Polymers: Molecular Exciton versus Semiconductor Band Model*, edited by N.S. Sariciftci (World Scientific Publishing, Singapore, 1997).
39. D. Mukhopadhyay, G.W. Hayden and Z.G. Soos, Phys. Rev. **B51**, 9476 (1995).
40. S. Ramasesha, S.K. Pati, H.R. Krishnamurthy, Z. Shuai and J.L. Bredas, Phys. Rev. **B54**, 7598 (1996).
41. W.T. Simpson, J. Am. Chem. Soc. **77**, 6164 (1955).
42. M.A. Martín-Delgado, G. Sierra, S. Pleutin and E. Jeckelmann, Phys. Rev. **B61**, 1841 (2000).
43. S. Mazumdar and D.K. Campbell, Phys. Rev. Lett. **55**, 2067 (1985).

EFFECTS OF SOLVATION FOR (R,R) TARTARIC-ACID AMIDES

MARCIN HOFFMANN
Quantum Chemistry Group, Department of Chemistry,
A. Mickiewicz University, Grunwaldzka 6, 60-780 Poznan, Poland

AND

JACEK RYCHLEWSKI
Institute of Bioorganic Chemistry, Polish Academy of Sciences,
Noskowskiego 12/14, 61-713 Poznan, Poland

Abstract. A review of the most recent applications of the semi-empirical solvation method *(SM)* as well as results of calculations for (*R*,*R*) tartaric-acid amides are presented. Solvation models *(SMx)* have shown to be efficient in predicting solvation effects for various organic molecules. Effects of solvation, as calculated with the use of *SM5.4*, alter conformational preferences of isolated molecules of (*R*,*R*) tartaric-acid amides, so that in water solution the lowest-energy conformer of the primary amide with respect to the main carbon chain is the *trans* (T) conformer while for the tertiary amide it is the *gauche minus* (G-) conformer. The lowest-energy structures of isolated molecules for these amides are *gauche plus* (G+), stabilized by intramolecular hydrogen bonds. The calculated results are in agreement with the experimental data, which indicate that both in crystal structure and in polar solution the T- and G- conformers are observed for the primary and tertiary amides, respectively. In addition, in non-polar chloroform solvent the G+ structure of the tertiary amide can be observed.

1. Introduction

(*R*,*R*)-tartaric acid (see Figure 1) and its salts hold a crucial role in the history of stereochemistry. Pasteur [1] discovered the enantiomers and Bijvoet et al. [2] assigned an absolute configuration to the very first organic molecule while studying (*R*,*R*)-tartaric acid salts. Currently, (*R*,*R*)-tartaric acid and its derivatives are widely used in resolution of chiral amines [3-5] and as chiral auxiliaries in many asymmetric syntheses [6-15]. Moreover, (*R*,*R*)-tartaric acid amides have been successfully used in designing biodegradable polymers [16, 17].

J. Maruani et al. (eds.), New Trends in Quantum Systems in Chemistry and Physics, Volume 2, 189–210.

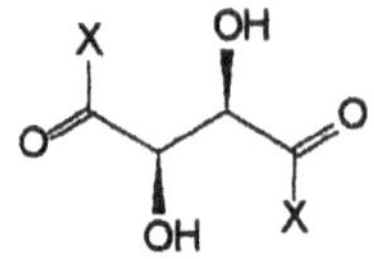

Figure 1. Zig-zag formula of (R,R)-tartaric acid and its derivatives. X = NH2 for the diamide and X = N(CH3)2 for the N,N,N',N'-tetramethyldiamide.

Computational efforts to describe the conformational preferences of (R,R)-tartaric acid and its derivatives - mainly for isolated molecules - were made recently [18-25]. The conformations of these molecules also attracted attention from experimental chemists [22-40]. (*R*,*R*)-tartaric acid and its dimethyl diester were observed in crystals, in conformations with extended carbon chain and planar α-hydroxy-carboxylic moieties (T*ss* and T*as* for the acid and the ester, respectively) [25-28] (see Figure 2). The predominance of the T-structure was also shown by studies of optical rotation [31], vibrational circular dichroism (VCD) [23], Raman optical activity [32, 35], and nuclear magnetic resonance (NMR) [22, 33, 34]. The results of ab-initio and semiempirical calculations indicated that for the isolated molecules the T*ss* and T*as* conformers were those of lowest energy [22, 21, 23, 25]. It should be noted, however, that early interpretations of NMR and VCD studies indicated that for the dimethyl diester of (*R*,*R*)-tartaric acid the G+ conformation is favored [36-38].

Conformations of primary, secondary, and tertiary amides of (R,R)-tartaric acid, both symmetrically and asymmetrically substituted, have been studied crystallographically [22, 24, 29, 30-40]. Moreover, ab initio studies up to MP2 / 6-31G*//RHF/6-31G* level [41] for both the diamide and *N,N,N',N'*-tetramethyldiamide of (*R*,*R*)-tartaric acid have been carried out [20, 22]. X-ray results have shown that primary and secondary amides of (*R*,*R*)-tartaric acid tend to adopt a conformation with the extended carbon chain – the T*aa* structure. In this T*aa* conformation both the α-hydroxy-amide moieties form planes and the structure gains stabilization from hydrogen bonding between donors, the NH, and acceptors, the proximal OH groups. Moreover, the T*aa* structure is favorably stabilized by the attraction of antiparallel local dipoles formed along distal $C_{sp3}H$ and C_{sp2}=O bonds [18, 21, 22].

In the case of tertiary amides, the planar arrangement of α-hydroxy-amide moieties is not favored due to bulky substituents. The repulsion between the *N*-alkyl substituents and the hydrogen atom attached to the distal C_{sp3} atom destabilizes the T structures of tertiary amides of (*R*,*R*)-tartaric acid [18, 21, 22]. As a result, the *N,N,N',N'*-tetramethyldiamide of (*R*,*R*)-tartaric acid is found in crystals in the G-*p*+*p*+ conformation, in which the main carbon chain is bent and the α-hydroxy-amide moieties do not form planes but the O-C_{sp3}-C_{sp2}=O torsion angle is about 90° [29, 39].

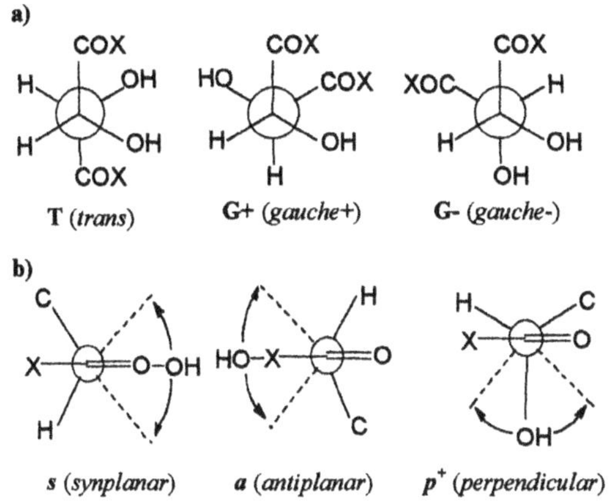

Figure 2. Rotational profiles of (*R,R*)-tartaric acid and its derivatives. (a) Rotation around the $C_{sp}3$–$C_{sp}3$ bond and (b) rotation around the $C_{sp}3$–$C_{sp}2$ bond. X = NH_2 or $N(CH_3)_2$.

In contrast to the X-ray results, ab initio calculations of the isolated molecules of primary, and tertiary amides indicated that for these molecules the G+*aa* structure is favored. The driving force being hydrogen bonds between distal OH and C=O groups and in the case of the primary amide hydrogen bonds with NH groups as donors and proximal OH as acceptors [20, 22].

The differences between the crystallographic results and ab-initio calculations appear to be in line with NMR studies. NMR analyses indicated that in a polar solution of *N,N,N',N'*-tetramethyldiamide of (*R,R*)-tartaric acid the G-conformer predominates, whereas in a nonpolar solution in chloroform there is a significant amount of the G+ conformers [22].

These differences stimulated our interest and prompted us to study the effects of solvation on (*R,R*)-tartaric acid amides. Encouraged by the widespread usage of AMSOL [42] (*vide infra*) we decided to use it to calculate Gibbs free energies of hydratation. We performed the calculations with the use of the solvation model SM5.4 [43-46] and hamiltonian PM3 [47] for all structures, optimized at the RHF/6-31G* level [20].

2. Computational methods

Solvation models *(SMx)* are semi-empirical models that introduce into calculations the effects of solvents, for example water [43, 48-52], alkanes [53, 54], chloroform [46], and others [44]. They were developed by Donald G. Truhlar,

Christopher J. Cramer and their co-workers during the last decade [61, 100, 55, 56]. In *SMx*, terms responsible for cavity formation, dispersion, solvent structure and local field polarization are present [51, 57]. The solvation energy is obtained via the usual approximation that the solute, treated at the quantum mechanical level, is immersed in an isotropic polarizable continuum representing the solvent. Therefore the standard free energy of the solute in solution can be expressed as:

$$G^0(sol) = G^0(gas) + \Delta G_S^0 \tag{1}$$

where $G^0(gas)$ is the gas phase solute energy and ΔG_S^0 is a free energy of solvation.

The solvation energy is written as:

$$\Delta G_S^0 = \Delta G_{ENP} + G_{CDS} \tag{2}$$

where G_{CDS} (cavitation-dispersion-solvent structure) is the contribution of the first-shell solvation effects to the standard-state transfer free energy, and ΔG_{ENP} (electronic-nuclear-polarization) includes the change in electronic and nuclear internal energy of the solute and the electron polarization free energy of the solute-solvent system upon insertion of the solute into the solvent. Furthermore, the first-shell solvation term is expressed as:

$$G_{CDS} = \sum_k \sigma_k A_k \tag{3}$$

so that the assumption of proportionality of the solvent-accessible surface and of the cavity energy is explicitly shown. In this equation, σ_k is the atomic surface tension of atom k and A_k, which is the solvent-accessible area of atom k, is a function that depends on the van der Waals radius for the k-th atom and on the radius of the sphere encompassing an explicit solvent molecule. Obviously, the solvent-accessible area does not include regions that overlap with the cognate spheres computed for all other atoms surrounding the k-th atom and must therefore be reevaluated during each geometry optimization step in order to reflect changes in the molecular conformation.

The ΔG_{ENP} term, called the electrostatic term, can be written as:

$$\Delta G_{ENP} = \Delta E_{EN} + G_P \tag{4}$$

where G_P represents the polarization free energy, and ΔE_{EN} is the change in the solute's internal free energy upon insertion in solution - approximated, as usual, as the change in the sum of electronic total energy and nuclear repulsion energy of the solute in going from the gas phase to solution.

The polarization free energy of inserting a molecule in a medium of dielectric constant ε is expressed in the generalized Born approximation for multi-centered systems as:

$$G_P = -\frac{1}{2}\left(1-\frac{1}{\varepsilon}\right)\sum_{k,k'} q_k q_{k'} \gamma_{kk'} \qquad (5)$$

where k and k' are atomic centers, q_k is the partial charge on atom k, and $\gamma_{kk'}$ is a Coulomb integral. The Coulomb integrals $\gamma_{kk'}$ ($k \neq k'$) are modeled such that they reduce to the self Coulomb integrals γ_{kk} when the distance $R_{kk'}$ between identical atoms tends to zero, and at large distances they are asymptotic to $1/R_{kk'}$ where $R_{kk'}$ is the interatomic distance between k and k'. The expression for Coulomb integrals $\gamma_{kk'}$ is:

$$\gamma_{kk'} = \frac{1}{\sqrt{\left[R_{kk'}^2 + \alpha_k \alpha_{k'} \left(\exp\left(\frac{-R_{kk'}^2}{d_{kk'}\alpha_k \alpha_{k'}}\right) + C_{kk'}\right)\right]}} \qquad (6)$$

where α_k is an effective atomic radius and $d_{kk'}$ is an empirically optimized constant. $C_{kk'}$ was introduced in the SM1–SM3 models to correct empirically certain trends in experimental data. However, in the SM5 solvation models $C_{kk'}$ is equal to zero.

In this work we were using SM5.4 aqueous model, [43] in which the solute Hamiltonian is modeled using NDDO molecular orbital theory – AM1 or PM3 – with class IV atomic charges from the CM1A and CM1P charge models [61]. In these CM1A and CM1P charge models, charges are obtained from one electron density matrix by a semiempirical linear mapping. These charges have two major advantages. Firstly, they make up for errors intrinsic to replacing a continuous charge distribution by a set of distributed point charges because the mapping from which they are obtained is chosen to minimize errors in the physical observables predicted from point charges. Secondly, they make up for deficiencies in the semiempirical wave function from which they are obtained because the parameterizations are chosen to minimize deviations from experiment.

All calculations were performed with the AMSOL 6.5.3 package [42] on the Cray J916 in Poznan Supercomputing and Networking Center.

3. Use of *SMx* solvation models

The SM2/AM1 model was used to examine anomeric and reverse anomeric effects and allowed to state that aqueous solvation tends to reduce anomeric stabilization [58]. Moreover, SM2/AM1 and SM3/PM3 models were accounted for in calculations of the aqueous solvation effects on the anomeric and conformational equilibria of D-glucopy-ranose. The solvation models put the relative ordering of the hydroxymethyl conformers in line with the experimentally determined ordering of populations. The calculations indicated that the anomeric equilibrium is controlled primarily by effects that the *gauche/trans* O-C6-C5-O hydroxymethyl conformational equilibrium is dominated by favorable solute-solvent hydrogen bonding interactions, and that the rotameric equilibria were controlled mainly by dielectric polarization of the solvent [59]. On the other hand, Monte Carlo results for the effects of solvation on the anomeric equilibrium for 2-methoxytetrahydropyran indicated that the AM1/SM2 method tends to underestimate the hydration effects for this compound [60].

The SM4/AM1 method with the charge model CM1A [61] was employed to account for solvation effects during the study on relative stability of alternative chair forms and hydroxymethyl conformations of β-D-glucopyranose. The results indicated that the conformation with axial OH groups is significantly further destabilized in solution in relation to the conformer with equatorial OH groups. This finding was elucidated as the result of favorable electrostatic polarization and hydrogen bonding interactions with the first solvation shell for the equatorial conformer [62]. Similarly, the SM5.4/AM1 method was used in calculating free energies of solvation for exo-anomeric effects for different conformers of glucose and related systems. It was concluded that solvation effects reduced the energetic stabilization associated with the exo-anomeric effect in the tetrahydropyrans and that in glucose and glucoside systems solvation effects oppose the formation of intramolecular hydrogen bonds [63].

During the quantum mechanical conformational analysis of 1,2-ethanediol, aqueous solvation effects were taken into account via SM1a/AM1, SM2/AM1 and SM3/PM3 methods. By adding calculated free energies of solvation to gas-phase free energies it was found that the *trans* population increased from 2% (in gas phase) to 12% (in water solvent) and that the portion of conformers having no internal hydrogen bond increased from 17% to 25%. The calculated results were in reasonable agreement with experimental data both in the gas phase and in aqueous solution [64].

The SM2/AM1 model performed well in predicting free energies of hydration for the substituted benzenes [65]. The use of the SM2/AM1 model to assess solvation effects allowed to propose a mechanism for the *cis-trans* isomerization of diazene, which proved consistent with experimental observations [66]. The results obtained indicated that trace amounts of acid can rapidly protonate *trans*

N_2H_2 to form $N_2H_2^+$ and that $N_2H_2^+$ can lose a proton to form either the *cis* or *trans* isomer in an equilibrium ratio.

The SM2.1/AM1 model was used to calculate solvent effects for bioactive isoxazoline and oxime derivatives obtained in reaction from 7-ketolignans [67]. Solvent effects on the conformation of a synthetic *N*-methylpiperazine derivative which shows antihelminthic potency and binds to minor groove of B-DNA were estimated by Monte Carlo free energy simulations with explicit solvent representation and SM1/AM1 model. The authors concluded that the SM1/AM1 model provides a good qualitative description of the variation of the solvation free energy during conformational changes at a fraction of the cost of free energy simulations including explicit solvents [68].

A computational study of substituent and solvent effects for conformational preferences of 2-phenethylamines and on the intramolecular amine-aryl interactions in charged and neutral molecules was carried out up to MP2/6-311++G(d,p)//MP2/6-31G(d,p) level and solvent effects were accounted for using SM2/AM1 methods. The best agreement with experiment was obtained when the high level ab initio results were combined with the SM2/AM1 calculated effect [69]. The effects of aqueous solvation on the conformational properties of a conformationally restricted analogue of a dipeptide – phenylalanine – were studied by means of the SM2/AM1 method. Results indicated that the rotational barriers between the minima were lower in aqueous solution than in gas phase, which was in good agreement with experimental observations that most small peptides in aqueous solution consist of a large number of rapidly interconverting structures [70].

Similarly, the effects of solvation on the conformational properties of cocaine and its two potent substitutes were calculated with SM2/AM1 and, based on the calculated increased pK_a values, decreased conformational flexibility, and decreased aqueous solvation, the authors proposed a model explaining the increased binding affinity of the studied cocaine substitutes to the cocaine receptor [71]. SM2/AM1 and SM3/PM3 models were exploited during the comparison of experimental and theoretical structures of a transition state analogue used for the induction of anti-cocaine catalitic antibodies. Results implied that accurate modeling of the potential energy surfaces for the hydrolysis of cocaine requires solvation methods like *SMx* [72]. The SM2/AM1 computational analysis gave insight into the role of solubility and the positioning of the 3β-substituents in ecgonine methyl esters for inhibition of cocaine binding and dopamine uptake. The results indicated that more favorable aqueous solubility contributes to a decrease in experimental binding values due to additional energy required to desolvate the analogues as they interact with the cocaine binding site [73].

Moreover, the SM2/AM1 calculations indicated that free energy terms associated with cavity formation, solvent reorganization, and dispersion interactions can be correlated to binding affinity, activation and neural response at the L-alanine receptor. From the consideration of this model, molecular features of the analogues that were important for binding and neural response were deduced, and other analogues or ligands were developed and tested [74].

The SM2.1/AM1 solvation model was used to calculate the aqueous potential energies in modeling hydrogen bonds during calculations of the potential energy surface for intermolecular amide hydrogen bonds [75]. In a study on relative stability of trimethylamine – formic acid versus trimethylammonium formate SM2 /AM1 as well as SCIPCM method [124] were used to assess solvation free energies for the two complexes. On the basis of high level ab initio calculations it is concluded that *in vacuo* the trimethylamine-formic acid complex is favored by 7.0 kcal/mol, whereas in water the trimethylammonium formate is favored by 6.2 kcal/mol and 8.4 kcal/mol, according to SM2/AM1 and SCI-PCM methods, respectively [76].

Rabinowitz and Little concluded during their computations of the biologically relevant reactivities of cyclopentapolycyclic aromatic hydrocarbons that the use of SM2/AM1 provide important insights for modeling molecular processes in biological systems [77]. Harris and Loew demonstrated in their comparative study of free energies of solvation of phenylimidazole inhibitors of cytochrome P450*cam* that SM2/AM1 and free energy simulations give very similar results, whereas the Poisson-Boltzmann method is inferior to them [78].

Correlation and solvation effects on heterocyclic equilibria in aqueous solution were analyzed with the use of SM2/AM1 and Onsager models. It was found that the Onsager model was inferior to SM2/AM1, because it underestimates the solvation of the *syn*-form. Local bond moments, as shown by SM2/AM1, had significant effects on the bulk electric polarizaton, even when they largely cancel in the net dipole moment. Moreover, the equilibrium shifts calculated with SM2 /AM1, due to the effects of methyl substitution on the isoxazole ring, were consistent with the available experimental data [79].

The SM2/AM1 model was also used to correct, to aqueous solution, gas free energies of oxidation of tryptophan to dioexethane, in studying isotope effects and alternative substrate reactivities for tryptophan 2,3-dioxygenase [80], and also in studying the mechanism of the decomposition of pyridinyltriazenes in aqueous buffer and helping to chose the correct mechanism for the experimental data [81]. Moreover, the SM2/AM1 model was exploited during a study of the interaction of substrates and inhibitors with xanthine oxidase [82].

The study on the dehalogenation mechanism *in vacuo* and in solution shed some light on the dehalogenation of 4-chlorobenzoyl CoA by 4-chlorobenzoyl CoA dehalogenase. As shown from the influence of solvation effects on the nonenzymatic reaction profile determined with SM3/PM3, there was a large reaction barrier in solution, of 37.4 kcal/mol [83]. Similarly, ab initio computations up to MP2/6-31G(d)//HF/6-31G(d) and semiempirical SM3/PM3 calculations gave some insight on the catalytic mechanism of formate dehydrogenase and hydrid transfer from formate to NAD^+. In gas phase the reacton between formate and NAD^+ is extremely favorable, but due to severe solvation effects of the charged reactants there is a significant potential energy barrier for the hydrid transfer in solution so that the ester adduct is formed instead. The obstacle the enzyme must overcome is to prevent this side reaction – the formation of the ester adduct – from happening. Therefore the enzyme must orient the formate correctly in the active site to keep the carboxylate oxygens away from the C4 position of NAD^+ [84].

Free energies of solvation calculated with the SM3/PM3 model were used in examining the factors controlling the catalytic efficiency of a transmethylation enzyme – catechol *O*-methyltransferase – in aqueous solution. The good agreement between the calculated kinetic isotope effects for the model reaction and the measured kinetic isotope effects for the enzymatic reaction suggest that the structure of the enzymatic transition state is very similar to that of the nonenzymatic reaction. The authors, on the basis of experiments on similar compounds, suggested that the barrier in aqueous solution for transmethylation obtained with SM3/PM3 is probably a good estimate of the actual barrier in solution [85].

The SM2/AM1 model was successful in studies on alkaline hydrolysis of clavulanic acid – β-lactam antibiotic and β-lactamase inhibitor [86]. The SM3/PM3 model was used to study the influence of the solvent on the basic hydrolysis of the β-lactame ring, and calculations for a supramolecular complex with 20 molecules of water in the solvation sphere around the solute yielded a potential barrier very close to the experimental value [87, 88].

Free energies of solvation in water calculated with the use of the SM2/AM1 method were incorporated as independent variables in a predictive model of the structure-function relationship of polyamine transport inhibitors, which affect the maintenance of the intracellular polyamine concentrations necessary for cell growth and proliferation [89].

Free energies of solvation in water and in *n*-hexadecane calculated with the SM2.1 /AM1 and SM4/AM1 models were used to correlate the logarithm of the ratio of brain-blood partitioning with the computed free energy of solvation in water. According to the authors, the power of this approach is in practical applications such as the discovery of therapeutic agents that penetrate to brain, and it can be used as a powerful tool for rank-ordering compounds prior to synthesis [90].

The SM5.4/AM1 model was used in explaining what controls partitioning of the nucleic acid bases between chloroform and water. The results obtained were in excellent agreement with experimental data, showing the high predictive accuracy that can be obtained for chloroform-water partitioning coefficients from the SM5.4/AM1 model. The model took into account nonlinear interactions between different solute functional groups that were not typically included in partitioning models based on summing fragment contributions. The authors proudly stated: "the excellence of this approach means that it may be possible to use it for rational design of novel nucleic-acid bases so as to influence such oligonucleotide properties as helix stability, base pairing fidelity, binding constants, tendency towards triple-helix formation, and other biological properties and functions sensitive to solvation and desolvation effects" [91].

Calculations of solvation and binding free energy differences between (3S)-tetrahydro-3-furyl-*N*-((1S,2R)-3-(4-amino-*N*-isobutylbenzenesulfonamido)-1-benzyl-2-hydroxypropyl) carbamate – a potent low molecular weight and orally bioactive inhibitor of the HIV-1 protease – and its analogues were carried out with free energy perturbation, SM2/AM1 and SM3/PM3 methods. Results showed how physicochemical properties like aqueous solubilities and bioavailabilities of these HIV-1 protease inhibitors were improved while maintaining their inhibitory potency [92].

Free energies of solvation in water and chloroform were calculated with the SM5.4/AM1 and SM5.4/PM3 models during a study on the effects of substitution of the hydroxyl group by the fluorine atom for (R,R)-tartaric acid derivatives. The results indicated that the substitution of OH by F results in greater conformational freedom of these compounds, and that solvation tends to decrease conformational diversity [18].

Computational studies with the use of the SM2/AM1 and SM3/PM3 models were performed to investigate the effects of fluorine substitution on phenol acidities. These solvation models produced free energies that were in good agreement with available experimental data and appeared quite useful in the prediction of solvent effects on the relative acidities of fluorophenols with respect to phenol [93]. Similarly, SM2/AM1 was used in modeling pK_a for 16 carboxylic acids and 15 chlorinated phenols, and it was stated that semiempirical solvation models predict solution phase acidity in qualitative agreement with experiment, but are not sufficiently accurate to yield absolute pK_a values. However, scaling through linear regression for specific compound classes led to pK_a estimations with standard deviations below or equal to 0.4. It was concluded that bulk water polarization is more important than solute-solvent coupling effects for pK_a estimates [94]. On the other hand, early studies with the use of the SM1/AM1 method to account for aqueous solvation effects on chemical equilibria suggested that a useful accuracy is obtained and that errors occur in treating the electronic structure of the solute, not the hydratation effect [95].

Results from SM2/AM1 calculations were used in predicting from neural networks one-electron potentials for bioreductive agents to an accuracy of about 67 mV. According to the authors, the success of such a procedure is very promising for future use in the design of molecules where the electrode potential is of crucial importance, like in anticancer drugs [96]. Density functional calculations in combination with calculations of free energy of hydratation were shown to be a powerful tool for calculating two-electron electrode potentials for quinones and naphtoquinones, and it was seen that the computationally inexpensive SM3/PM3 method performs well for those systems where no internal hydrogen bond is formed [97].

SM2/AM1 and SM3/PM3 models were used to study the hydrolysis of pyrophosphate, which is coupled to virtually all biosynthetic reactions. However, the authors concluded that extreme care must be taken when applying semiempirical methods to compounds containing second-row atoms, since they may produce anomalously high atomic charges [98]. On the other hand, a study on *syn* and *anti* conformations of solvated cyclic 3',5'-adenosine monophosphate indicated that SM3/PM3 and SM2/AM1 models are inexpensive yet accurate approaches for determining the effects of bulk solvent, but they are not very useful for the representation of specific solute solvent interactions, which are better described by the supermolecule approach [99]. SM3/PM3 method was exploited during examining of the phosphate ester hydrolysis and confirmed that the cyclic transition state is better solvated and indicated that the effect of solvation is dominated by the electric polarization associated with the hydroxy oxygen [100].

In addition, the SM2/AM1 model together with the SCRF method [101-103] was used to assess solvent effects on relative *N*- and *O*- acidity of 2- and 4-[(2,4, 6-trinitrophenyl)amino] benzioic acids. The authors stated that SCRF appears superior to SM2/AM1 but that the poorer performance of the latter might be due to limitations of the underlying quantum mechanical – semiempirical – method rather than to the approximate treatment of the solvent [104].

The SM2/AM1 and SM3/PM3 methods were used to compare gas phase and aqueous alkaline hydrolysis of methyl acetate and predicted an energy barrier to hydrolysis in relatively good agreement with experiment [105]. The SM3/PM3 method was used for the determination of activation energies in water through a computationally efficient procedure for modeling the first step of the alkaline hydrolysis of esters proposed by Turner et al. [106]. The SM2.1/AM1 model was used in theoretical studies of the base-catalysed rearrangement of 4,4'-disubstituted benzils in gas phase and in aqueous solution and the results, in line with experimental trends, indicated that substituent effects are much larger in water, and that the contribution of both the migration step and the migrating ring become much more important in water than in the gas phase [107].

The SM1/AM1 and SM2/AM1 methods were used in the examination of solvent effects on prototropic tautomerism of 6-thiopurine, and they showed that the thione tautomer was greatly stabilized in aqueous solution, in agreement with NMR and UV experimental data [108]. Similarly, the SM2/AM1 solvation model was used in a study on tautomerism and protonation of 2-thiopurine, and the results indicated that in aqueous solution the thione forms are more stabilized than the thiol ones [109]. The biochemical and pharmacological importance of the tautomerism and protonation of 7-aminopyrazolopyrimidine was studied in water solution with the use of SM2/AM1 and other models accounting for solvation effects [110]. SM3/PM3 and SM4/PM3 methods were used in calculations of solvation effects for tautomeric equilibria of 2-pyridone and 2-hydroxypyridine dimeric forms, and the results led to a possible mechanism suggesting that tautomerization of a specific DNA base may be induced by forming a complex with an intruder – a specific molecule with multiple hydrogen bonding capability, which triggers the mutation process [111].

Yet another example of the use of SM2/AM1 for nucleic acid bases comes from studies on solvation effects on tautomerizm of 8-azapurine [112]. Proton affinities for different atoms of 3*N*-, 7*N*- and 9*N*-[(1'R,3'R)-*trans*-3'-hydroxycyclopentanyl]adenine were calculated with the use of SM2/AM1 model and the obtained theoretical results together with experimental ones showed that the 9*N*-analogue can be separated from a mixture via acid wash [113]. Aqueous phase calculations were carried out with the SM2/AM1 model during a study on mutagenic significance of proton acidities in methylated guanine and thymine bases and deoxynucleosides. The calculated proton acidities correlate well with experimental data. In line with experimental data the calculations predicted that O^6-alkyldeoxyguanosines, and O^2- and O^4-alkyldeoxythymidines would be mutagenic, while N^7-alkyldeoxy-guanosines would not. The experimentally untested N^3-methyldeoxyguanosine was predicted to be mutagenic [114].

The effects of solvation on the conformational and tautomeric equilibria of formohydoxamic acid were calculated with the use of SM2/AM1 and SM3/PM3 models, and the results suggested that N-oxide tautomers were preferentially stabilized in water. But not to the extent of making them competitive with the keto tautomer corresponding to the global minimum [115]. Similarly, SM2/AM1 and SM3/PM3 methods were used to account for solvent effects in water during the study on tautomerism of hydroxy-pyridazines. The calculated properties were generally in good agreement with existing experimental data [116].

SM1/AM1, in agreement with experimental observations, predicted that the *anti* conformer of the neurotransmitter dopamine, at neutral pH, is an important contributor to the conformer population [117]. SM1/AM1 calculations provided support for the proposed solvent induced twisting mechanism of the biphenyl linkage, which offered a useful building block toward the design of molecular switches, where the rate of electron transport through the molecule, and therefore solvatochromic behavior might be modified by the degree of twist between the two phenyl groups [118].

In an exceptionally interesting way, Pietro demonstrated the usefulness of the SM1/AM1 solvation method in teaching computational chemistry. Only the inclusion of solvation effects led to the expected energy barrier (in agreement with experiment) in the reaction profile of deprotonation of nitroalkanes [119].

The SM2/AM1 model was used to elucidate the source of the aqueous acceleration of the Claisen rearrangement. The calculations allowed to state that this acceleration is caused by electronic polarization and first-shell hydration hydrophillic effects, with the relative magnitudes and even the signs of these effects being quite sensitive to the substitution pattern [120].

Reaction coordinate calculations on phenylthiocyanates of fungicidal activity at SM1/AM1 level in water strongly suggested the position and mechanism for nucleophilic attack, which were in line with experimental data [121]. The mechanism for the hydrolysis of 2,2'-dichlorodiethyl sulfide was studied in water solution with the use of the SM3/PM3 model and the results obtained showed the reaction in aqueous solution is substantially faster, with the neighboring group mechanism preferred over the conventional S_N2 one [122].

Factors controlling regioselectivity in the reduction of polynitroaromatics in aqueous solution were examined with the use of SM2/AM1 model. The analysis supported the reliability of aqueous geometries predicted by SM2/AM1. Moreover, successful predictions for the regioselectivity of reduction further supported the validity of the modeling approach. The authors stated that "the ability of the *SMx* series of solvation models to predict the electronic structure of reagents in solution should open the door for the study of many environmentally important reactions taking place in aqueous or other liquid media" [123].

The mechanism of catalysis of ester aminolysis by weak amine bases of substituted phenylquinoline-6- and -8-carboxylates were evaluated, and the solvent effects were examined by semiempirical SM2.1/AM1, ab initio SCIPCM [124] and PS-Solv [125]. All these models predicted that a zwitterionic tetrahedral intermediate should be formed in this reaction. The results did not support the proposed earlier proton slide mechanism and allowed to propose a mechanism which was fully consistent with the experimental observations [126].

Effects of solvation on zwitterion formation between methylamine and formaldehyde were studied by various solvation methods. The SM2/AM1 model predicted the expected zwitterionc minimum while SM3/PM3 failed to do so [127]. Calculations were performed with the use of AMSOL to account for solvation effects in the study of molecular properties and pharmacokinetic behavior of cetirizine, a zwitterionic third-generation antihistaminic. Results indicated that the folded conformation remains of low energy not only *in vacuo* but also in water solution [128].

In a computational investigation of substituent and solvent effects on acyloxy shifts in radicals, SM2/AM1 and SCIPCM models were exploited with explicit consideration of water molecules. All solvent models utilized agreed in predicting the [3,2] acyloxy shift to proceed more readily in aqueous solution than in gas phase, but in spite of the qualitative agreement with experimental data the calculated substituent and solvent effects were smaller than those observed experimentally [129].

SM2/AM1 and SM3/PM3 calculations in water as well as SM4/AM1 and SM4/PM3 calculations were performed on cyclohexane to study proton transfer reactions in 1-methylindene with two bases, ammonia and trimethylamine. The calculations confirmed predictions that the proton moves relatively freely over the indene ring once it is abstracted from the original location by the base [130].

Solvation effects in *n*-heptane and acetonitrile were accounted for by using the SM5.4/AM1 method, in which the CM1A charge model was used during studies on singlet-triplet splittings and 1,2-hydrogen shift barriers for methylphenylborenide, methylphenylcarbene and methylphenylnitrenium. The calculations allowed to rationalize the trends observed in these compounds as being due to different charge distributions and polarizabilities for singlets, triplets, and hydrogen-shift transition state structures [131].

In a test of the quality of solvation models proposed by Lankin et al. [132], AMSOL aqueous models provided semiquantitative agreement with the NMR data for protonated 3-fluoropiperidines in water solution. AMSOL calculations were used to estimate the solvation energies during conformational studies on methyl 4-*O*-(*β*-D-galactopyranosyl)- *β*-D-xylopyranside [133]. Solvation free energies in water and cyclohexane were utilized in conformational analysis and quantitative size and shape affinity relationship studies on 24 arylpiperazines (ligands of the serotoninergic receptor) in their neutral and protonated forms [134]. Moreover, the AMSOL package was utilized to calculate solvent effects during studies on enolization of molecular models representing D-ribulose-1,5-bisphosphate, which studies allowed to propose a possible molceular mechanism for the enolization step in Rubisco enzyme [135]. Similarly, AMSOL calculations were performed to estimate solvation effects in studies on the enzymatic reaction of interconversion of lactate and pyruvate catalysed by lactate dehydrogenase with nicotinamide adenine dinucleotide (NAD) as a cofactor [136].

It is worth mentioning that a new quantum mechanical and molecular mechanical method was developed by Kaminski and Jorgensen, based on the CM1A charges computed by AMSOL [61]. Results obtained with this method demonstrated the proposed approach provides results competitive with other methods for estimating solvent effects on tautomerization [137].

4. Results and Discussion

Tables 1 and 2 present results of ab initio calculations for isolated molecules of (*R*,*R*)-tartaric acid diamide and *N*,*N*,*N'*,*N'*-tetramethyldiamide, respectively [20].

TABLE 1. Ab initio results for the diamide of (*R*,*R*)-tartaric acid (ref. 20)

Conformer	CC*C*C [°]	O=CC*O [°]	O=CC*O [°]	μ [D]	ΔE RHF [kcal/mol]	ΔE MP2 [kcal/mol]
T*aa*	-149.5	161.0	161.0	3.78	11.97	13.11
G-*aa*	-27.0	-157.7	-157.7	3.15	15.62	17.10
G+*as*	48.3	155.2	-4.1	7.19	11.45	11.66
T*as*	-160.4	175.9	2.2	2.07	7.21	7.89
T*as*	-175.5	166.1	11.0	2.95	10.48	12.00
T*aa*	-175.1	141.6	176.7	3.76	5.79	7.23
G+*aa*	53.4	143.4	173.5	1.69	8.18	9.03
G+*aa*	62.6	160.5	160.5	0.83	0.00	0.00
G+*ss*	48.3	-10.0	-10.0	0.06	3.16	1.83
T*aa*	168.5	154.5	154.5	1.41	3.64	5.17

T*ss*	161.6	-24.0	-24.0	2.57	10.29	10.46
G-*as*	-93.9	172.2	1.3	2.68	3.27	2.68

TABLE 2. Ab initio results for *N,N,N',N'*-tetramethyldiamide of (*R,R*)-tartaric acid (ref. 20)

Conformer	CC*C*C [°]	O=CC*O [°]	O=CC*O [°]	μ [D]	ΔE RHF [kcal/mol]	ΔE MP2 [kcal/mol]
G-*p*+*p*+	-58.5	62.8	62.8	4.89	6.91	6.21
G-*aa*	-38.5	-140.5	-140.5	3.24	11.14	11.07
E+*ss*	132.2	15.1	15.1	5.24	4.95	4.07
G+*ss*	49.7	25.2	25.2	0.96	2.72	0.92
T*as*	-152.7	155.6	8.7	3.40	5.27	5.19
G+*aa*	65.1	140.5	140.5	0.17	0.00	0.00
G-*as*	-69.6	-149.3	31.9	4.40	7.26	6.05
T*aa*	-170.6	-159.2	-159.2	4.90	11.92	13.40

Tables 3 and 4 present results of SM5.4 calculations with PM3 hamiltonian for molecules in water solution of (*R,R*)-tartaric acid diamide and *N,N,N',N'*-tetramethyldiamide, respectively.

TABLE 3. AMSOL calculations for the diamide of (*R,R*)-tartaric acid in water solution

Conformer	Solvation energy [kcal/mol]	ΔE SM5.4 [kcal/mol]
T*aa*	-24.37	0.00
G-*aa*	-21.96	5.08
G+*as*	-21.93	5.42
T*as*	-20.52	2.72
T*as*	-20.43	2.97
T*aa*	-20.04	3.51
G+*aa*	-19.26	5.29
G+*aa*	-17.62	1.38
G+*ss*	-17.27	4.74
T*aa*	-17.03	7.55
T*ss*	-15.89	7.24
G-*as*	-14.89	5.59

TABLE 4. AMSOL calculations for *N,N,N',N'*-tetramethyldiamide of (*R,R*)-tartaric acid in water solution

Conformer	Solvation energy [kcal/mol]	ΔE SM5.4 [kcal/mol]
G-*p+p+*	-10.23	0.00
G-*aa*	-12.50	2.54
E+*ss*	-7.92	6.42
G+*ss*	-6.74	8.06
T*as*	-6.94	8.45
G+*aa*	-4.97	9.17
G-*as*	-8.12	9.90
T*aa*	-9.49	11.52

For the isolated molecules of both the primary and tertiary amides studied, the lowest-energy conformers are the G+*aa* ones. These structures, shown in Figs 3a (diamide) and 4a (tetramethyldiamide), are stabilized by intramolecular hydrogen bonds, each closing a six-membered ring with OH as donor and distal C=O as acceptor. In the case of the diamide, hydrogen bonds (each closing a five-membered ring with NH as donor and proximal OH as acceptor) are also present.

As it can easily be noticed from a comparison between Tables 3 and 4, the energy of hydratation of (*R,R*)-tartaric acid diamide is greater than that of (*R,R*)-tartaric acid *N,N,N',N'*-tetramethyldiamide. This is understandable because the diamide has NH_2 groups and is more polar than the tetramethyldiamide, which has $N(CH_3)_2$ groups.

In water solution, as shown by AMSOL calculations, the conformational preferences of the studied amides show up. Gibbs free energies of hydratation calculated with the SM5.4 model and PM3 hamiltonian for structures optimized at the RHF/6-31G* level indicated that those most favored by hydratation are the T- and G-conformers for the diamide and *N,N,N',N'*-tetramethyldiamide of (*R,R*)-tartaric acid, respectively.

The lowest-energy conformer in water solution of (*R,R*)-tartaric acid diamide – the T*aa* one (see Figure 3b) – closely resembles the structure found in crystals (see Figure 3c) and in polar solvents. It is stabilized by hydrogen bonds, each closing a five-membered ring with NH as donor and proximal OH as acceptor. Moreover, this T*aa* structure is stabilized by the attraction of antiparallel local dipoles formed along C_{sp3}-H and distal C=O bonds.

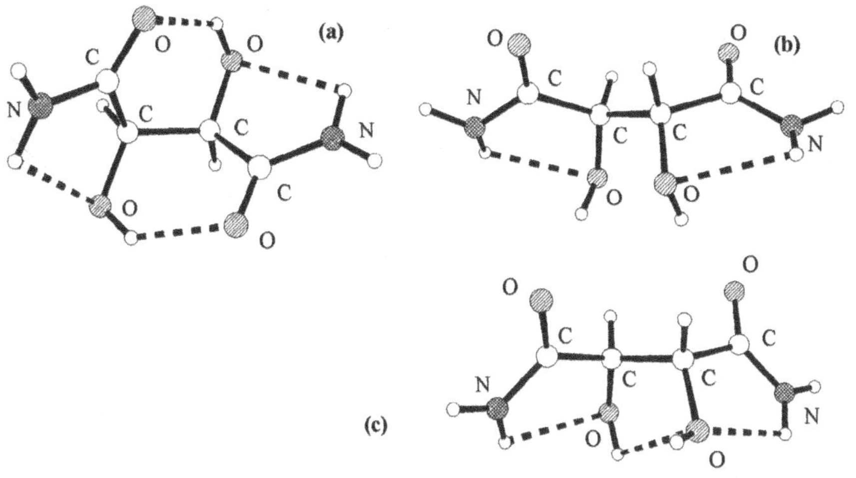

Figure 3. Structures of (*R*,*R*)-tartaric acid diamide: (a) lowest-energy conformer for isolated molecule; (b) lowest-energy conformer in aqueous solution; (c) crystal structure. Figs (a), (c) adapted from ref. 22.

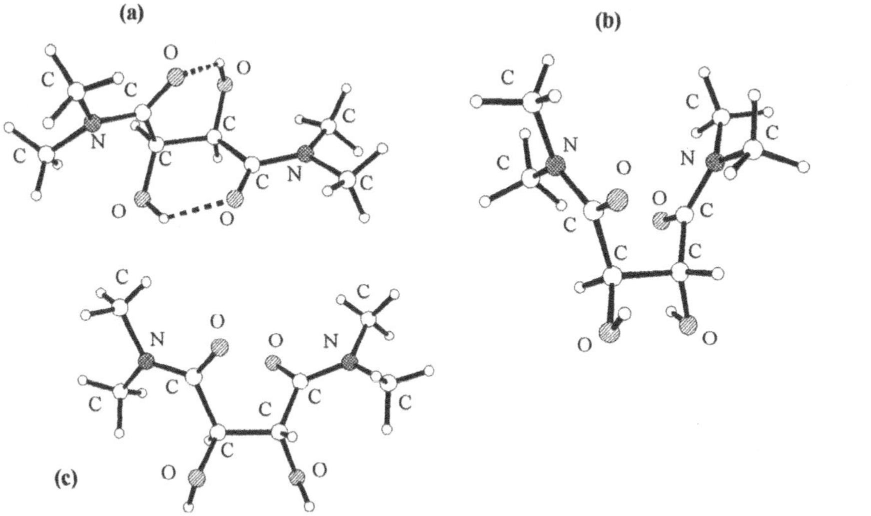

Figure 4. Structures of (*R*,*R*)-tartaric acid *N*,*N*,*N'*,*N'*-tetramethyl diamide: (a) lowest-energy conformer for isolated molecule; (b) lowest-energy conformer in aqueous solution; (c) crystal structure. Fig. (a) adapted from ref. 22; fig. (c) adapted from ref. 29.

In the case of *N,N,N',N'*-tetramethyldiamide of (*R,R*)-tartaric acid the lowest-energy conformer in water solution is the G-*p*+*p*+ one (Figure 4b). This structure also closely resembles that observed in crystals (Figure 4c) and in polar solvents.

5. Conclusions

Solvation effects, as calculated with the use of the AMSOL method, affect the conformational preferences of (*R,R*)-tartaric acid amides, so that their lowest-energy structures are different from the G+*aa* conformers favored in isolated molecules. Hydratation favores the T*aa* conformation for the primary amides and the G-*p*+*p*+ conformation for the tertiary amides of (*R,R*)-tartaric acid.

The calculated results are in a very good agreement with the experimental data obtained from X-ray and NMR analyses [22, 29, 39]. This seems to indicate that the relatively simple and computationally economical semiempirical SM5.4 model with PM3 hamiltonian is efficient enough in predicting conformational preferences, even for more difficult cases like (R,R)-tartaric acid amides in water solution.

Acknowledgements

The authors are grateful to Prof. D. G. Truhlar and Prof. C. J. Cramer for providing an academic license for AMSOL-6.5.3. They also express their gratitude to Prof. U. Rychlewska for a careful reading of the manuscript and suggestions and support in preparing Figures 3 and 4. Finally the authors gratefully acknowledge financial support from Polish Committee for Scientific Research (KBN), grant 8 T11F 028 15, and access to the computer resources of the Poznan Supercomputing and Networking Center.

References

1. Pasteur, L. (1848) *Ann. Chim. Phys.* **24**, 442.
2. Bijvoet, J. M.; Peerdman, A. F.; Van Bommel, A. J. (1951) *Nature* **168**, 271-272.
3. Wilen, S. H. (1971) *Top. Stereochem.* **6**, 107-176.
4. Wilen, S. H.; Collet, A.; Jacques, J. (1977) *Tetrahedron* **33**, 2725-2736.
5. Jacques, J.; Collet, A.; Wilen, S. H. (1981) *Enantiomers, Racemates and Resolutions*, John Wiley, New York.
6. Katuski, T.; Sharpless, K. B. (1980) *J. Am. Chem. Soc.* **102**, 5974-5976.
7. Haruta, R.; Ishiguro, M.; Ikeda, N.; Yamamoto, H. (1982) *J. Am. Chem. Soc.* **104**, 7667-7669.
8. Toda, F.; Tanaka, K.; Nassimbeni, L.; Niven, M. (1988) *Chem. Lett.* **1988**, 1371-1374.
9. Roush, W. R.; Banfi, L. (1988) *J. Am. Chem. Soc.* **110**, 3979-3982.
10. Furuta, K.; Maruyama, T.; Yamamoto, H. (1991) *Synlett.* 439-440.
11. Furuta, K.; Maruyama, T.; Yamamoto, H. (1991) *J. Am. Chem. Soc.* **113**, 1041-1042.

12. Ishihara, K.; Gao, Q.; Yamamoto, H. (1993) *J. Org. Chem.* **58**, 6917-6919.
13. Gao, Q.; Ishihara, K.; Maruyama, T.; Mouri, M.; Yamamoto, H. (1994) *Tetrahedron* **50**, 979-988.
14. Roush, W. R.; Grover, P. T. (1995) *J. Org. Chem.* **60**, 3806-3813.
15. Gawronski, J.; Gawronska, K. (1999) *Tartaric and Malic Acids in Synthesis – A Source Book of Building Blocks, Ligands, Auxiliaries, and Resolving Agents*, John Wiley, New York.
16. Iribarren, I.; Aleman, C.; Bou, J. J.; Munoz-Guerra, S. (1996) *Macromolecules* **29**, 4397-4405.
17. Iribarren, I.; Aleman, C.; Regano, C.; Martinez de Ilarduya, A.; Bou, J. J.; Munoz-Guerra, S. (1996) *Macromolecules* **29**, 8413-8424.
18. Hoffmann, M.; Rychlewski, J.; Rychlewska, U. (1999) *J. Am. Chem. Soc.* **121**, 1912-1921.
19. Aleman, C.; Galembeck, S. E. (1997) *J. Org. Chem.* **62**, 6562-6567.
20. Hoffmann, M.; Szarecka, A.; Rychlewski, J. (1998) *Adv. Quant. Chem.* **32**, 109-125.
21. Hoffmann, M.; Rychlewski, J.; Rychlewska, U. (1996) *Comput. Methods Sci. Technol.* **2**, 51-63.
22. Gawronski, J.; Gawronska, K.; Skowronek, P.; Rychlewska, U.; Warzajtis, B.; Rychlewski, J.; Hoffmann, M.; Szarecka, A. (1997) *Tetrahedron* **53**, 6113-6144.
23. Polovarapu, P. L.; Ewing, C. S.; Chandramouly, T. (1987) *J. Am. Chem. Soc.* **109**, 7382-7386.
24. Szarecka, A.; Hoffmann, M.; Rychlewski, J.; Rychlewska, U. (1996) *J. Mol. Struct.* **374**, 363-372.
25. Rychlewska, U.; Warzajtis, B.; Hoffmann, M.; Rychlewski, J. (1997) *Molecules* **2**, 106-113.
26. Stern, F.; Beevers, C. A. (1950) *Acta Crystallogr.* **3**, 341-346.
27. Okaya, Y.; Stemple, N. R.; Kay, M. I. (1966) *Acta Crystallogr.* **21**, 237-243.
28. Albertson, J.; Oskarsson, A.; Stahl, K. (1979) *J. Appl. Crystallogr.* **12**, 537-544.
29. Rychlewska, U. (1992) *Acta Crystallogr.* **C48**, 965-969.
30. Szczepanska, B.; Rychlewska, U. (1994), in D. Jones and A. Katrusiak (eds.), *Correlations, Transformations and Interactions in Organic Chemistry*, Oxford University Press, Oxford, pp. 23-244.
31. Hargreaves, M. K.; Richardson, P. J. (1957) *J. Chem. Soc.* 2260-2271.
32. Barron, L. D.; Gargano, A. R.; Hecht, L.; Polvarapu, P. L.; Sugeta, H. (1992) *Spectrochim. Acta* **48A**, 1051-1066.
33. Ascenso, J.; Gil, V. M. S. (1980) *Can. J. Chem.* **58**, 1376-1379.
34. Hasan, M. (1980) *Org. Magn. Res.* **14**, 309-312.
35. Barron, L. D. (1978) *Tetrahedron* **34**, 607-610.
36. Keiderling, T. A.; Stephens, P. J. (1977) *J. Am. Chem. Soc.* **99**, 8061-8062.
37. Su, C. N.; Keiderling, T. A. (1980) *J. Am. Chem. Soc.* **102**, 511-515.
38. Freedman, T. B.; Balukjian, G. A.; Nafie, L. A. (1985) *J. Am. Chem. Soc.* **107**, 6213-6222.
39. Gawronski, J.; Gawronska, K.; Rychlewska, U. (1989) *Tetrahedron Lett.* **30**, 6071-6074.
40. Rychlewska, U.; Szarecka, A.; Rychlewski, J.; Motola, R. (1999) *Acta Crystallogr.* **B55**, 617-625.
41. We follow throughout the standard notational convention that x/y//z/w denotes a calculation with method x and basis y at geometry optimized with method z and basis w. See Hehre, W.J.; Radom, L.; Schleyer, P. v. R.; Pople, J. A. (1986) *Ab initio Molecular Orbital Theory*, Wiley, New York.
42. Hawkins, G. D.; Giesen, D. J.; Lynch, G. C.; Chambers, C. C.; Rossi, I.; Storer, J. W.; Li, J.; Rinaldi, T. Z. D.; Liotard, D. A.; Cramer, C. J.; Truhlar, D. G.: AMSOL version 6.5.3 (1998), based in part on AMPAC, version 2.1 by Liotard, D. A.; Healy, E. F.; Ruiz, J. M.; Dewar, M. J. S., and on EF by Jensen, F., University of Minnesota.
43. Chambers, C. C.; Hawkins, G. D.; Cramer, C. J.; Truhlar, D. G. (1996) *J. Phys. Chem.* **100**, 16385-16398.
44. Giesen, D. J.; Gu, M. Z.; Cramer, C. J.; Truhlar, D. G. (1996) *J. Org. Chem.* **61**, 8720-8721.

45. Giesen, D. J.; Hawkins, G. D.; Cramer, C. J.; Truhlar, D. G. (1997) *Theor. Chem. Acc.* **98**, 85-109.
46. Giesen, D. J.; Chambers, C. C.; Cramer, C. J.; Truhlar, D. G. (1997) *J. Phys. Chem.* **101**, 2061-2069.
47. Stewart, J. J. P. (1989) *J. Comput. Chem.* **10**, 221-264.
48. Cramer, C. J.; Truhlar, D. G. (1991) *J. Am. Chem. Soc.* **113**, 8305-8311.
49. Cramer, C. J.; Truhlar, D. G. (1992) *Science* **256**, 213-217.
50. Cramer, C. J.; Truhlar, D. G. (1991) *J. Comput. Chem.* **13**, 1089-1097.
51. Cramer, C. J.; Truhlar, D. G. (1992) *J. Comput.-Aided Mol. Design* **6**, 629-666.
52. Hawkins, G. D.; Cramer, C. J.; Truhlar, D. G. (1996) *J. Phys Chem.* **100**, 19824-19839.
53. Giesen, D. J.; Storer, J. W.; Cramer, C. J.; Truhlar, D. G. (1995) *J. Am. Chem. Soc.* **117**, 1057-1068.
54. Giesen, D. J.; Cramer, C. J.; Truhlar, D. G. (1995) *J. Phys Chem.* **99**, 7137-7146.
55. Giesen, D. J.; Cramer, C. J.; Truhlar, D. G. (1994) *J. Phys Chem.* **98**, 4141-4147.
56. Liotard, D. A.; Hawkins, G. D.; Lynch, G. C.; Cramer, C. J.; Truhlar, D. G. (1995) *J. Comput. Chem.* **16**, 422-440.
57. Richards, N. G. J. (1998), in Sapse, A.-M. (ed.), *Molecular Orbital Calculations for Biological Systems*, Oxford University Press, New York, pp. 11-48.
58. Cramer, C. J. (1992) *J. Org. Chem.* **57**, 7034-7043.
59. Cramer, C. J.; Truhlar, D. G. (1993) *J. Am. Chem. Soc.* **115**, 5745-5753.
60. Jorgensen, W. L.; Morales de Tirado, P. I.; Severance, D. L. (1994) *J. Am. Chem. Soc.* **116**, 2199-2200.
61. Storer, J. W.; Giesen, D. J.; Cramer, C. J.; Truhlar, D. G. (1995) *J. Comput.-Aided Mol. Design* **9**, 87-110.
62. Cramer, C. J.; Truhlar, D. G. (1995) *Carbohydr. Res.* **276**, 219-251.
63. Cramer, C. J.; Truhlar, D. G. (1997) *Carbohydr. Res.* **298**, 1-14.
64. Cramer, C. J.; Truhlar, D. G. (1994) *J. Am. Chem. Soc.* **116**, 3892-3900.
65. Jorgensen, W. L.; Nguygen, T. B. (1993) *J. Comput. Chem.* **14**, 195-205.
66. McKee, M. L. (1993) *J. Phys. Chem.* **97**, 13608-13614.
67. Corral, J. M. M. del; Gordaliza, M.; Castro, M. A.; Lopez-Vasquez, M. L.; Garcia-Gravalos, M. D.; Broughton, H. B.; Feliciano, A. S. (1997) *Tetrahedron* **53**, 6555-6564.
68. Sapse, A. M.; Mezei, M.; Karakhanov, I.; Jain, D. C. (1997) *J. Mol. Struct. (Theochem)* **393**, 25-30.
69. Urban, J. J.; Cronin, C. W.; Roberts, R. R.; Famini, G. R. (1997) *J. Am. Chem. Soc.* **119**, 12292-12299.
70. Lovas, S.; Murphy, R. H. (1994) *J. Mol. Struct. (Theochem)* **311**, 297-304.
71. Yang, B.; Wright, J.; Eldefrawi, M. E.; Pou, S.; MacKerell, A. D. Jr. (1994) *J. Am. Chem. Soc.* **116**, 8722- 8732.
72. Sherer, E. C.; Yang, G.; Turner, G. M.; Shields, G. C.; Landry, D. W. (1997) *J. Phys. Chem A* **101**, 8526- 8529.
73. Lieske, S. F.; Yang, B.; Eldefrawi, M. E.; MacKerell, A. D. Jr.; Wright, J. (1998) *J. Med. Chem.* **41**, 864- 876.
74. Venanzi, T. J.; Bryant, B. P.; Venanzi, C. A. (1995) *J. Comput.-Aided Mol. Design* **9**, 439-447.
75. Adalsteinsson, N., Maulitz, A. H.; Bruice, T. C. (1996) *J. Am. Chem. Soc.* **118**, 7689-7693.
76. Liljefors, T.; Norrby, P.-O. (1997) *J. Am. Chem. Soc.* **119**, 1052-1058.
77. Rabinowitz, J. R.; Little, S. B. (1994) *Int. J. Quantum Chem.* **52**, 681-691.
78. Harris, D.; Loew, G. (1996) *J. Comput. Chem.* **17**, 273-288.
79. Cramer, C. J.; Truhlar, D. G. (1993) *J. Am. Chem. Soc.* **115**, 8810-8817.
80. Leeds, J. M.; Brown, P. J.; McGeehan, G. M.; Brown, F. K.; Wiseman, J. S. (1993) *J. Biol.*

Chem. **268**, 17781-17786.
81. Fransworth, D. W.; Wink, D. A.; Roscher, N. M.; Michejda, C. J.; Smith, R. H. Jr. (1994) *J. Org. Chem.* **59**, 5942-5950.
82. Rastelli, G.; Costantino, L.; Albasini, A. (1997) *J. Am. Chem. Soc.* **119**, 3007-3016.
83. Zheng, Y.-J.; Bruice, T. C. (1997) *J. Am. Chem. Soc.* **119**, 3868-3877.
84. Schiott, B.; Zheng, Y.-J.; Briuce, T. C. (1998) *J. Am. Chem. Soc.* **120**, 7192-7200.
85. Zheng, Y.-J.; Bruice, T. C. (1997) *J. Am. Chem. Soc.* **119**, 8137-8145.
86. Frau, J.; Donoso, J.; Munoz, F; Blanco, F. G. (1994) *Helv. Chim Acta* **77**, 1557-1569.
87. Frau, J.; Donoso, J.; Munoz, F; Garcia-Blanco, F. (1996) *Helv. Chim Acta* **79**, 353-362.
88. Frau, J.; Donoso, J.; Munoz, F; Garcia-Blanco, F. (1997) *J. Mol. Struct. (Theochem)* **390**, 247-254.
89. Li, Y.; Mackerell, A. D. Jr.; Egorin, M. J.; Ballesteros, M. F.; Rosen, D. M.; Wu, Y.-Y.; Blamble, D. A.; Callery, P. S. (1997) *Cancer Res.* **57**, 234-239.
90. Lombardo, F.; Blake, J. F.; Curatolo, W. J. (1996) *J. Med. Chem.* **39**, 4750-4755.
91. Giesen, D. J.; Chambers, C. C.; Cramer, C. J.; Truhlar, D. G. (1997) *J. Phys. Chem. B* **101**, 5084-5088.
92. Rao, B. G.; Kim, E. E.; Murcko, M. A. (1996) *J. Comput.-Aided Mol. Design* **10**, 23-30.
93. Urban, J. J.; von Tersch, R. L.; Famini, G. R. (1994) *J. Org. Chem.* **59**, 5239-5245.
94. Schuurmann, G. (1996) *Quant. Struct.-Activ. Relat.* **15**, 121-132.
95. Cramer, C. J.; Truhlar, D. G. (1991) *J. Am. Chem. Soc.* **113**, 8552-8554.
96. Wolfe, J. J.; Wright, J. D.; Reynolds, C. A.; Saunders A.C.G. (1994) *Anti-Cancer Drug Design* **9**, 85-102.
97. Reynolds, C. A. (1995) *Int. J. Quantum Chem.* **56**, 677-687.
98. Colvin, M. E.; Evleth, E.; Akacem, Y. (1995) *J. Am. Chem. Soc.* **117**, 4357-4362.
99. Krischner, K. N.; Sherer, E. C.; Shields, G. C. (1996) *J. Phys. Chem.* **100**, 3293-3298.
100. Cramer, C. J.; Truhlar, D. G. (1994) *J. Chem. Soc. Perkin Trans.* **90**, 1799-1810.
101. Wong, M. W.; Frisch, M. J.; Wiberg, K. B. (1991) *J. Am. Chem. Soc.* **113**, 4776-4782.
102. Wong, M. W.; Frisch, M. J.; Wiberg, K. B. (1992) *J. Am. Chem. Soc.* **114**, 523-529.
103. Wong, M. W.; Frisch, M. J.; Wiberg, K. B. (1992) *J. Am. Chem. Soc.* **114**, 1645-1652.
104. Bagno, A.; Dorgio, F.; McCrae, P.; Scorrano, G. (1996) *J. Chem. Soc. Perkin Trans.* **2**, 2163-2168.
105. Sherer, E. C.; Turner, G. M.; Shields, G. C. (1995) *Int. J. Quantum Chem. Quantum Biol. Symp.* **22**, 83-93.
106. Turner, G. M.; Sherer, E. C.; Shields, G. C. (1995) *Int. J. Quantum Chem. Quantum Biol. Symp.* **22**, 103-112.
107. Lee, I.; Lee, D.; Lee, J. K.; Kim, C. K.; Lee, B.-S. (1996) *J. Chem. Soc. Perkin Trans* **2**, 2519-2523.
108. Contreras, J. G.; Alderete, J. B. (1994) *J. Mol. Struct. (Theochem)* **309**, 137-141.
109. Contreras, J. G.; Alderete, J. B. (1995) *J. Mol. Struct. (Theochem)* **334**, 223-228.
110. Orozco, M.; Luque, F. J. (1995) *J. Am. Chem. Soc.* **117**, 1378-1386.
111. Chou, P.-T.; Wei, C.-Y.; Hung, F.-T. (1997) *J. Phys. Chem. B* **101**, 9119-9126.
112. Contreras, J. G.; Madariaga, S. T.; Alderete, J. B. (1996) *J. Mol. Struct. (Theochem)* **365**, 63-69.
113. Barbuch, R.; Curran, T. T.; Hay, D. A.; Vaz, R. J. (1997) *Tetrahedron* **53**, 7181-7190.
114. Venkateswarlu, D.; Lyngdoh, R. H. D. (1997) *J. Mol. Struct. (Theochem)* **393**, 151-161.
115. Stinchcomb, D. M.; Pranata, J. (1996) *J. Mol. Struct. (Theochem)* **370**, 25-32.
116. Greenwood, J. R.; Capper, H. R.; Allan, R. D.; Johnston, G. A. R. (1997) *J. Mol. Struct. (Theochem)* **419**, 97-111.
117. Urban, J. J.; Cramer, C. J.; Famini, G. R. (1992) *J. Am. Chem. Soc.* **114**, 8226-8231.
118. Metcalfe, R. A.; Dodsworth, E. S.; Lever, A.B.P.; Pietro, W. L.; Stufkens, D. J. (1993) *Inorg.*

Chem. **32**, 3581-3582.
119. Pietro, W. J. (1994) *J. Chem. Educ.* **71**, 416-420.
120. Cramer, C. J.; Truhlar, D. G. (1992) *J. Am. Chem. Soc.* **114**, 8794-8799.
121. Morley, J. O.; Naji, M. (1995) *J. Chem. Soc. Perkin Trans.* **2**, 1301-1304.
122. Donovan, W. H.; White, W. E. (1996) *J. Mol. Struct. (Theochem)* **370**, 209-220.
123. Barrows, S. E.; Cramer, C. J.; Truhlar, D. G.; Elovitz, M. S.; Weber, E. J. (1996) *Environ. Sci. Technol.* **30**, 3028-3038.
124. Foresman, J. B.; Frisch, Æ. (1996) *Exploring Chemistry with Electronic Structure Methods*; Gaussian Inc.; Pittsburgh, PA, USA.
125. Tannor, D. J.; Marten, B.; Murphy, R.; Friesner, R. A.; Sitkoff, D.; Nicholls, A.; Ringnalda, M.; Goddard, W. A. III; Honig, B. (1994) *J. Am. Chem. Soc.* **116**, 11875-11882.
126. Adalsteinsson, H.; Briuce, T. C. (1998) *J. Am. Chem. Soc.* **120**, 3440-3447.
127. Hall, N. E.; Smith, B. J. (1998) *J. Phys. Chem. A* **102**, 3985-3990.
128. Pagliara, A.; Testa, B.; Carrupt, P.-A.; Jolliet, P.; Morin, C.; Morin, D.; Urien, S.; Tillement, J.-P.; Rihoux, J.-P. (1998) *J. Med. Chem.* **41**, 853-863.
129. Zipse, H. (1997) *J. Am. Chem. Soc.* **119**, 1087-1093.
130. Agback, M.; Lunell, S.; Hussenius, A.; Matsson, O. (1998) *Acta Chem. Scand.* **52**, 541-548.
131. Cramer, J. C.; Truhlar, D. G.; Falvey, D. E. (1997) *J. Am. Chem. Soc.* **119**, 12338-12342.
132. Lankin, D. C.; Chandrakumar, N. S.; Rao, S. N.; Spangler, D. P.; Snyder, J. P. (1993) *J. Am. Chem. Soc.* **115**, 3356-3357.
133. Asensio, J. L.; Lopez, R.; Fernandes-Mayoralas, A.; Jimenez-Barbero, J. (1994) *Tetrahedron* **50**, 6417-6432.
134. Cocchi, M.; Fanelli, F.; Menziani, M. C.; Benedetti, P. G. (1997) *J. Mol. Struct. (Theochem)* **397**, 129-145.
135. Tapia, O.; Andres, J.; Safont, V. S. (1994) *J. Phys. Chem.* **98**, 4821-4830.
136. Andres, J.; Moliner, V.; Krechl, J.; Silla, E. (1995) *J. Chem. Soc. Perkin Trans.* **2**, 1551-1558.
137. Kaminski, G. A.; Jorgensen, W. L. (1998) *J. Phys Chem. B* **102**, 1787-1796.

INTERPRETATION OF VIBRATIONAL SPECTRA IN ELECTRO-CHEMICAL ENVIRONMENTS FROM FIRST-PRINCIPLE CALCULATIONS: COMPUTATIONAL STRATEGIES

M. GARCÍA-HERNÁNDEZ[1], A. MARKOVITS[1,2,3],
A. CLOTET[3], J. M. RICART[3] AND F. ILLAS[1]

1) Departament de Química Física i Centre de Recerca en Química Teòrica, Universitat de Barcelona, C/ Martí i Franqués 1, 08028 Barcelona, Spain
2) Laboratoire de Chimie Théorique (UMR 7616 CNRS), UPMC, B.P. 137, 4 place Jussieu, 75252 Paris Cedex 05, France
3) Departament de Química Física i Inorgànica, Universitat Rovira i Virgili, Pl. Imperial Tarraco 1, 43005 Tarragona, Spain

Abstract. The vibrational frequencies of carbonate adsorbed on Pt (111) electrodes have been investigated by the cluster-model approach within the hybrid B3LYP DFT-based method. A rather large cluster model was used to represent the Pt (111) surface and a fast and efficient procedure based on the normal coordinate approach is proposed to compute the vibrational frequencies of adsorbed carbonate. This normal coordinate approach avoids the expensive computation of the hessian matrix of the cluster plus adsorbate supersystem. In addition, the vibrational analysis has also been carried out for carbonate in mono- and bi-dentate complexes that are often used as reference to assign the vibrational spectra of adsorbed carbonate. First, it is shown that for the mono- and bi-dentate complexes the B3LYP frequencies are comparable to the experimental data. The comparison to computed vibrational frequencies for adsorbed carbonate does not support the Surface vs Organometallic Chemistry analogy usually invoked to assign the infrared bands of adsorbed species. Finally, to better represent the electrochemical environment, the effect of a uniform external electric field on the adsorbed carbonate frequencies has been analysed.

1. Introduction

Surface Electrochemistry is an important field in Surface Science, which is undergoing a very important development. In spite of the complexity of the systems, the experimental measurements have acquired a very high degree of sophistication and atomic resolution has almost been reached. After the development of special techniques that allowed the preparation of well defined single crystal electrodes [1, 2], attempts have been made to use surface analysis techniques in an electrochemical environment. However, one must

J. Maruani et al. (eds.), New Trends in Quantum Systems in Chemistry and Physics, Volume 2, 211–226.

realise that such a direct transfer from Surface Science to Electrochemistry is not so simple because these techniques have been developed to carry out the experimental measurements under ultra high vacuum conditions. A huge effort has been necessary to accommodate these surface science techniques to electrochemical environments. In particular, "in situ" infrared spectroscopy [3] has been one of the pioneering surface analysis techniques applied in electrochemical conditions.

The main goal of in-situ infrared spectroscopy is to provide information about the adsorption sites and coordination modes of molecules on electrodes. In principle, the specific bands in the spectra of species adsorbed at electrodes should allow to distinguish between different adsorption modes and to obtain structural information about the adsorbate-substrate interaction. Infrared spectroscopy has become more and more popular among electrochemists although band assignment is far from being straightforward. In gas phase IR experiments, well-defined patterns based on comparison to spectra of highly purified compounds always exist. On the contrary, the interpretation of IR bands in Surface Science and Electrochemistry suffers from the lack of good references to be used to assign the IR bands of the adsorbed species. Therefore, spectroscopists attempt to add some analogies to chemical intuition. A very tempting possibility is to compare the IR spectra of a given molecule in different but related chemical environments. Think, for example, of an adsorbate on a metal surface and of the same species as a ligand in a metal complex [4]. The IR spectra of a large number of inorganic and organometallic complexes are well established and could provide a reasonable first-order approach to interpret the spectra of the adsorbed species. However, the analogy between Surface Science and Organometallic Chemistry has been strongly questioned [5,6]. It has been shown that the assignment of adsorption sites based on the comparison of the vibrational frequencies of adsorbates and of coordination compounds is not univocal at all, even using single-crystal well-defined surfaces [5, 6].

The use of quantum mechanical models for adsorbates on surfaces provides a useful and alternative procedure to assign the experimental IR bands. Theoretical models allow to include some key features of the electrochemical systems as the influence of an external electric field on the vibrational frequencies, the effect of solvent or that of coadsorbed species. A particular, very convenient, peculiarity of theoretical models is the ability to separate the influence of each one of these effects. This feature permits to quantify the importance of different mechanisms and provides further understanding of Electrochemistry at a molecular level. The main goal of this paper is to propose an efficient strategy for the computation of the vibrational frequencies of adsorbed species. In particular, we will consider the adsorption of carbonate on Pt (111), a system extensively studied in electrochemistry from experiment [7] and theory [8]. Moreover, the assignment of the IR bands was precisely carried out by comparing the "in situ" spectra to that of mono- or bi-dentate carbonato-cobalt complexes identified long time ago [9, 10]. This choice permits a comparison of the vibrational frequencies of carbonate in two different chemical environments. From this comparison it will appear that some of the previous assignments are not supported. Furthermore, the analysis of the vibrational modes of adsorbed carbonate permits to understand the origin of this failure of the Surface Science and Organometallic Chemistry analogy.

2. Computational methods

The electronic structure of the cobalt carbonato complexes and of the Pt (111) surface cluster models interacting with the carbonate species in the different coordination modes have been studied by means of an ab-initio hybrid Density Functional Theory (DFT) based method. In particular, the widely used B3LYP exchange-correlation functional [11] has been chosen. The B3LYP functional was originally designed to correctly reproduce the thermochemical data of a series of organic molecules. However, it has been shown that this hybrid DFT approach is also adequate for transition metals containing system as well as to describe metal-oxide [12-14] or adsorbate-metal [15, 16] interactions.

The Kohn-Sham molecular orbitals used to define the electron density have been expressed as a linear combination of Gaussian Type Orbitals. The C and O atoms of the carbonate anion are described with a standard split-valence plus polarisation basis set, 6-31G*. Previous studies have shown that for the carbonate-Pt(111) interaction, these basis sets and the B3LYP method provide results with a level of accuracy that is accurate enough for the present purposes [8]. The description of the surface cluster models is based in a mixed approach where the cluster is divided into a "local" and an "outer" region. The "local" region contains the atom directly involved with the adsorbate, whereas the "outer" region is mainly devoted to provide an adequate environment to the "local" one and to improve the representation of the metal conduction band. For the atoms on the "local" region, the effect of the inner core electrons on the cluster electronic structure is replaced by a Relativistic, small core, Effective Core Potential, RECP [17]. Therefore, the $5s^2 5p^6 5d^{10}$ electrons of each Pt atom in the "local regions" are explicitly included and are described by a standard double-ζ basis set [17].

The same approach, basis set and ECP have been used to describe the electronic density of Co in the carbonato complexes. For the Pt atoms in the "outer" region we used recently developed one-electron pseudopotentials [18,19] especially appropriate for embedding purposes [20-23]. This one-electron pseudo-potential was initially derived in the form of Durand and Barthelat [24] and later fitted to the Hay-Wadt form [17] to allow its use in most of the avail-able computer codes. A basis [5s3p/2s1p] set for the one-electron Pt pseudo-atoms has been used. In all systems the search for the optimum geometrical parameters has been carried out employing analytical gradients. The frequency calculations depend on whether complexes or surface cluster models are used. In the first case the full hessian matrix is computed, the second derivatives being obtained by finite difference of analytical first derivatives. In the case of surface cluster models a two-step procedure was used: this procedure will be described in detail in Section III.c. All calculations were carried out using the Gaussian 94 suite of computer codes [25].

3. Models for the interpretation of infrared spectra

The cluster model approach has long been used to model surfaces or chemisorption phenomena. The successes and limitations of this approach have recently been extensively reviewed [26-28]. This approach assumes that the properties of interest are of local nature, and hence, can be described by means of finite representations of the extended surface. It must be pointed out that for moderately large cluster models, ~25 atoms, the computation of the IR vibrational modes is computationally costly because the energy gradients calculation is a very time consuming step even when analytical first derivatives are used. However, comparison to experiment requires to consider a limited number of vibrational modes only, those corresponding to nuclear displacements of the adsorbate. Therefore, the approach that is often used for gas phase molecules is here highly inefficient, it requires a huge computational effort and most of it is worthless. Even if computers are becoming more and more powerful, a computational efficient strategy is required to compute the vibrational spectra of adsorbed molecules. Unfortunately, general methods able to describe the vibrational modes of a reduced part of the system are not available although one must mention the procedure recently proposed by Head [29]. The normal coordinate approach outlined below is a cheap alternative to Head's method.

3.1. SURFACE CLUSTER

Reasonably large cluster models are used to simulate the various interactions that are expected to occur between the carbonate species and the Pt (111) surface. Different clusters are employed to represent different coordination modes, although some of the different chemisorption sites may be studied using the same cluster. For the interactions directly above the atop site, the cluster model used is $Pt_{13}(7,3,3)$, where numbers in parenthesis indicate that it contains seven Pt atoms in the first layer, three in the second and three in the third, Fig. 1a. Upon carbonate adsorption, the C_{3v} symmetry point group of this bare surface cluster is reduced to C_s. A $Pt_{18}(12,6)$ cluster, Fig. 1b, is used to represent the bridge sites. In principle, this cluster model also exhibits C_{3v} point symmetry but due to the definition of the "local" region its symmetry is lowered to C_s and to C_s or C_1 upon carbonate adsorption on the long and short bridge, respectively. For each coordination mode and adsorption site described above, geometry optimisations have been performed. The internal geometry of the adsorbed carbonate and the distance to the surface were optimised. These optimisations were carried out maintain the C_{2v} symmetry of the adsorbed molecule and its orientation perpendicular to the surface. The Pt-Pt distance in the cluster was kept fixed at the bulk value, 2.77Å, in all the calculations.

Next, we describe the Pt atoms included in the "local" and "outer" regions of each cluster. For Pt_{13}, the seven atoms on the first layer are included in the "local" region although we have also explored the case where only the first layer central atom is included in the "local" region. For Pt_{18} the "local" region is defined by the four cluster atoms surrounding the bridge site, shadowed in Fig. 1.

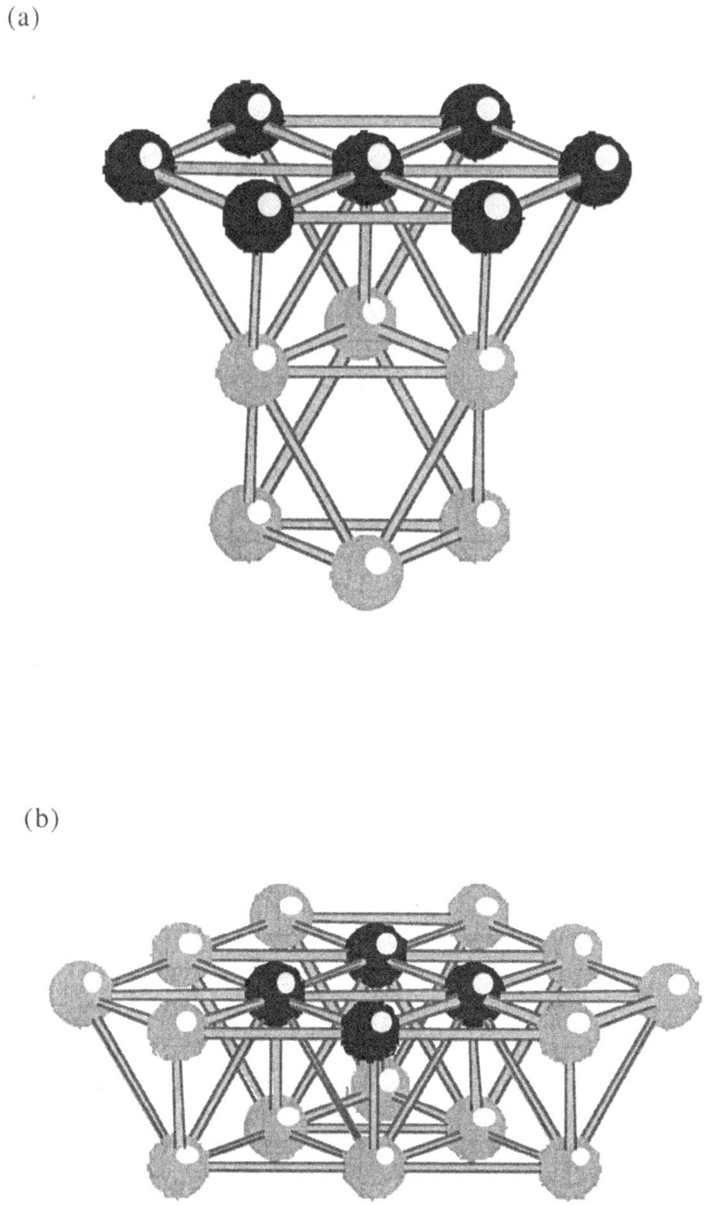

Figure 1. Cluster models employed for the study of chemisorption of carbonate: (a) $Pt_{13}(7,3,3)$; (b)$Pt_{18}(12,6)$. Darker atoms define the "local" region, see text.

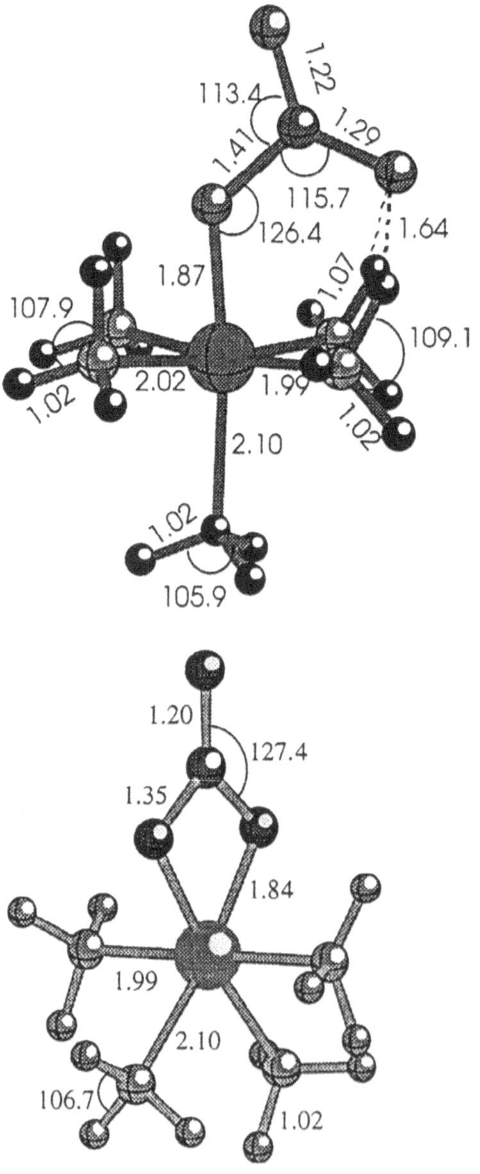

Figure 2. The bidentate and monodentate complexes.

3.2. CARBONATO COMPLEXES

The vibrational frequencies of carbonato complexes have been used to assign some of the frequencies experimentally observed by Iwasita et al. [7] in an electrochemical environment. These are hexacoordinated Co (III) complexes with five or six ligands, one carbonato interacting with the metal either through one or two oxygen atoms and either five or four NH_3 molecules respectively, see Fig. 2.

Within the approximation of considering the complexes in the gas phase, the optimum geometry and vibrational frequencies of bi- and mono-dentate complexes have been obtained from a full geometry optimisation. To the best of our knowledge, no precise geometrical parameters are available for these complexes. Next, the vibrational frequencies have been computed as commented above and comparison with the experimental data is used to validate the particular theoretical approach chosen to carry out the present study.

3.3. THE NORMAL COORDINATES APPROACH

The brute force calculation of the vibrational frequencies of an adsorbate on a metal surface is computationally demanding because one is constrained to obtain and diagonalise the hessian matrix of the whole system. The first step is strongly time-consuming and it would be useful to have some procedure to compute directly the vibrational frequencies of interest while avoiding the construction of the whole hessian matrix. A possibility that has been used recently consists in computing just the block of the hessian matrix that corresponds to the nuclear displacements of the adsorbate alone [30]. Though this procedure results in considerable time saving it still requires several evaluations of energy gradients, at least one for each displacement of each coordinate of the adsorbate. The result of such a procedure is the complete set of frequencies of the adsorbate. In many cases this information is surabondant because one is often concerned with a limited number of adsorbate normal modes. A direct approach to these normal modes is clearly welcome.

In the case of a simple diatomic molecule like CO chemisorbed on metal surfaces the computation of its stretching frequency is straightforward. The usual procedure is to vary the CO length maintaining the centre of mass at the equilibrium distance [21, 31-34]. This stratagem permits to obtain different values for the energy at different internuclear distances that are then fitted into a polynomial curve and the second derivative of the polynomial at the minimum leads to the force constant and by using the CO reduced mass the stretching vibrational frequency is obtained. This is the easiest possible case given that the internuclear distance coincides with the normal coordinate of the vibration. In principle the above procedure can be applied to any normal mode of any molecule, but the recipe to be used is not so clear when the target of the study is a polyatomic molecule vibrational mode. Then the motion of the nuclei in the molecule is approached by a series of uncoupled oscillators, or normal modes. Each oscillator has a normal coordinate specific of this particular normal mode and indicates the motion of the atoms of the system in that specific uncoupled vibration. The normal coordinates are given by the eigenvectors of the full hessian matrix. Later, we will outline a procedure that permits to obtain a good approximation of the normal coordinates bypassing the construction of the hessian matrix of the full system.

Once the normal coordinate of interest is known, it is possible to simultaneously move all atoms of the system, following the displacements indicated by the appropriate eigenvector. The total energy of the system as a function of the normal displacements results in potential curve similar to that described for the case of the chemisorbed CO. From this curve the force constant of the vibration can be easily obtained. In practice it is common to express the displacements defining each normal coordinate, Q_i, as

$$Q_i = \sum_{k=1}^{3N} L_{ki} q_k \tag{1}$$

i.e., a linear combination of mass-weighted cartesian coordinates q_k ($k = 1, 3N$; N = number of atoms). The motion of all atoms along Q_i requires a simultaneous displacement of each mass-weighted coordinate q_k proportional to L_{ki}. Because of the units of q_k, a further transformation in the $\mathbf{L}_i$ vector is necessary to produce displacements along Q_i that effectively represent a displacement in units of length and, hence, can be directly added to $\mathbf{r}$, the vector containing the equilibrium coordinates of all the atoms of the molecule. The transformed $\mathbf{L}_i$, hereafter referred to as $\Delta\mathbf{r}_i$, displaces all atoms from the equilibrium position according to the normal vibration mode, maintains the coordinates of the centre of mass fixed at the equilibrium position, and leads to displacements that have the proper length units. Displacements of all nuclei proportional to $\Delta\mathbf{r}_i$ lead to geometries that are shifted from equilibrium along Q_i to an extent that is precisely given by the proportionality constant. In principle, values of the proportionality constant leading to changes in the total energy of ~1mH are appropriate and guarantee that the displacements are within the harmonic region.

The transformation from $\mathbf{L}_i$ to $\Delta\mathbf{r}_i$ simply consists in dividing each component of $\mathbf{L}_i$ by the square root of the mass of its corresponding atom. Defining λ as the second derivative of the curve of the energy versus these normal coordinates permits to obtain the vibrational frequency by

$$\nu = \sqrt{\lambda} / 2\pi \tag{2}$$

However, it must be pointed out that $\Delta\mathbf{r}_i$, expressed in cartesian coordinates, is not normalised. Gaussian94 first normalises $\Delta\mathbf{r}_i$ before printing and also prints the former norm or, more precisely, its inverse. This last quantity has mass units and is referred to as the *normal mode reduced mass*, which is not to be confused with the *reduced mass* defined in classical mechanics. The *normal mode reduced mass* appears because of the fact that, after transforming to the new normalised coordinates, one is no longer representing the energy versus Q_i but versus Q_i divided by the square root of this *reduced mass*. As a result, the resulting *force constant* is multiplied by this mass and the actual frequency of the normal mode is obtained by dividing the result by this *normal mode reduced mass*. This is the origin of the *reduced mass* term, formally it divides the force constant and permits to obtain the frequency.

The point that remains to be explained is how to obtain each $\mathbf{L}_i$ without carrying out a total or partial hessian matrix calculation for the full system. The procedure suggested in the present paper is approximated yet very efficient. The basic idea relies on the fact that functional groups in chemistry have representative vibrational frequencies rather insensitive on the particular environment. This fact suggests that it is possible to transfer the normal modes from one system to another. Here, a full hessian matrix calculation for the adsorbate-cluster model supersystem is carried out but using a rather reduced surface model. This simple calculation allows to obtain the normal coordinate and the normal mode reduced mass of interest. The present procedure assumes that the adsorbate geometry and vibrational frequencies are not largely affected by the cluster size. There is strong evidence that this is indeed the case [26-28].

Following the above arguments, the geometry of the molecule above the reduced cluster model is optimised and the vibrational frequencies are computed by explicit diagonalization of the hessian matrix. In this diagonalization, the masses of the atoms of the substrate are considered to be infinite to decouple the internal modes from the surface phonons. Once the normal modes are determined and the normal coordinates Q_i of the mode are obtained in cartesian coordinates, $\Delta\mathbf{r}_i$, the vibrational frequency of a given mode can be easily calculated for the same molecule interacting with a larger cluster model. For the Pt_{18}-carbonate on the short bridge coordination mode, the calculation of the vibrational frequencies from the explicit computation of the full hessian matrix takes about 520 CPU hours on a superescalar parallel machine. The calculation of a single frequency on the same cluster requires only 6 CPU hours on the same machine plus the time required to obtain the normal modes on the small cluster. This is, however, a small fraction of the total time. The reason for carrying out both calculations is to check the hypothesis made on the transferability of the normal modes. The error of the normal approach with respect to the usual method based in the construction of the full hessian is less than 0.3%. Here, we must point out that the gain in CPU time is much larger if one wishes to compute the dependence of the vibrational frequencies with respect to a uniform external electric field. This follows from the fact that the normal modes of the smaller cluster are also transferred to cases where the electric field is present, with the additional hypothesis that the intensity of the electric field is small enough not to induce noticeable changes in the adsorbate geometry.

4. Results and Discussion

4.1. RESULTS FOR COMPLEXES

Before starting the discussion concerning the Co (III) carbonato complexes it is worth to recall that the purpose of this section is two-fold. First, to show that the present computational approach properly describes the vibrational frequencies of these metals complexes supporting the calculated values for adsorbed carbonate. Second, to provide an adequate comparison between carbonato as a ligand and as an adsorbate.

First, a full geometry optimisation of the monodentate and bidentate Co (III) complexes has been carried out, a summary of the optimised geometrical parameters is given in Fig 2. For the monodentate complex, the carbonate ligand C_2 axis tilts to allow a favourable interaction between the carbonate and two of the NH_3 ligands. The two NH bonds of the NH_3 molecules interacting with the carbonate are somehow longer than the rest and the three CO bond lengths of the carbonate are significantly different. The longest CO bond distance connects the carbon atom with the oxygen atom interacting with the metal atom. This is easy to understand because this interaction is stronger than that existing between the oxygen atom and the NH_3 ligands. Accordingly, the CO bond involving the O interacting with the hydrogen atoms is longer than the other one. However, this bond length, 1.29 Å, is shorter than the one of the free carbonate CO bond length calculated at the same level of theory. This is probably because the interaction between the carbonate and the metal weakens the CO bond directly involved in this interaction; as a consequence the two other CO bonds are strengthened. The internal geometry of all the four remaining NH_3 ligands of the bidentate complex is almost the same. In addition, the equatorial and axial N-Co(III) bond lengths are also almost equal, 2.10Å vs. 1.99Å. One must caution that solvation effects, not considered here, can affect the final geometry.

Contrary to the monodentate complex, the carbonato ligand of the bidentate complex has a C_{2v} symmetry, with no direct interaction between the carbonate and any other ligand. Surprisingly, the O-Co(III) bond length is almost the same in the bidentate and monodentate complexes, 1.84 vs. 1.87Å. This may be originating from steric factors between the carbonato and NH_3 ligands in the monodentate complex. This result contradicts the assumption of Fujita et al. [9], who performed a normal coordinate analysis and computed several C-O force constants assuming that the oxygen-metal bond should be stronger in the monodentate case and, as a consequence, the bond length shorter.

To the best of our knowledge, no experimental data are available for direct comparison with our computed structures. Fortunately, experimental infrared spectra are available [9, 10, 35] that can be directly compared to the frequencies obtained from our calculations. In general the present calculated values are in a rather good agreement with the experiment. However, there is disagreement concerning the assignment of a few peaks. A simulated spectrum has been obtained (Fig. 3) by fitting the data to lorentzian functions, using the SYNSPEK program [36].

First, we comment the band involving the carbonate ligand and computed at 1214 cm^{-1}. This is in agreement with the assignment proposed by Gatehouse et al. [10] for the band appearing experimentally at 1297 cm^{-1}, and indicates that the calculated values are within 6% close to the experiment. Here, we must note that calculations refer to the gas phase complex whereas experiments are carried out for a crystal form of the complex. Therefore the agreement between experiment and theory is reasonable. However, some of our calculations disagree with assignments made either by Gatehouse et al. [10] or by Fujita et al. [9]. According to the present calculations, the band at

1440 cm^{-1} is due to displacements of the NH_3 ligands interacting with the carbonato ligands, not to the carbonate as suggested in [10]. Likewise, the mode appearing at 924 cm^{-1} in the calculation and experimentally measured at 1050 cm^{-1} corresponds to NH_3 ligands, not to the carbonato ligand.

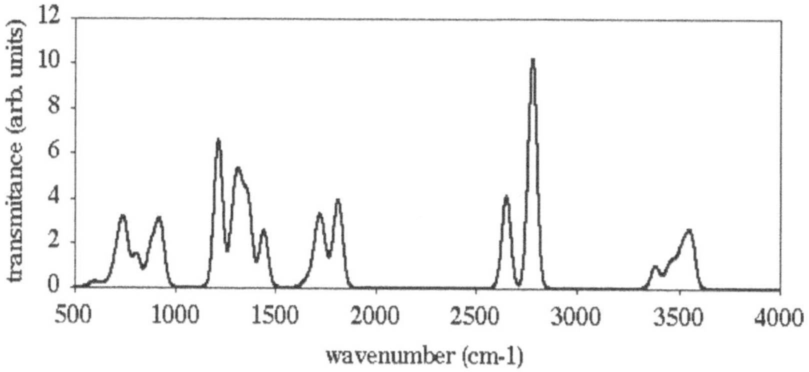

Figure 3. Calculated infrared spectrum of the monodentate $[Co(III)(NH_3)_5CO_3]^{2+}$ complex.

A similar description is obtained for the bidentate complex. The experimental highest frequencies are there 1590-1640 cm^{-1} (A_1), 1260-1290 cm^{-1} (B_2) and 1300 cm^{-1} (A_1) [37], and are very close to those yielded by the calculations. Therefore, the comparison between experimental and calculated frequencies for the two complexes is fairly good and validates the theoretical approach used, giving confidence to the predictions made for the adsorbed species.

4.2. RESULTS FOR THE ADSORBED CARBONATE

In a previous study [8] two different adsorbed species were considered: the carbonate anion, CO_3^{2-}, and the carbon trioxide, CO_3 (the corresponding neutral molecule). This was a necessary test to avoid possible cluster artefacts arising from the use of a charged adsorbate. The result of these calculations showed that, once adsorbed, the carbon trioxide and the carbonate become the same species. This is supported by the fact that the two adsorbates have the same equilibrium geometry, the same final Mulliken charge distribution - the carbonate gives one electron to the cluster and the carbon trioxide takes one electron from the cluster - and the distance between the adsorbates and the substrate are identical. The only observed difference concerned the adsorption energy and this is due to a well-known artefact of the cluster model [38, 39]. While the absolute values of the Ionisation Potential (IP) and Electron Affinity (EA) for a real metal surface are the same, the computed values for a given cluster model representing a metal surface are somewhat different showing one of the limitations of the cluster model approach.

Using the strategy described in Sec. 3, we have calculated the two highest frequencies of the adsorbed carbonate. As a first step, the frequencies for the adsorbed molecule were computed following the standard procedure, i.e., explicit construction and diagonalization of the hessian matrix assuming infinite substrate mass. Indeed, the same calculation has been carried out without and with the presence of a uniform external electric field with the idea to validate the normal coordinates computational strategy outlined in the previous section. A summary of results is given in Table 1.

TABLE 1. The two highest frequencies of the short bridge carbonate adsorbed on Pt_4 and Pt_{18} surface cluster models: (a) frequencies obtained from explicit diagonalization of the full hessian matrix; (b) frequencies obtained using the normal coordinate approach

Mode	Field = 0.00 a.u.			Field = 0.01 a.u.		
	Pt_4 (a)	Pt_{18} (a)	Pt_{18} (b)	Pt_4 (a)	Pt_{18} (a)	Pt_{18} (b)
A_1	1678	1657	1653	1624	1592	1589
B_2	1233	1250	1250	1250	1275	1274

The first important result is that the frequencies obtained from the diagonalization of the hessian matrix for carbonate adsorbed on Pt_4 or Pt_{18} clusters are almost the same. This shows that the adsorbate vibrational properties do not largely depend on the cluster model and can be considered as a local property. Moreover, it allows one to trust the results obtained on moderately large clusters such as Pt_{18} without needing to consider more extended models. In addition, results on Table 1 clearly show that the agreement between the vibrational frequencies obtained from the hessian and from the normal cooordinates approach is outstanding. This holds in the case where there is no external electric field and, also, in the case where a uniform external field of 0.01 a.u. is applied in the direction perpendicular to the surface.

Next, we describe the influence of the electric field on the frequencies by using the normal coordinate approach. A summary of results is given in Fig. 4. The optimised geometry of adsorbed carbonate is not largely changed by the electric field although the distance from the adsorbate to the surface increases with its intensity, till desorption occurs when the intensity of the field reaches 0.03 a.u. This result shows that the energetics of carbonate adsorption is controlled by the electric field whereas the internal geometry is not. In any case, the comparison with the experimental data of Iwasita et al. [7] is not straightforward because of the particularities of the electrochemical measurements. The information obtained from the electrochemical experiments concerns the variation of the frequencies versus the potential and not versus the electric field. Therefore, a direct comparison is not possible. A detailed description of the influence of electric fields in the vibrational frequencies of adsorbed carbonate will be reported elsewhere [40].

4.3. COMPARING CARBONATE ADSORBED ON PT (111) TO THE CO (III) CARBONATO COMPLEX

The first important remark is that, as a result of these two different chemical environments, large differences are found for the distance of the carbonate oxygen to the metal atom. In the monodentate complex, the metal-oxygen atom of the carbonate distance is 1.87 Å while in the cases of the atop, fcc and hcp monodentate adsorption mode, the optimised distances are 2.37, 1.63 and 1.71 Å, respectively [8]. In the bidentate complex, the optimised oxygen-metal distance is 1.84 Å. On Pt_{18}, the optimised O-Pt bond lengths are also quite different for the short bridge, long bridge and chelating adsorption modes, respectively 2.07, 2.37 and 2.53 Å [8]. The differences in the optimised bond lengths show that the comparison is speculative and add further arguments to the claim that the nature of the bond in the complex and at the surface is not the same. Indeed, the vibrational normal modes may be different in each case.

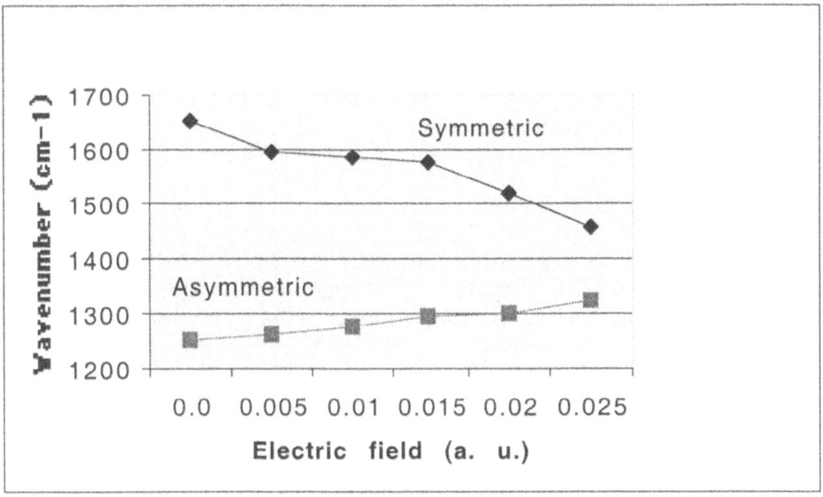

Figure 4. Symmetric and asymmetric adsorbed carbonate frequencies versus a uniform external electric field applied perpendicularly to the surface.

In fact, in a previous paper [8] it was shown that the comparison between the frequencies of carbonate in a Co (III) complex and adsorbed on a Pt (111) surface was quite misleading. In particular, the splitting of the two degenerate highest frequencies of free carbonate was different for the species in the complex and on the surface. Hence, the use of the complex spectrum as a reference leads to a wrong assignment of the frequencies obtained with the carbonate adsorbed on a Pt (111) electrode. A detailed analysis of the different vibrational modes shows that only some of these modes are transferable from the complex to the surface. The transferability depending on which part of the adsorbate is involved in the normal mode and whether the motion of the atoms in this normal mode imply a strong interaction with the surface.

Several arguments can be invoked to show that a comparison to the complex should be avoided. First, the nature of the chemical bonding between the carbonate and a metal atom is different in a complex than in an adsorbate / surface system. In our case, the metal atom of the complex is a Co (III) while the metal of the surface is Pt, a meaningful comparison will require at least to use Pt complexes. It is well established that, in different carbonate-containing species, the splitting of the degenerate frequencies depends on the metal, its oxidation number, and on the particular compound [10, 37]. A second reason why a comparison to complexes often leads to untrustful conclusions is simply the fact that the carbonate environment is quite different. In fact, some ligands of the complex can give rise to steric factors not existing on a surface. An additional point is that in the bidentate complex the ligand interacts with one metal atom only while when adsorbed in a bridging mode it interacts with two platinum atoms. A similar situation occurs in the cases on-top, fcc-hollow and hcp hollow-site. It clearly follows that a comparison of the frequencies of carbonate adsorbed on these three different adsorption sites with those of a single complex can be used for qualitative purposes only. For instance, to recognise carbonate adsorbed on the surface but neither to assign the adsorption site nor to assign the different IR bands to particular vibrational modes.

5. Conclusions

A new and efficient computational strategy has been presented, that simplifies the calculation of the vibrational frequencies of a molecular system adsorbed on moderate to large cluster models. This procedure is based on a certain hypothesis and assumption. Nevertheless, present results show that these do not affect the numerical accuracy of the calculated frequencies. An important consequence of this strategy is that largely simplifies the study of the effect of a uniform electric field on the frequencies of an adsorbed species. This is because it is not necessary to recalculate the normal coordinates at each value of the electric field. The method has been presented in connection to a cluster model representation of the surface, but it can be directly applied to periodical approaches without further modification.

In the case of carbonate it has been shown that the B3LYP calculated frequencies of the mono- and bi-dentate complexes give results that are comparable to available experimental data. A more important conclusion follows from the examination of the normal modes. This analysis has permitted to reinterpret the assignments of a few infrared bands. The present study does also show that it is not possible to use these carbonato complexes to assign frequencies of adsorbate carbonate on Pt (111). The difference between the metal in complexes and at a surface, the different topologies and the steric factors due to ligands of complexes make the analogy between Surface Chemistry and Organometallic Chemistry extremely dangerous. In particular, the different metal-adsorbate bond lengths are a univocal proof of the difference in the nature of the chemical bond at a surface and in a complex.

Alternatively, it is proposed that a comparison between theoretical frequencies obtained from cluster model calculations and experimental spectra is a worthwhile approach to study and assign the vibrational frequencies of oxyanions adsorbed on single-crystal electrodes.

Acknowledgements

Financial support from the Spanish "Ministerio de Educación y Ciencia", projects CICyT PB95-0847-C02-01 and -02, and partial support from the "Generalitat de Catalunya", project 1997SGR00167, are gratefully acknowledged. Part of the computer time was provided by the "Centre de Supercomputació de Catalunya", C^4-CESCA, through a research grant from the University of Barcelona. A.M. thanks the Researchers Training and Mobility program of the European Commission, under contract ERB FMGE CT95 0062, for making possible his stays in Barcelona and Tarragona. M.G.H. is grateful to the "Generalitat de Catalunya" for a predoctoral fellowship.

References

1. Clavilier, J. (1980) *J. Electroanal. Chem.* **107**, 211.
2. Clavilier, J., Faure, R., Guinet, G. and Durand, R. (1980) *J. Electroanal. Chem.* **107**, 205.
3. Ashley, K. and Pons, S. (1988) *Chem. Rev.* **88**, 673.
4. Albert, M. R. and J. J. T. Yates (1987). *The Surface Scientist's Guide to Organometallic Chemistry*, ACS, Washington, DC.
5. Schindler, K.-M., Hofmann, Ph., Weiß, K. U., Dippel, R., Gardner, P., Fritzche, V., Bradshaw, A. M., Woodruff, D. P., Davila, M. E., Asensio, M. C., Conesa, J. C. and González-Elipe, A. R. (1993) *J. Elect. Spect. & Rel. Phen.* **64/65**, 75.
6. Davila, M. E., Asensio, M. C., Woodruff, D. P., Schindler, K.-M., Hofmann, Ph., Weiss, K. U., Dippel, R., Gardner, P., Fritzsche, V., Bradshaw, A. M., Conesa, J. C. and González-Elipe, A. R. (1994) *Surf. Sci.* **311**, 337.
7. Iwasita, T., Rodes, A. and Pastor, E. (1995) *J. Electroanal. Chem.* **383**, 181.
8. Markovits, A., Garcia-Hernandez, M., Ricart, J. M. and Illas, F. (1999) *J. Phys. Chem. B.* **103**, 509.
9. Fujita, J., Martell, A. E. and Nakamoto K. (1962) *J. Chem. Phys.* **36**, 339.
10. Gatehouse, B. M., Livingstone, S. E. and Nyholm, R. S. (1958) *J. Chem. Soc.*, 3137.
11. Becke, A. D. (1993) *J. Chem. Phys* **98**, 5468. See also Stephens, P. J., Deulin, F. J., Chabalowski, C. F. and Frisch, M. J. (1994) *J. Phys. Chem.* **99**, 11623.
12. Ricca, A. and Bauschlicher, C. W. (1994) *J. Phys. Chem.* **98**, 12899.
13. Russo, T. V., Martin, R. L. and Hay, P. J. (1995) *J. Chem. Phys.* **102**, 8023.
14. Siegbahn, P. E. M. and Crabtree, R. H. (1997) *J. Am. Chem. Soc.* **119**, 3103.
15. Illas, F., Mele, F., Curulla, D., Clotet, A. and Ricart, J. M. (1998) *Electrochim. Acta*, **44**, 1213.
16. Lopez, N. and Illas, F. (1998) *J. Phys. Chem. B* **102**,1430.
17. Hay, P. J. and Wadt, W. R. (1985) *J. Chem. Phys.* **82**, 299.
18. Zurita, S., Rubio, J., Illas, F. and Barthelat, J. C. (1996) *J. Chem. Phys.* **104**, 8500.
19. Rubio, J., Zurita, S., Barthelat, J. C. and Illas, F. (1994) *Chem. Phys. Lett.* **217**, 283.
20. Zurita, S., Rubio, J. and Illas, F. (1996) *Electrochim. Acta* **41**, 2275.
21. Illas, F., Márquez, A., Zurita, S. and Rubio, J. (1995) *Phys. Rev. B* **52**, 12372.
22. Illas, F., Zurita, S., Márquez, A. and Rubio, J. (1997) *Surf. Sci.* **376**, 279.
23. Illas, F., Clotet, A. and Ricart, J. M. (1997) *J. Phys. Chem.* **101**, 9732.
24. Durand, P. and Barthelat, J. C. (1975) *Theoret. Chim. Acta* **38**, 283.

25. Gaussian 94, Revision D2, Frisch, M. J., Trucks, G. W., Schlegel, H. B., Gill, P. M. W., Johnson, B. G., Robb, M. A., Cheeseman, J. R., Keith, T., Petersson, G. A., Montgomery, J. A., Raghavachari, K., Al-Laham, M. A., Zakrzewski, V. G., Ortiz, J. V., Foresman, J. B., Peng, C. Y., Ayala, P. Y., Chen, W., Wong, M. W., Andres, J. L., Replogle, E. S., Gomperts, R., Martin, R. L., Fox, D. J., Binkley, J. S., Defrees, D. J., Baker, J., Stewart, J. P., Head-Gordon, M., Gonzalez, C. and Pople, J. A. Gaussian, Inc., Pittsburgh PA (1995).
26. Bagus, P.S. and Illas, F. (1998) *The Surface Chemical Bond,* in Encyclopedia of Computational Chemistry, Schleyer, P. v. R., Allinger, N. L., Clark, T., Gasteiger, J., Kollman, P. A., Schaefer III, H. F. and Schreiner P. R., eds. John Wiley & Sons, Chichester UK, Vol. 4, 2870.
27. Whitten, J.L. and Yang, H. (1996) *Surf. Sci. Rep.* **24**, 59.
28. Pacchioni, G. (1996) *Heterogenous Chemistry Reviews* **2**, 213.
29. Head, J.D. (1997) *International Journal of Quantum Chemistry* **65**, 827
30. Habas, M.P., Mele, F., Sodupe, M. and Illas, F. (1999) *Surf. Sci.* **431**, 208.
31. Bagus, P.S. and Müller, W. (1985) *Chem. Phys. Lett.* **115**, 540.
32. Müller, W. and Bagus, P.S. (1985) *J. Vac. Sci. & Technol. A* **3**, 1623.
33. Bagus, P.S. and Pacchioni, G. (1990) *Surf. Sci.* **236**, 233.
34. Illas, F., Zurita, S., Márquez, A. and Rubio, J. (1997) *Surf. Sci.* **376**, 279.
35. Nakamoto, K., Fujita, J., Tanaka, S. and Kobayashi, M. (1957) *J. Am. Chem. Soc.* **79**, 4904.
36. SYNSPEK program obtained via Internet from Irikura, K.K., National Institute of Standards and Technology, Gaithersburg, MD 20899, USA.
37. Little, L. H. (1986) *Infrared Spectra of Adsorbed Species*. London, Academic Press.
38. Pacchioni, G. and Bagus, P. S. (1993) *Surf. Sci.* **286**, 317.
39. Bagus, P. S and Pacchioni, G. (1995) *J. Chem. Phys.* **102**, 879.
40. Markovits, A., García-Hernández, M., Ricart, J.M. and Illas, F., work in progress.

EXCITED STATES IN METAL OXIDES BY CONFIGURATION INTERACTION AND MULTIREFERENCE PERTURBATION THEORY

CARMEN SOUSA[1], COEN DE GRAAF[1], FRANCESC ILLAS[1] AND GIANFRANCO PACCHIONI[2]

1) Departament de Química Física i Centre de Recerca en Química Teòrica, Universitat de Barcelona, C/ Martí i Franquès 1, 08028 Barcelona, Spain
2) Istituto Nazionale di Fisica della Materia, Dipartimento di Scienza dei Materiali, Università di Milano-Bicocca, via R. Cozzi 53, 20125 Milano, Italy

Abstract. In this paper we address the study of optical and magnetic properties of metal oxides based on the cluster model approach and ab initio quantum chemistry techniques. Two different multireference approaches are used to compute the wave functions of the ground and the excited electronic states involved in optical transitions and magnetic interactions. The Difference Dedicated Configuration Interaction (DDCI) approach is a variational method based on the direct computation of energy differences between states. The Complete Active Space Self-Consistent Field (CASSCF) followed by Complete Active Space Second-Order Perturbation Theory (CASPT2) approach combines the Configuration Interaction (CI) techniques and many-body perturbation theory to account for electron correlation effects. Both methods turn out to be very suitable to estimate the differential electronic correlation and compute excitation energies. To illustrate the suitability of both methods, four applications are presented in this work: the study of d-d transitions and magnetic interactions in fcc transition-metal oxides, and the study of excitonic and vacancy states in MgO.

1. Introduction

The combination of the cluster model approach and modern powerful quantum chemistry techniques can provide useful information about the electronic structure of local phenomena in metal oxides. The theoretical description of the electronic states involved in local optical transitions and magnetic phenomena, for example, in these oxides needs very accurate computational schemes, because of the generally very large differential electron correlation effects. Recently, two very promising methods have become available, that allow to study optical and magnetic phenomena with a high degree of precision. The first one, the Differ-

J. Maruani et al. (eds.), New Trends in Quantum Systems in Chemistry and Physics, Volume 2, 227–245.

ence Dedicated Configuration Interaction (DDCI) method [1, 2], is based on the direct computation of the energy difference between the electronic states of interest by Configuration Interaction (CI) techniques, whereas the second method, referred to as CASSCF / CASPT2 [3, 4], includes part of the electron correlation effects in a variational way and estimates the rest by many body perturbation theory. Here, we present four applications in which both methods are explored for their accuracy and efficiency to describe electronic states involved in local electronic properties.

The paper is organized as follows: in the subsequent section we give a short description of the two computational schemes. Thereafter, Section III discusses the four applications in separate subsections. Each subsection first gives some information about the material model, basis sets and the approximations made in the calculation of the wave functions. The first application concerns the d-d transitions in the transition metal (TM) oxides NiO, CoO and MnO. These transitions are very localized and it has been shown before that an accurate description of the states involved can be obtained within the cluster model approach [5-13]. Secondly, the theoretical description of the transitions accompanying an oxygen vacancy (F center) in MgO is discussed. Not only bulk vacancies are treated but attention is also focused on the F centers on the regular surface at a step and corner site. In the third place, we intent to give an accurate estimate of the excitation energy of a Frenkel exciton in MgO. This exciton is created when an electron is excited form a core level on Mg into the conduction band [14]. There is theoretical evidence that this excitation leads to a localized exciton because of the strong hole-particle interaction [15]. In the last place, the magnetic interactions in the cubic transition metal oxides —CuO, NiO, CoO, FeO and MnO— are studied. The unpaired 3d-electrons on the transition metals give rise to a net spin moment. The strength of the interaction between the spin moments on neighboring metal sites can be accurately calculated within the cluster model approach provided that electron correlation effects are accounted for [16-19]. Section IV summarizes the conclusions.

2. Computational methods

2.1. DDCI

The most straightforward way to improve the mean-field approximation of the Hartree-Fock approach is to extent the N-electron wave function with more determinants and variationally optimize the expansion coefficients. For systems where the Hartree-Fock wave function is a good starting point, the inclusion of determinants connected with single and double excitations usually recovers a large part of the electron correlation effects. However, for more complicated systems a multireference wave function is required, which implies a much faster increase in size of the computational problem than for a single reference wave function. Several very efficient implementations exist nowadays to reduce the number of parameters in the calculation without losing accuracy. One way is to impose additional restrictions on the configuration expansion coefficients as is done in the internally contracted CI method [20, 21]. In this approach, the configuration expansion is constructed by applying excitation operators to the reference wave function as a whole. On the other hand, the Difference Dedicated CI scheme is based on the understanding that many external determinants contribute equally to the correlation energy of the electronic states involved in the process under study. Therefore, a selection is made and only those determinants are included that contribute to the energy difference between the states. First, an uncontracted list of determinants is constructed from the multireference wave function by single and double replacements from the determinants in the reference space, usually a Complete Active Space (CAS) that reflects the essential physical mechanisms of the system under study. Thereafter, a selection is made by second-order perturbation theory. For a system with two unpaired electrons, it has been proven that in case of a degenerate reference space only the determinants $|k\rangle$ that fulfill the condition:

$$\frac{\langle i|\hat{H}^{(0)}|k\rangle\langle k|\hat{H}^{(0)}|j\rangle}{E_0 - E_k} \neq 0 \tag{1}$$

(where $|i\rangle$ and $|j\rangle$ represent two different determinants in the reference space) contribute to the energy difference of the states involved [22]. This condition selects those determinants with at most two holes in the inactive orbitals, one hole in the inactive orbitals plus one particle in the secondary orbitals, or two particles in the secondary orbitals. The complete diagonalization of the resulting CI matrix is usually referred to as DDCI2, making reference to the two degrees of freedom in the generation of the external determinants. Although, for almost any real system the selection condition does not apply strictly - normally the system has a non-degenerate reference space and often more than two unpaired electrons - the DDCI2 method gives very reasonable results for magnetic interaction problems. However, it has been shown recently that adding some well defined set of determinants to the wave function gives an important contribution to the energy difference between the states of interest. These extra determinants are characterized by at most one hole in the inactive orbitals plus two particles in the secondary orbitals or two holes in the inactive plus one particle in the secondary orbitals. The resulting method is labeled DDCI3 because of the three degrees of freedom in the generation of the determinants. DDCI3 is also the appropriate DDCI variant to compute energy differences involving electronic excitations [23, 24]. Figure 1 schematically depicts the determinants treated in the DDCI2 and the DDCI3 approaches. The DDCI3 calculations reported in this paper have been performed with the CASDI suite of programs [25].

2.2. CASSCF / CASPT2

Another possibility to improve the Hartree-Fock wave function is to estimate electron correlation effects by many-body perturbation theory. The division of the

Hamiltonian proposed by Møller and Plesset gives rise to a very successful and efficient method to treat electron correlation effects in systems that can be described by a single reference wave function. However, for a multireference wave function the Møller-Plesset division can no longer be made and an alternative choice of $\hat{H}^{(0)}$ is needed. One such scheme is the Complete Active Space Second-Order Perturbation Theory (CASPT2) developed by Anderson and Roos [3, 4]. We will briefly resume the most important definitions of the theory; one is referred to the original articles for a more extensive description of the method. The reference wave function is a CASSCF wave function that accounts for the largest part of the non-dynamical electron correlation. The zeroth-order Hamiltonian is defined as follows and reduces to the Møller-Plesset operator in the limit of zero active orbitals:

$$\hat{H}^{(0)} = \sum_{pq\sigma} f_{pq\sigma} \hat{E}_{pq} \quad with\ f_{pq\sigma} = -\langle 0|[[H, a^{\dagger}_{q\sigma}], a_{p\sigma}]_{+}|0\rangle \tag{2}$$

where $|0\rangle$ is the CASSCF reference wave function and the E_{pq} are excitation operators. In contrast to the Møller-Plesset $\hat{H}^{(0)}$ the one-electron operator in the present zeroth-order Hamiltonian has non-zero matrix elements between the different orbital subspaces and is only diagonal within these subspaces. For this reason a diagonalization step is required in order to obtain the second-order correction to the energy. The first-order wave function is written as an internally contracted expansion of all configurations connected to single and double replacements from the CASSCF wave function:

$$|\Psi^{(1)}\rangle = \sum_{pqrs} C_{pqrs} \hat{E}_{pq} \hat{E}_{rs} |0\rangle \tag{3}$$

As in any perturbational scheme based on Fock-type zeroth-order Hamiltonians, configurations can appear in the first-order wave function with an expectation value of $\hat{H}^{(0)}$ very close to (or even lower than) the expectation value of the reference wave function. This can lead to very small (or even negative) energy denominators both in the expressions for the second-order correction to the energy, $E^{(2)}$, and in the expressions for the coefficients of the configurations in the first-order wave function, $C^{(1)}$, and cause a breakdown of the perturbation treatment of the electron correlation. The best way to solve this problem is to include the trouble causing configuration in the CASSCF wave function, however, it often happens that the state that causes the breakdown does not have a large interaction matrix element with the reference state and therefore not very important to obtain a good estimate of the second-order energy. Roos and Anderson [26] introduced a so-called level-shift technique for the CASPT2 that shifts up in energy all configurations in first-order wave function by an arbitrary amount μ, avoiding in this way the destructive influence of the intruder state without being forced to extend the active space in the CASSCF step. Once the zeroth-order

Hamiltonian has been diagonalized, a correction is applied for the appearance of the level shift in the denominators of the expressions for $E^{(2)}$. This level shift method has been applied successfully to a wide variety of problems in the field of spectroscopy and can be considered as a pragmatic solution to the intruder state problem inherent to perturbation theory. All CASSCF / CASPT2 calculations presented here have been done with MOLCAS-4 [27].

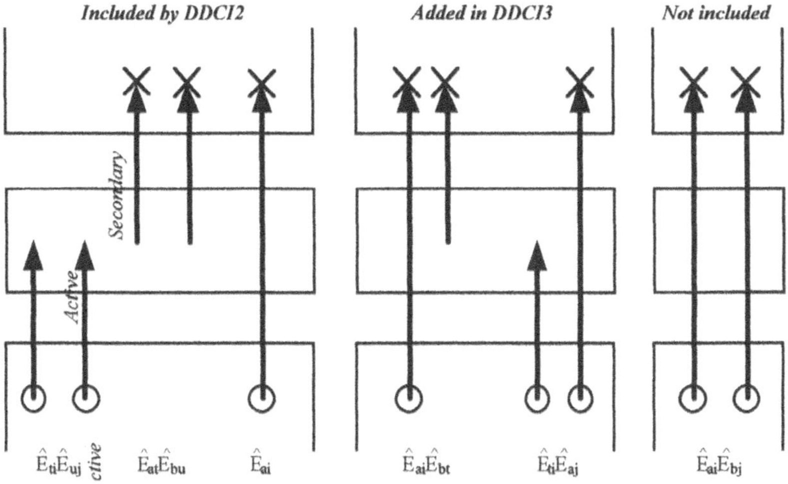

Figure 1. Schematic representation of the determinants treated in the different DDCI approaches. The different classes of external determinants are labeled by the excitation operator, $\hat{E}_{pq} = a_p^{\dagger} a_q$, that by acting on a determinant in the reference space generates a determinant of this class. The following labels are used: i, j for inactive orbitals; t, u for active orbitals; a, b for secondary orbitals. The open circles denote the creation of a hole in the inactive orbitals, whereas the crosses indicate the creation of an electron in the secondary space.

3. Applications

3.1. d-d EXCITATIONS IN NiO, CoO AND MnO

All three transition metal oxides considered crystallize in an fcc cubic structure with a lattice parameter of 4.16 Å, 4.26 Å, and 4.43 Å, for NiO, CoO and MnO, respectively [28]. The materials are modeled with a TMO_6 cluster embedded in optimized point charges [10] to account for the long-range electrostatic potential. These charges reproduce the Madelung field in the cluster region with high accuracy (less than 1 mHa in the cluster atoms). We also investigate the effect of the Pauli repulsion between the cluster and the ions in the direct surrounding in NiO and MnO. This short range repulsion is either included by adding ab initio embedding model potentials (AIEMP) [29] to the cluster or by embedding the cluster in frozen ions. The basis sets are constructed from atomic natural orbitals (ANO), for the transition metal ions a [21s, 15p, 10d, 6f] / (7s, 6p, 4d, 2f) basis set is employed and for the oxygen anions, we use a [14s, 9p, 4d] / (6s, 5p, 1d) basis set [30, 31]. The formal ionic charges of the transition metal cations of 2+ gives rise to a $(\ldots t_{2g}^{6} e_{g}^{2})$ ground state configuration for Ni, a $(\ldots t_{2g}^{5} e_{g}^{2})$ configuration for Co and a $(\ldots t_{2g}^{3} e_{g}^{2})$ configuration for Mn. The electrons in open-shell orbitals are coupled to a $^{3}A_{2g}$, $^{4}T_{1g}$, and $^{6}A_{1g}$ ground state, respectively. In this section we discuss the relative energies of the $3d^n$ states arising from the rearrangement of the electrons in the 3d-shell. For this purpose, we construct CASSCF wave functions for each $3d^n$ state with an active space that consists of the orbitals with mainly TM-3d character and a set of correlating virtuals, the so-called d' orbitals. In the subsequent CASPT2 step, we use as zeroth-order this multireference wave function to account for the remaining electron correlation effects arising from the TM-3s, 3p and 3d electrons and the O-2s and 2p electrons. A previous systematic study has shown that this division between variational and perturbational treatments of the electron correlation effects is adequate to obtain excitation energies within 0.15 eV of the experimental values [10].

Table 1 lists CASSCF and CASPT2 energies of the lowest $3d^n$ states of NiO, CoO, and MnO. The results are compared with experimental data obtained from optical absorption experiments for NiO and CoO [32, 33] and electron-energy-loss spectroscopy for MnO [34]. Note that results listed for CoO are obtained for a CoO_6 cluster embedded in point charges only and, therefore, do not have included the effect of the Pauli repulsion. In NiO, the Pauli repulsion due to the 18 Ni^{2+} ions coordinating the six oxygen ions in the cluster is included by representing these ions by AIEMPs (1s, 1p) [29], and in MnO, we included the Pauli repulsion by embedding the MnO_6 cluster in 18 frozen ions. To circumvent the problem of freezing open-shell ions, we modeled the extra Mn^{2+} ions by frozen Mg^{2+} ions, which have a rather similar ionic radius [35]. For MnO, we only list the sextet-quartet transitions because transitions to doublet states are expected to have a very low intensity, which prevents the experimental observation of these transitions. It is seen in Table 1 that the CASSCF wave function already gives a rather satisfactory description of the relative energies of the $3d^n$ states, the ordering is, with a few exceptions, similar as observed in the experiments. However, CASPT2 introduces important changes in the relative energies of the $3d^n$ states in all three compounds, ranging from an increase of the excitation energy of ~ 0.15 eV for the first excited $3d^n$ state in NiO to a decrease of more than 0.5 eV for several higher lying $3d^n$ states in all three materials.

Comparison of the experimental data to the CASPT2 values shows that for the large majority of states the calculated value does not deviate more than 0.2 eV from the experimentally determined one. Exceptions are the $^{4}A_{2g}$ state in Co-O (1.79 vs. 2.14) and the lower d-d transitions in MnO, for which the calculated transition energies are too high. In none of these cases are there indications from the perturbation theory that intruder states artificially affect the calculated transition energies. Hence, it is unlikely that the discrepancy between calculated and experimental values can be attributed to an incorrect or incomplete treatment of the electron correlation effects.

TABLE 1. Relative energies (in eV) of the lowest $3d^n$ states in bulk TMO (TM=Ni, Co, Mn) obtained by CASSCF and CASPT2. The active space is formed by the TM-3d orbitals and a set of correlating virtuals with the same symmetry character. CASPT2 correlates the TM-3s, 3p, 3d electrons and the O-2s, 2p electrons

NiO	CASSCF	CASPT2	Exp. [32-34]
$^3A_{2g}$	0.00	0.00	0.00
$^3T_{2g}$	0.95	1.12	1.13
a^3T_{1g}	1.71	1.73[(a)]	1.75
a^1E_g	2.20	1.85	1.95
a^1T_{2g}	3.05	2.86	2.75
a^1A_{1g}	3.64	3.04	2.94
b^3T_{1g}	3.53	3.11[(a)]	3.25
$^1T_{1g}$	3.87	3.68	3.52
CoO			
a^4T_{1g}	0.00	0.00	0.00
$^4T_{2g}$	0.72	0.76	0.95
a^2E_g	1.79	1.81	1.61
a^2T_{1g}	2.45	2.22	2.03
a^2T_{2g}	2.50	2.16	2.05
$^4A_{2g}$	1.49	1.79	2.14
b^4T_{1g}	2.70	2.27	2.30
b^2T_{1g}	3.16	2.47	2.52
$^2A_{1g}$	2.99	2.64	2.61
MnO			
$^6A_{1g}$	0.00	0.00	0.00
a^4T_{1g}	2.91	2.82	2.13
a^4T_{2g}	3.45	3.09	2.40
$^4A_{1g}$	3.68	3.29	2.82
a^4E_g	3.70	3.30	
b^4T_{2g}	4.28	3.66	3.31
b^4E_g	4.62	3.91	3.82
b^4T_{1g}	4.78	4.48	4.57
$^4A_{2g}$	6.16	5.19	5.08

(a) with a level shift of 0.20 au

On the other hand, a previous study showed that the lack of Pauli repulsion between the cluster atoms and the direct surrounding introduces a small but not negligible error in the calculated values of the d-d transitions [36]. It was found that the Pauli repulsion increases the splitting of the e_g and t_{2g} TM-3d orbitals by ~ 0.15 eV in NiO. Assuming a similar effect for CoO and recalling that the excitation energy of the $^4A_{2g}$ state is proportional to $2x10D_q$, we may conclude that the excitation energy for this state listed in Table 1 will increase by about 0.3 eV when Pauli repulsion is accounted for. This increase will bring the calculated excitation energy within the typical accuracy of the CASSCF / CASPT2 method, ~ 0.2 eV.

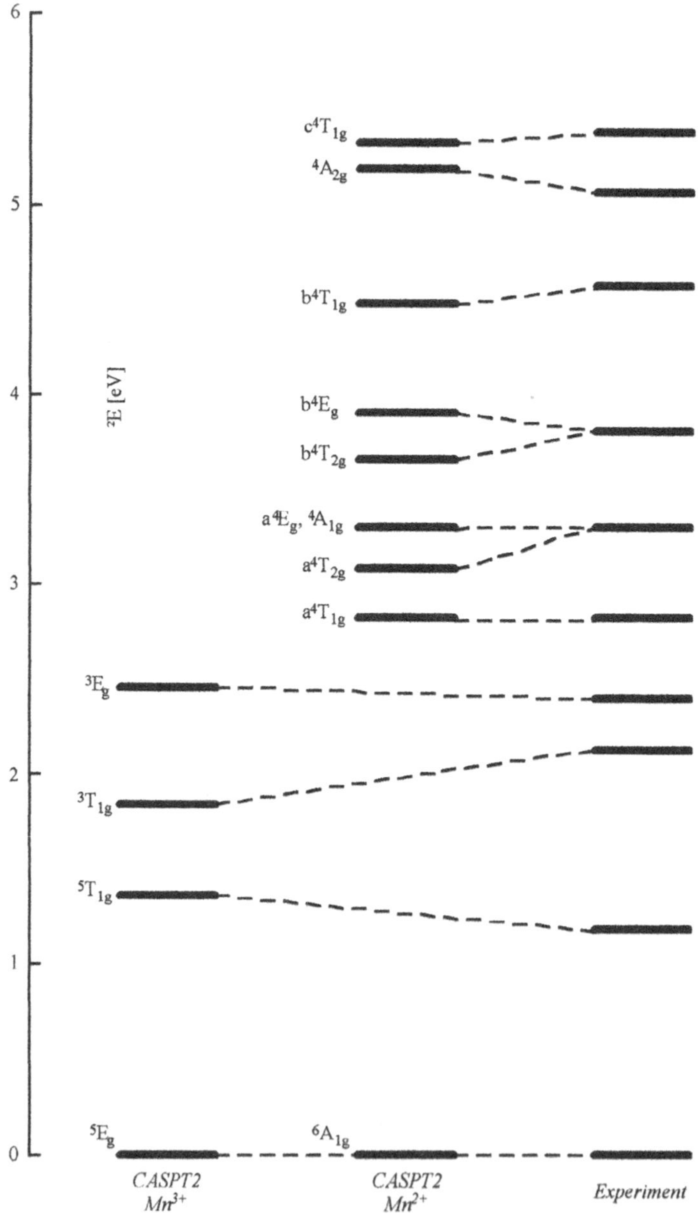

Figure 2. Level diagram of the d-d transitions in bulk MnO. The CASPT2 values obtained for the Mn^{2+} and the Mn^{3+} ions are compared with the EELS data of Ref. [34].

For MnO the situation is less clear. The calculated transition energies of the lower d-d transitions are significantly higher than the experimental ones: ~ 0.7 eV for the a^4T_{1g} and the a^4T_{2g} states, and ~0.5 eV for the $^4A_{1g}$, ^4E_g and b^4T_{2g} states. The remaining d-d transitions are calculated within the accuracy of the computational method of ±0.15 eV as observed for NiO and CoO. In order to explain the differences between theory and experiment, we tested two different hypotheses. The first one is that the lower d-d transitions arise from Mn^{2+} ions located at the (100) surface. In NiO, a surface state has been observed with a transition energy of 0.6 eV, i.e., 0.5 eV lower than the first bulk transition [37-39]. The second possibility is that these lower d-d transitions arise from Mn^{3+} impurities in the crystal. In fact, Fromme et al. [34] assign a small shoulder observed at 1.18 eV to the $^5E_g \rightarrow {}^3T_{1g}$ transition on a Mn^{3+} impurity. Calculations on a MnO_5 cluster embedded in Evjen charges show that the symmetry lowering at the surface from O_h to C_{4v} does not cause the appearance of surface states at lower transition energies than those for the bulk. The calculated $^6A_1 \rightarrow {}^4A_2$, ^4E and $^4B_2$ transition energies are larger than 3 eV. On the other hand, the calculated transitions for a Mn^{3+} ion impurity in MnO appear at energies significantly lower than the ones calculated for the Mn^{2+} ion; the lowest Mn^{3+} transition, $^5E_g \rightarrow {}^5T_{1g}$, appears at 1.36 eV, the transition to the $^3T_{1g}$ state at 1.85 eV and the ^3E_g transition is calculated at 2.46 eV. Figure 2 summarizes the calculated and observed transition energies. The level diagram suggests an alternative interpretation for the peaks observed in the EELS experiments [34]. The lower three peaks might arise from a Mn^{3+} impurity, while the peak at 3.0 eV may be assigned to the first d-d transition of the Mn^{2+} ion. The study of the d-d transitions in MnO with X-ray fluorescence spectroscopy [40] gives additional support for this hypothesis. In these spectra the lowest peak that carries significant intensity only arises at 3 eV. Nevertheless it is obvious that a more comprehensive analysis of the transition energies in MnO is needed.

3.2. F CENTERS IN MgO

The oxygen vacancies in solid MgO give rise to interesting features in the optical spectra. The so called F centers are characterized by the trapped electrons in the cavity left by removing the anion. The removal of an O^- ion or a neutral O atom results in F^+ or F centers with one or two electrons trapped, respectively. It is widely accepted that these electrons are localized at the center of the vacancy [41] and therefore, the combination of cluster model approach and high accurate ab initio methodologies is a reliable technique to study the spectroscopy of these vacancies. It is well known that both F and F^+ centers give rise to an intense absorption band around 5 eV which is composed of two bands, one at 4.96 eV due to F^+ centers and one due to neutral F centers at 5.0 eV [42, 43]. In a previous work, Illas and Pacchioni [41] studied the F centers in bulk and surface MgO by means of cluster models and MRCI wave functions. The computed excitation energies for bulk F and F^+ centers were 6.0 and 5.8 eV respectively, with an error of 0.8-1 eV with respect to the experiment, due to limitations in the level of calculation, mainly the basis sets size. Applying the same error estimation, these authors could assess the surface optical transitions at 2.2-2.5 eV for F and F^+ defects. The interpretation of the optical measurements of MgO surface is much less clear. The bands assigned in the literature to surface F centers vary from 2-2.3 eV to 1.15 eV while other bands remain unassigned (see reference [44] and references therein).

In this work we will present the results of the optical transitions associated to F centers formed at low-coordinated sites of the MgO (100) surface. As these excitations will appear at lower energy than those in the regular surface we will check whether they can be responsible of some of the unassigned features of the spectrum.

To model a regular surface, a step and a corner, we use the following cluster models: a Vac $(Mg^{2+})_5$ $(O^{2-})_{12}$ $(Mg^{2+})_{12}$ $(TIP)_{12}$ cluster embedded in an array of 634 point charges for the surface, a Vac $(Mg^{2+})_4$ $(O^{2-})_{10}$ $(Mg^{2+})_{20}$ cluster embedded in an array of 563 point charges for a step and a Vac $(Mg^{2+})_3$ $(O^{2-})_6$ $(Mg^{2+})_{10}$ cluster embedded in 323 point charges for a corner, where Vac refers to the vacancy left by an O anion and TIP is the total ion potential [45] without basis set to model the 12 outermost Mg^{2+} cations. All atoms are treated as effective core potentials (ECP) that include the 1s, 2s and 2p electrons for the Mg cations and the 1s core orbital for the O anions. The basis sets used are the following: an optimized (3s, 2p, 1d) uncontracted basis set for the vacancy, a contracted [4s, 1p] / (2s, 1p) basis for the next Mg^{2+} neighbors, a [4s, 4p] / (2s, 2p) for O^{2-} and a [4s] / (1s) contraction for the outermost Mg^{2+} ions (for more details concerning basis set and cluster models see ref. [44]). Since the creation of a vacancy can induce non-negligible geometrical relaxation, the positions of the Mg nearest neighbors and the O next nearest neighbors of the vacancy are optimized at the Hartree-Fock level for the ground state of the F and F^+ centers for each cluster. The rest of the atoms and point charges are fixed at the ideal crystal position. DDCI calculations have been performed for the lowest excited states for both F and F^+ centers in the three clusters. The model space is a CAS which involves two orbitals and either two or one electron for F and F^+ centers, respectively. To construct the CAS a set of molecular orbitals has to be chosen. However, excitation energies computed by DDCI3 are not very dependent upon the molecular orbitals used [44].

In Table 2 we summarize the results of the DDCI calculations on the lowest excitation energies for terrace, step and corner F and F^+ centers. As expected, for a given transition the excitation energy decreases as the coordination of the defect decreases. For F centers, the allowed singlet to singlet lowest transition occurs at 3.4 eV for the surface, 2.9 eV for the step and 2.6 eV for the corner. The same trend is found for the F^+ centers, where the first doublet to doublet transition, goes from 3.6 eV for the surface, to 2.6 eV for the step and 2.4 eV for the corner. As observed previously for the bulk calculations, excitations for the F^+ centers appear around 0.2 eV below the excitations due to F centers. The excitations at the step and corner sites are shifted about 0.5-1.2 eV compared to the surface transitions indicating that, if these defects are present on the surface, two different optical bands should be detected in the spectrum.

TABLE 2. Summary of excitation energies (in eV) of F and F^+ centers on the MgO surface step and corner sites [(a)]

	Surface		Step		Corner	
Center	Transition	DDCI	Transition	DDCI	Transition	DDCI
F	$^1A_1 \rightarrow {}^3A_1$ (1s →2p$_z$)	1.98	$^1A' \rightarrow {}^3A''$ (1s →2p$_z$)	1.63	$^1A \rightarrow {}^3E$ (1s →2p$_{x,y}$)	1.39
	$^1A_1 \rightarrow {}^1A_1$ (1s →2p$_z$)	3.39	$^1A' \rightarrow {}^1A''$ (1s →2p$_z$)	2.92	$^1A \rightarrow {}^1E$ (1s →2p$_{x,y}$)	2.60
	$^1A_1 \rightarrow {}^1E$ (1s →2p$_{x,y}$)	4.72	$^1A' \rightarrow {}^1A'$ (1s →2p$_{x,y}$)	3.82		
F^+	$^2A_1 \rightarrow {}^2A_1$ (1s →2p$_z$)	3.56	$^2A' \rightarrow {}^2A''$ (1s →2p$_z$)	2.60	$^2A \rightarrow {}^2E$ (1s →2p$_{x,y}$)	2.43

(a) DDCI calculations for F centers were performed using the orbitals from the $^3A_1$ state for the surface and those from the $^3A'$ and ^3A states for the step and corner, respectively. For F^+ centers the orbitals of the $^2A_1$ state were used for the surface and $^2A'$ and ^2A for the step and corner.

Comparison of our results with experiment is difficult because, except for the band appearing at 5 eV, the rest of the bands are not unambiguously assigned. For bulk MgO previous calculations showed an overestimation of 0.8-1 eV of the excitation energy, that is, an error of ~15%. Applying a similar correction to our present calculations, the singlet to singlet allowed transitions are predicted to occur at ~2.9 and ~3.0 eV for F and F^+ surface centers, at ~2.5 and ~2.2 eV for F and F^+ step centers and at ~2.2 and ~2.1 eV for F and F^+ corner centers. These values allow us to tentatively assign the bands observed around 2-2.3 eV to low-coordinated (step and corner) F centers while the transitions due to surface vacancies should be seen around 3 eV. Recent EELS measurements on MgO ultrathin films grown on Ag(100) have shown bands at 3.3 eV and 2.1 eV, that were assigned to surface F centers [46]. These new experimental data give additional support to our theoretical assignments. The assignment we propose implies that the oxygen vacancies detected by optical spectroscopy are localized both at regular surface and at low-coordinated sites.

A final point concerns the comparison of the DDCI method and the CASSCF / CASPT2 approach. We have performed some test calculations for the step cluster using all electron basis sets of moderate size for all cluster atoms. These calculations not only permit a direct comparison of the two approaches for the inclusion of electron correlation but also provide a test of the ECP and basis used in the previously commented calculations. Compared to the results listed in Table 2 for the surface, the DDCI excitation energies obtained with all electron basis sets change less than 0.1 eV for all states considered. Moreover,

CASPT2 excitation energies are also very similar to the DDCI energies obtained with the all electron basis sets. The perturbative scheme gives energies which are about 0.2 eV lower. Hence we conclude that the present choice of ECP and basis set are sufficient to compute excitation energies of F centers in the MgO (100) surface.

3.3. FRENKEL EXCITONS IN MgO

In a previous paper, we studied the character of the excited states of bulk MgO [47]. This type of excitations involves the creation of a hole in an occupied orbital and a particle in an empty orbital. When the hole-particle interaction is strong enough a localized excitonic state arises. Our previous study showed that there are no excitonic states when the hole is made on the ligand orbitals, either in the valence O(2p) or the core O(1s) orbitals. However, we found evidence for the existence of an excitonic state arising from an excitation from either an inner Mg(1s) orbital or a Mg(2p) orbital [15, 47]. In these cases, the localization of the hole on a Mg^{2+} center, leading to an effective Mg^{3+} center, results in an excited orbital localized inside the cluster region which extends up to the first shell of oxygen neighbors only. The electron in the excited state showed a strong Mg-3s character and we concluded that the excitonic states coming from an excitation in the Mg levels are indeed localized. These results confirm the assignment arising from the measurements of O'Brien et al. [14]. These authors measured the absorption spectrum of bulk MgO and reported a peak around 0.3 eV below the MgO conduction band. This peak was assigned to an excitonic state due to a Mg $2p^6 \rightarrow 2p^5\ 3s^1$ transition with a transition energy of 53.4 eV.

In our previous work, we focused on the character of the excited states by means of the analysis of SCF wave functions. At this level of calculation, the computed excitation energy for the Mg $2p^6 \rightarrow 2p^5\ 3s^1$ transition was 54.6 eV, that is, more than 1 eV above the experimental result. In the present study we attempt the calculation of this excitation by means of more accurate multiconfigurational methods, particularly, the CASSCF / CASPT2 approach commented before. To model the MgO crystal, a $Mg_{13}O_{14}$ cluster is used. This cluster model is arranged in a cube and contains a central Mg, the six O nearest neighbors, the next 12 Mg neighbors and 8 O occupying the vertices of the cube. The cluster is embedded in a set of optimized point charges to ensure a good representation of the long-range electrostatic potential in the cluster region. All electrons of the cluster are explicitly considered, and the following ANO basis sets are used for all atoms: for the central Mg, a [17s, 12p, 5d] / (5s, 4p, 1d) contracted basis set; for the six O's, a [14s, 9p, 4d] / (4s, 3p, 1d); for the 12 Mg's, a [17s, 12p, 5d] / (4s, 3p), and for the remaining 8 O's a [14s, 9p, 4d] / (3s, 2p) contracted basis set. The system contains 270 electrons and 358 basis functions.

CASSCF wave functions based on an active space formed by two orbitals (a central Mg-2p orbital and an orbital with strong central Mg-3s character) and by two electrons are constructed for a singlet closed-shell ground state and singlet and triplet excited states. The CASSCF result is improved by perturbation theory performing a CASPT2 calculation in which all Mg and O 2s and 2p electrons are correlated. The presence of a hole in a Mg(2p) orbital in the reference wave function, makes the perturbative treatment of the correlation energy extra sensitive for intruder state problems, since many configurations will appear in the first-order wave function with similar (and, of course, also much lower) energy expectation values of $\hat{H}^{(0)}$ as the reference wave function. For this reason, we observe rather low weights of the CASSCF wave function in the first-order corrected one. However, these low weights do not significantly affect the second-order estimate of the total correlation energy, because the individual terms have very small interaction matrix elements with the CASSCF wave function. In test calculations on a MgO_6 cluster embedded in point charges, we observe a very similar correction of the excitation energy by CASPT2 (around 1 eV in both cases) without the appearance of intruder states in the wave function, giving additional evidence that the perturbative correction to the excitation energy is not significantly influenced by the intruder states encountered in the larger cluster.

The CASSCF excitation energies for the excited triplet state is 54.5 eV, similar to the SCF value reported in a previous work. The Mulliken population analysis of the active orbitals indicate a high degree of localization of the excited state, in agreement with previous results. The inclusion of electron correlation by CASPT2 decreases the CASSCF excitation energies around 1 eV, leading to a final value for the excitation of 53.4 eV. For the excited singlet state the CASSCF value is 54.7 eV and the CASPT2, 53.6 eV. The experimental transition energy, 53.4 eV, has to be compared with the excitation energy to the excited singlet state, which corresponds to the spin allowed transition, at 53.6 eV. This result is converged for the size of the basis set, as the excit-ation energies obtained with smaller basis sets on the first layer of O ions and the second shell of Mg ions are virtually the same. Furthermore, the extension of the present cluster with the next shell of Mg atoms, $Mg_{19}O_{14}$, does not change the values of the excitation energies. Agreement with the experiment is very satisfactory showing that the CASSCF / CASPT2 method is a good approach to study these localized excited states in ionic materials.

3.4. MAGNETIC COUPLING IN TMO (TM=Cu, Ni, Co, Fe, Mn)

The localized nature of magnetic moments in ionic transition-metal compounds permits an accurate study of the interactions between the moments located on neighboring TM ions. The extraction of the magnitude of the interaction from experimental data is commonly done by fitting magnetic susceptibility data to a Heisenberg Hamiltonian that contains two center interactions only and is usually limited to neighboring TM ions. For this reason the cluster model approach emerges as a natural choice to study the magnetic interactions in ionic compounds. More specifically, recent cluster model studies have established the appropriateness of the approach by quantitatively reproducing experimental values of the magnetic coupling in a variety of ionic materials with a two-center cluster only [19]. It has also been demonstrated that the inclusion of more magnetic centers in the cluster do not significantly change the calculated values [48]. Here, we report the results of CASSCF / CASPT2calculations on the magnetic interactions in the series of late transition metal oxides TMO (TM=Cu, Ni, Co, Fe, Mn). It is expected that the magnetic interaction in this series decreases from Cu to Mn, since the decrease of the nuclear charge on the metal causes an increase in the charge transfer energy and, hence, a decrease in the importance of the covalent interactions.

To facilitate the comparison between the different compounds, we use idealized structures for CuO and FeO, whose real crystal structures show defects and / or orthorhombic distortions. The lattice parameters are as follows: 4.087 and 4.332 Å for CuO and FeO (extrapolation to x=0 of FeO_{1-x}), respectively. Those for the other compounds are given in subsection A of this section. The clusters used in the calculations contain two TM ions and a bridging oxygen between them. The 10 oxygen anions directly surrounding the TM_2O unit are also included in the cluster. The short range repulsion due to the 4 TM ions surrounding the bridging oxygen is taken into account by including Mg^{2+} AIEMPs provided with a (1s,1p) basis set to ensure the strong-orthogonality between cluster and embedding ions [49]. In addition, calculations are performed for the magnetic interactions at the TMO(100) surface. For this purpose, TM_2O_9 clusters are constructed. The three TM ions around the bridging oxygen are described in the same way as for the bulk, and the effect of the rest of the crystal is included by a set of Evjen charges. We use the following ANO basis sets to describe the one-electron space: for the TM ions, [21s, 15p, 10d, 6f] / (6s, 5p, 3d, 1f); for the bridging oxygen, [14s, 9p, 4d] / (5s, 4p, 1d), and for the edge oxygens [14s, 9p] / (3s, 2p). The active space in the CASSCF calculation is formed by the open-shell orbitals and all the unpaired electrons, i.e., (2/2) for CuO, (4/4) for NiO, (10/14) for CoO since the ground state configuration of Co^{2+} in an octahedral field has contributions from both the $e_g^2 t_{2g}^5$ and the $e_g^3 t_{2g}^4$ configurations, (8/8) for FeO, and (10/10) for MnO. CASPT2 correlates the TM-3s, 3p, 3d and the O-2s, 2p electrons.

Table 3 summarizes results obtained for the bulk and the surface. First, we observe that the trend of decreasing magnetic interaction strength, parameterized by J in the Heisenberg Hamiltonian, is reproduced both by CASSCF and CASPT2. The analysis of the one-electron functions and multireference N-electron wave function indicates that indeed the covalent interaction - appearing either through a mixing in the one-electron functions of the O-2p and TM-3d atomic orbitals or as configurations connected to charge transfer excitations in the N-electron wave function - is strongest for CuO and a gradual decrease is observed towards MnO. Secondly, the results in Table 3 show that the treatment of external electron correlation increases the calculated J by a factor of about 4. This has been observed in many applications before and illustrates once more the necessity to account for this electron correlation in order to obtain reasonable estimates of the magnetic coupling parameter. In third place, it must be noted that in some cases CASPT2 fails to estimate the electron correlation effects. For MnO (bulk), CoO (100) and MnO (100) the zeroth-order Hamiltonian cannot be diagonalized. More specifically, this occurs for the lower spin coupled (singlet and / or triplet) states, which all have a very long reference wave function expansion; typically between 10^3 and 10^4 CSFs, none of them with coefficients > 0.1. This prevents the diagonalization of the zeroth-order Hamiltonian in CASPT2 within a reasonable number of iterations. However, the comparison of the estimate of the magnetic interaction parameters derived from the higher spin coupled states with the materials that do not show convergence problems indicate that the effect of CASPT2 is the same in these troublesome cases.

TABLE 3. CASSCF and CASPT2 values of the magnetic coupling parameter J (in meV) for bulk TMO and for the TMO (100) surface. The active space in the CASSCF is formed by the open shell orbitals located on the TM ions and the unpaired electrons. CASPT2 correlates the TM-3s, 3p, 3d and the O-2s, 2p electrons

Bulk	CASSCF	CASPT2
CuO	-22.65	-78.20
NiO	-4.15	-13.65
CoO	-1.16	-4.45
FeO	-0.68	-2.28
MnO	-0.34	–
Surface		
CuO	-14.24	-51.79
NiO	-3.17	-10.68
CoO	-0.86	–
FeO	-0.45	-1.98
MnO	-0.23	–

The value calculated for NiO (bulk), 13.7 meV, compares very reasonably with the experimental value - 19 meV - obtained from magnetic susceptibility [50] and Raman scattering [51]. The agreement can even be slightly improved by extending the active space with the 2p orbital of the bridging oxygen and a set of virtuals with the same symmetry character as the TM-3d and O-2p orbitals. This active space allows for a more precise treatment of the charge transfer configurations in the wave function and leads to an improvement of the calculated J-value. These calculations have only been performed for CuO and NiO since for the other compounds the active space would increase to an unmanageable size, e.g. for CoO an active space with 22 active orbitals needs to be evaluated. Both for CuO and for NiO an increase of about 30% of the CASSCF value is obtained with the larger active space, -30.7 meV and -5.2 meV respectively. The CASPT2 value increases to -90.7 meV for CuO and to -14.8 meV for NiO, that is an increase of 16% and 8% respectively. Comparison of the computed magnetic coupling constant for bulk MnO with experimental data is not as direct as for bulk NiO, because of the lack of a CASPT2 estimate. However, assuming a similar increase as in the other oxides with respect to the CASSCF value (approximately a factor of 3.5), a value of -1.2 meV can be extrapolated for MnO.

This is in fair agreement with the experimental estimate of -1.7 meV obtained from inelastic neutron scattering [52] and analysis of thermodynamic data [53].

Finally, the comparison of bulk and surface reveals the decrease of the magnetic interaction with the coordination number of the TM ion. In all cases, the calculated J value decreases by approximately 30% at the CASSCF level and between 30% (CuO) and 10% (FeO) for CASPT2. This effect has been studied in more detail for NiO previously [17]. It was shown that the decrease of J is completely determined by the lower coordination of the Ni ions at the surface and that the lower Madelung potential at the surface only has a minor influence on the magnetic coupling. This is in contrast with another theoretical prediction based on the Hubbard model Hamiltonian, which predicts an increase of the coupling at the surface by ~ 50% [54]. Very recently, a helium atom diffraction study has been published of the antiferromagnetic transition on the NiO(100) surface [55]. The interpretation of the experimental data suggest that the magnetic coupling at the surface is lower than in the bulk.

4. Conclusions

We have performed electronic structure calculations to study local physical processes in metal oxides. The energy scale at which the processes occur is rather wide, it ranges from ~50 eV for the excitonic state in MgO to less than 1 meV involved in the magnetic interaction in MnO. Multireference wave functions are constructed that include the essential ingredients of the physical phenomena of interest and thereafter, the remaining electron correlation effects are accounted for either by second-order perturbation theory (CASPT2) or in variational manner (DDCI). Both approaches give accurate results and permit to analyze the underlying physical mechanisms.

The study of the d-d transitions in NiO, CoO and MnO with the CASSCF / CASPT2 method makes evident the importance of electron correlation effects on the relative energies of the different $3d^n$ states. For NiO and CoO we confirm the interpretation made in optical spectroscopy experiments, but the assignment of the lowest peaks in the EELS spectrum of MnO cannot be confirmed. The lowest transitions have been calculated at 3.0 eV, which is 0.8 eV higher than the lowest peak in the spectrum. Additional calculations suggest that these peaks might arise from Mn^{3+} impurities in the crystal.

The DDCI results for the F centers in MgO at different sites of the (100) surface show a decrease of the excitation energy with the decrease in the coordination of the vacancy. Assuming an error in the calculated transition energies of 15%, similar to the error observed for the transitions in bulk F centers, we can assign the peaks observed around 3 eV to regular surface F centers, the band observed around 2.4 eV to step sites and the band around 2.2 eV can be assigned to corner sites.

The excitonic state in MgO (bulk) arising from the creation of a hole in the Mg-2p core level appears as an impurity level within the band gap of MgO. Previous cluster model studies provided theoretical evidence of the local character of this excitonic state. In the present work we improved the estimated transition energy by accounting for electron correlation effects by means of CASSCF / CASPT2. Modeling MgO with a $Mg_{13}O_{14}$ cluster embedded in point charges, CASSCF gives a transition energy of 54.7 eV. This energy is lowered by about 1 eV at the CASPT2 level. The final estimate of 53.6 eV compares very well with the experimental value of 53.4 eV.

Finally, the magnetic interactions in the fcc TMO (TM=Cu, Ni, Co, Fe, Mn) have been analyzed. We find the magnetic interaction decreases with decreasing nuclear charge on the TM ion. A comparison of magnetic interactions between bulk atoms and between atoms located at the (100) surface shows that the interaction in all five compounds is lowered at the surface. The calculated value of the interaction in bulk NiO compares rather well with the measured one, especially when charge transfer effects are included in a more complete manner by extending the active space in CASSCF.

Acknowledgements

We thank Dr Ria Broer of the University of Groningen for her valuable help and discussions. This work was financed by the Spanish "Ministerio de Educación y Ciencia", under CICyT project PB95-0847-CO2-01, and by "Acción Integrada Hispano-Italiana", HI1998-0042. Part of the computer time was provided by the "Centre de Supercomputació de Catalunya" C^4-CESCA through a research grant from the University of Barcelona. C. de G. acknowledges financial help through the TMR action "Marie Curie research training grants", grant FMBICT983279, established by the European Community.

References

1. Miralles, J., Daudey, J.P., and Caballol, R. (1992) *Chem. Phys. Lett.* **198**, 555.
2. Miralles, J., Castell, O., Caballol, R., and Malrieu, J.-P. (1993) *Chem. Phys.* **172**, 33.
3. Andersson, K., Malmqvist, P.-Å., Roos, B.O., Sadlej, A.J., and Wolinski, K. (1990) *J. Phys. Chem.* **94**, 5483.
4. Andersson, K., Malmqvist, P.-Å., and Roos, B.O. (1992) *J. Chem. Phys.* **96**, 1218.
5. Bagus, P.S., and Wahlgren, U. (1977) *Mol. Phys.* **33**, 641.
6. Janssen, G.J.M., and Nieuwpoort, W.C. (1988) *Phys. Rev. B* **38**, 3449.
7. Pueyo, L., Luaña, V., Flórez, M., Francisco, E., Recio, J.M., and Bermejo, M. (1991) *Rev. Solid State Sci.* **5**, 137.
8. Martin, R.L. (1992), in G. Pacchioni, P. S. Bagus and F. Parmigiani (eds.), *Cluster Models for Surface and Bulk Phenomena*, Plenum Press, New York, p. 485.
9. Haßel, M., Kuhlenbeck, H., Freund, H.J., Shi, S., Freitag, A., Staemmler, V., Lütkehoff, S., and Neumann, M. (1995) *Chem. Phys. Lett.* **240**, 205.
10. de Graaf, C., Broer, R., and Nieuwpoort, W.C. (1996) *Chem. Phys.* **208**, 35.
11. Fujimori, A., and Minami, F. (1984) *Phys. Rev. B* **30**, 957.
12. van Elp, J., Eskes, H., Kuiper, P., and Sawatzky, G.A. (1992) *Phys. Rev. B* **45**, 1612.

13. Lorda, E., Illas, F., and Bagus, P.S. (1996) *Chem. Phys. Lett.* **256**, 377.
14. O'Brien, W.L., Jia, J., Dong, Q.-Y., Callcott, T.A., Rubensson, J.-E., Mueller, D.L., and Ederer, D.L. (1991) *Phys. Rev. B* **44**, 1013.
15. Sousa, C., Illas, F., Ricart, J.M., and Bagus, P.S. (1995) *Chem. Phys. Lett.* **239**, 263.
16. van Oosten, A.B., Broer, R., and Nieuwpoort, W.C. (1996) *Chem. Phys. Lett.* **257**, 207.
17. de Graaf, C., Broer, R., and Nieuwpoort, W.C. (1997) *Chem. Phys. Lett.* **271**, 372.
18. de P. R. Moreira, I., and Illas, F. (1997) *Phys. Rev. B* **55**, 4129.
19. de P. R. Moreira, I., Illas, F., Calzado, C.J., Sanz, J.F., Malrieu, J.-P., Ben-Amor, N., and Maynau, D. (1999) *Phys. Rev. B* **59**, 6593.
20. Werner, H.-J., and Knowles, P.J. (1988) *J. Chem. Phys.* **89**, 5803.
21. Knowles, P.J., and Werner, H.-J. (1992) *Theor. Chim. Acta* **84**, 95.
22. Malrieu, J.-P. (1967) *J. Chem. Phys.* **47**, 4555.
23. García, V.M., Castell, O., Caballol, R., and Malrieu, J.-P. (1995) *Chem. Phys. Lett.* **238**, 222.
24. García, V.M., Caballol, R., and Malrieu, J.P. (1998) *J. Chem. Phys.* **109**, 504.
25. D. Maynau and N. Ben Amor, CASDI suite of programs.
26. Roos, B.O., and Andersson, K. (1995) *Chem. Phys. Lett.* **245**, 215.
27. Andersson, K., Blomberg, M.R.A., Fülscher, M.P., Karlström, G., Lindh, R., Malmqvist, P.-Å., Neogrády, P., Olsen, J., Roos, B.O., Sadlej, A.J., Schütz, M., Seijo, L., Serrano-Andrés, L., Siegbahn, P.E.M., and Widmark, P.-O. (1997) MOLCAS 4, University of Lund, Sweden.
28. Wyckoff, R.W.G. (1963) *Crystal Structures, vol. 1*, Wiley, New York.
29. Barandiarán, Z., and Seijo, L. (1988) *J. Chem. Phys.* **89**, 5739.
30. Pou-Amérigo, R., Merchán, M., Nebot-Gil, I., Widmark, P.-O., and Roos, B.O. (1995) *Theor. Chim. Acta* **92**, 149.
31. Widmark, P.-O., Malmqvist, P.-Å., and Roos, B.O. (1990) *Theor. Chim. Acta* **77**, 291.
32. Newman, R., and Chrenko, R.M. (1959) *Phys. Rev.* **114**, 1507.
33. Pratt Jr., G.W., and Coelho, R. (1959) *Phys. Rev.* **116**, 281.
34. Fromme, B., Brunokowski, U., and Kisker, E. (1998) *Phys. Rev. B* **58**, 9783.
35. Shannon, R.D. (1976) *Acta Crystallografica* **A32**, 751.
36. de Graaf, C., Sousa, C., and Broer, R. (1999) *J. of Mol. Struct. (Theochem)* **458**, 53.
37. Freitag, A., Staemmler, V., Cappus, D., Ventrice Jr., C.A., Shamerey, K.A., Kuhlenbeck, H., and Freund, H.J. (1993) *Chem. Phys. Lett.* **210**, 10.
38. Gorschlüter, A., and Merz, H. (1994) *Phys. Rev. B* **49**, 17293.
39. Fromme, B., Möller, M., Anschütz, T., Bethke, C., and Kisker, E. (1996) *Phys. Rev. Lett.* **77**, 1548.
40. Butorin, S.M., Guo, J.-H., Magnuson, M., Kuiper, P., and Nordgren, J. (1996) *Phys. Rev. B* **54**, 4405.
41. Illas, F., and Pacchioni, G. (1998) *J. Chem. Phys.* **108**, 7835.
42. Chen, Y., Williams, R.T., and Sibley, W.A. (1969) *Phys. Rev.* **182**, 960.
43. Crawford Jr., J.H. (1983) *Semiconductors and Insulators* **5**, 599.
44. Sousa, C., Pacchioni, G., and Illas, F. (1999) *Surf. Sci.* **429**, 217.
45. Durand, P., and Barthelat, J.C. (1975) *Theor. Chim. Acta* **38**, 283.
46. Peterka, D., Tegenkamp, C., Schröder, K.-M., Ernst, W., and Pfnür, H. (1999) *Surf. Sci.* **431**, 146.
47. Bagus, P.S., Illas, F., and Sousa, C. (1994) *J. Chem. Phys.* **100**, 2943.

48. Illas, F., de P. R. Moreira, I., de Graaf, C., Castell, O., and Casanovas, J. (1997) *Phys. Rev. B* **56**, 5069.
49. Pascual, J.L., Seijo, L., and Barandiarán, Z. (1993) *J. Chem. Phys.* **98**, 9715.
50. Hutchings, M.T., and Samuelsen, E.J. (1972) *Phys. Rev. B* **6**, 3447.
51. Massey, M.J., Chen, N.H., Allen, J.W., and Merlin, R. (1990) *Phys. Rev. B* **42**, 8776.
52. Koghi, M., Ishikawa, Y., and Endoh, Y. (1972) *Solid Sate Comm.* **11**, 391.
53. Lines, M.E., and Jones, E.D. (1965) *Phys. Rev.* **139**, A1313.
54. Pothuizen, J.J.M., Cohen, O., and Sawatzky, G.A. (1996) *Mat. Res. Soc. Symp. Proc.* **401**, 501.
55. Marynowski, M., Franzen, W., El-Batanouny, M., and Staemmler, V. (1999) *Phys. Rev. B* **60**, 6053.

ELECTROSTATIC EFFECTS IN THE HETEROLYTIC DISSOCIATION OF HYDROGEN AT MAGNESIUM OXIDE

C. PISANI AND A. D'ERCOLE
Unità INFM di Torino and Dipartimento di Chimica IFM, Università di Torino, via Giuria 5, I-10125 Torino, Italy

Abstract. An embedded-cluster approximation is adopted for simulating the heterolytic dissociation of hydrogen at two intrinsic defects on the (001) surface of magnesium oxide: the isolated anion vacancy, and the "tub" divacancy. The dissociation process is shown to be critically dependent on the structure of the electrostatic field at the surface both as concerns energetics and final configuration.

1. Introduction

MgO powders activated at high temperatures have been known for many years to promote the heterolytic cleavage of hydrogen and methane molecules [1, 2]. The interest in this kind of reactions is motivated by the great technological relevance of the realization of heterogeneous catalysts capable of promoting in an efficient way the partial oxidation of methane [3]. While it is well established that the perfect (001) face, by far the most stable surface of crystalline MgO, is practically inert, it is not yet clear which types of defect at that face are responsible for its activity.

The present work reports the results of a theoretical investigation concerning the heterolytic dissociation of hydrogen at two intrinsic defects on the (001) surface of magnesium oxide: the isolated anion vacancy (the classical "F" defect) and the tub divacancy, the latter formed by removing two nearest neighbor ions at the MgO (001) surface (to be identified in the following with a T). In two parallel papers [4, 5] we discuss this problem in more detail with reference to experimental information derived from recent infrared and electron paramagnetic resonance experiments [6]. Here we concentrate on the dissociation process by stressing the importance of the structure of the electrostatic field at the defect.

J. Maruani et al. (eds.), New Trends in Quantum Systems in Chemistry and Physics, Volume 2, 247–255.

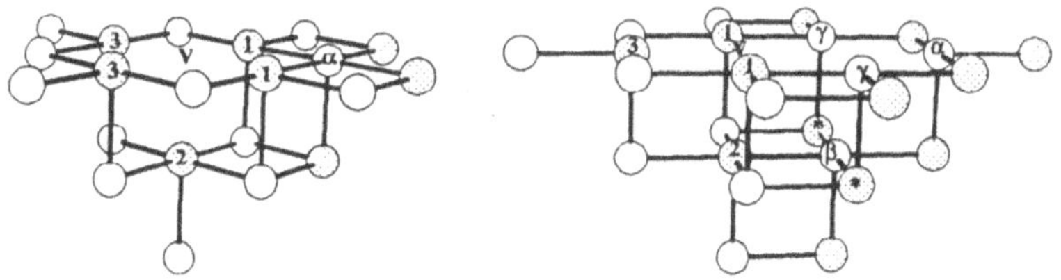

Figure 1. Embedded clusters used for simulating the anion vacancy (left) and the tub divacancy (right). Oxygen and magnesium ions are represented by white and light grey circles, respectively. The O ions at which the proton is bound are marked with Greek letters. The position of the missing oxygen is marked with V. The two Mg ions marked with an asterisk are included only when proton adsorption at O(β) is considered.

2. Method

We adopt for this study the perturbed-cluster technique [7] as implemented in the EMBED code in the frame of the Hartree-Fock (HF) approximation [8]. A detailed analysis of this method has been provided in a recent paper dealing with the characterization of different types of divacancy in MgO [9]. We only stress here an aspect which distinguishes the present from other embedding schemes (see for instance the contribution by Kantorovich in this volume). The general problem is to perform a quantum chemical calculation concerning a limited number of atoms (the *cluster*) within a complex structure, while keeping into account in an approximate way the effects of the rest of the system. In some special cases as those here considered, the cluster is part of a periodic system except for the presence at its interior of a local defect. The energy-dependent embedding potential to be added to the cluster hamiltonian can then be derived from the knowledge of the solution for the perfect host system. The latter is obtained by means of a periodic program which adopts the same techniques and approximations as for the treatment of the cluster: we use for this purpose the CRYSTAL code [10]. A consequence of this approach is observance of *self-embedding consistency* that is, in the absence of defects the one-electron density matrix within the cluster coincides with that of the host system. Most of the other embedding schemes are based on a combination of different approximations in the cluster region (where an ab-initio Hamiltonian is used) and in the environment (described with a semi-classical or semi-empirical Hamiltonian, as a set of point charges plus pseudo-potentials, etc.).

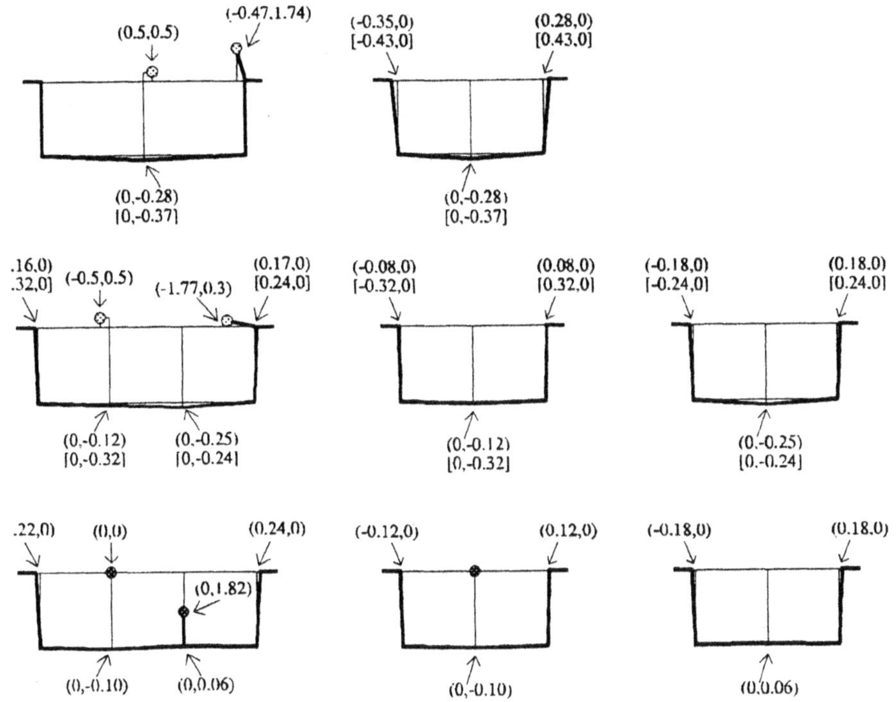

Figure 2. Equilibrium geometries for the three dissociated configurations. From top to bottom and from left to right, the following vertical sections are considered: Fα, in a plane through O(α) and Mg(2), and another through Mg(1) and Mg(2); Tα, in a plane through O(α) and Mg(2), and in two planes orthogonal to this one, through Mg(2) and O(β), respectively; Tβ, in the same planes as in the preceding case. Relaxations are given in Bohr with respect to the unperturbed lattice position: the two numbers in parentheses refer to the horizontal and vertical displacement. In square brackets, the corresponding data are provided for the bare defect prior to hydrogen adsorption. The position of H and H^+ is reported with reference to the missing oxygen site, and to the oxygen to which the proton is bound, respectively.

Figure 1 shows the embedded clusters adopted in the present study and introduces some conventions in the labelling of atoms; the two hydrogen atoms which are also part of the cluster are not represented. The three configurations of the dissociated H_2 molecule to be discussed in the following are reproduced in figure 2 along with the corresponding equilibrium geometry; they will be designed as Fα, Tα, and Tβ, respectively. The last two are distinguished from the oxygen where the hydroxyl is formed. The basis set is the same as in previous work [4]. Following the usual notation, it consists of 8-61 (Mg), 8-51 (O) and 31* (H) sets of contracted Gaussians; two sets of floating single Gaussian functions have been located at the site

TABLE 1. Calculated energies for the dissociation of the molecule (Hartree units). The HF results subsequently corrected for correlation (at an MP2 level of approximation) and for long range polarization are reported

Defect site	HF	HF+MP2	HF+MP2+pol
Fα	+0.032	+0.044	+0.041
Tα	-0.066	-0.033	-0.011
Tβ	-0.044	-0.014	-0.007

of the two removed surface ions. Corrections for the basis set superposition error have been included. The contributions to energy of electron correlation inside the cluster and of polarization of the surroundings have been estimated a posteriori [9].

3. Results and discussion

As shown in figure 2, local energy minima are found in the three cases corresponding to the formation of a hydride ion close to the position of the missing oxygen and of a hydroxyl near the defect; the Mulliken charges of the two hydrogens are 1.77, 1.97, 1.91, and 0.75, 0.57, 0.48, for Fα, Tα, and Tβ, respectively. In the presence of the dissociated molecule, the Mg and O atoms near the vacancy are closer to their position in the unperturbed crystal than in its absence. This is to be expected, since the molecular moieties partly compensate for the missing ion charges, more so for the T than for the F defect. The length of the hydroxyl bond is almost the same in the three cases: 1.80, 1.80, 1. 82 Bohr, in the same sequence as above.

Table 1 reports energy data for the dissociation reaction. At the HF level the two configurations at the tub are considerably more stable (41 and 28 kcal/mol) than the undissociated molecule; the opposite is true for the anion vacancy. The effect of the correlation correction is to destabilize appreciably the dissociated structures; in fact, its contribution to the strength of the H–H molecular bond and to the stability of the O^{2} ion is much higher than for O–H and $H^{\square}$. The polarization correction appears to act in the same direction, but to a rather different extent in the three cases. The explanation is similar to that we put forward when commenting on equilibrium geometries: the polarizing field acting on ions in the external zone is stronger in the absence than in the presence of the adsorbed molecule. Our estimate of this correction is rather rough [9] and can only be taken as indicative of the importance of long range polarization effects.

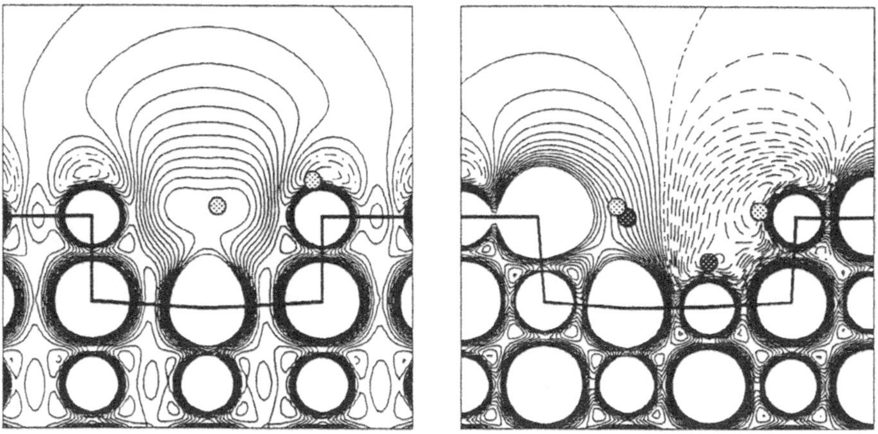

Figure 3. Electrostatic potential map for the two bare defects and identification of the position of the atoms of the dissociated H_2 molecule (H atoms in the Tβ configuration are in dark grey). The sections are in a vertical plane through the H atoms. Consecutive isopotential lines differ by 0.02 a.u. (0.54 V); continuous, dashed and dot-dashed curves refer to positive, negative, and zero potential, respectively. Lines corresponding to absolute values larger than 0.3 a.u. are not plotted.

In summary, dissociation is thermodynamically favored at the tub even if only by a few kcal/mol, while it is endothermic at the anion vacancy by as much as 25 kcal/mol. It is interesting to observe that the O(β) ion at the bottom of the tub, which is five-coordinated like all ions at the perfect (001) surface, is as reactive as the four-coordinated O(α) one.

The possibility for dissociation to take place at low temperatures as experimentally observed [6] also depends on the activation barrier. The reaction path at the anion vacancy has been identified in previous work [4]. The molecule approaches the surface in an upright position pointing to a location intermediate between a corner oxygen and the center of the vacancy; at about 2 Å from the surface, the lower H atom is attracted toward the corner oxygen, while the other points to the vacant site at the surface. At the HF level of approximation, the activation energy (25 kcal/mol) is close to the reaction energy (19 kcal/mol). We have tried to perform an analogous study for the case of the dissociation at the divacancy. In spite of our efforts, we found no path leading to the Tα configuration with an activation energy below 100 kcal/mol (note that the HF energy for the dissociation of the molecule into two separate atoms is about 80 kcal/mol). The results were quite different for Tβ, where the HF activation barrier is only 24 kcal/mol (see below).

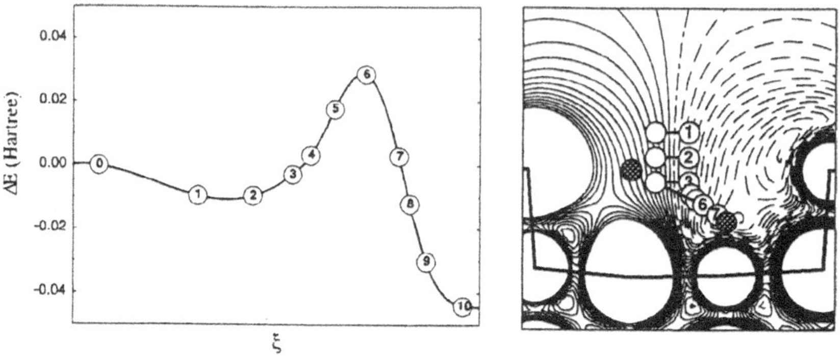

Figure 4. Energy versus ξ, the reaction coordinate (left plot), and subsequent configurations 1 to 7 of the molecule (right plot) for the process of H_2 dissociation leading to the $T\beta$ configuration. In the right plot the electrostatic potential is reproduced as in figure 3; the final configuration is represented by dark grey circles.

The difference between the three cases can be explained when considering the critical role played by the electrostatic field in these reactive processes. In order to discuss this point quantitatively, we report in figure 3 the electrostatic potential at the two defects. The potential difference between the location of the proton and the hydride for the dissociated $F\alpha$, $T\alpha$ and $T\beta$ configurations, is -0.252, -0.464, -0.403 a.u. In a naive picture, which takes the two species as possessing a charge of +1 and -1 $|e|$, these figures equal the net contribution (in Hartree) provided by electrostatic forces to the stabilization of the dissociated with respect to the undissociated configuration. To that, one must add the residual electrostatic interaction $1/d$, where d is the distance between the two species: -0.212, -0.145, -0.222 a.u., respectively. The electrostatic contribution to the stabilization of the ionic configuration is thus roughly estimated as 290, 390, 392 kcal/mol for the three cases. After adding the gain due to the formation of the O–H bond in the hydroxyl (around 80 kcal/mol), it is seen that the energy required for the heterolytic dissociation of H_2 in the gas phase (410 kcal/mol) is more than compensated for in the tub cases. While the above figures are undoubtedly exaggerated (in particular, there is a residual electronic charge at the proton), they may justify the much larger stability of the dissociated molecule at the tub than at the anion vacancy. Analogous considerations can be formulated to clarify the large difference in the activation barrier. For this purpose, we must consider the configuration of the transition state and the corresponding value of the electrostatic field. The transition state configuration for the case of the vacancy is not too dissimilar from both the undissociated and the dissociated configuration [4]: the molecule forms an

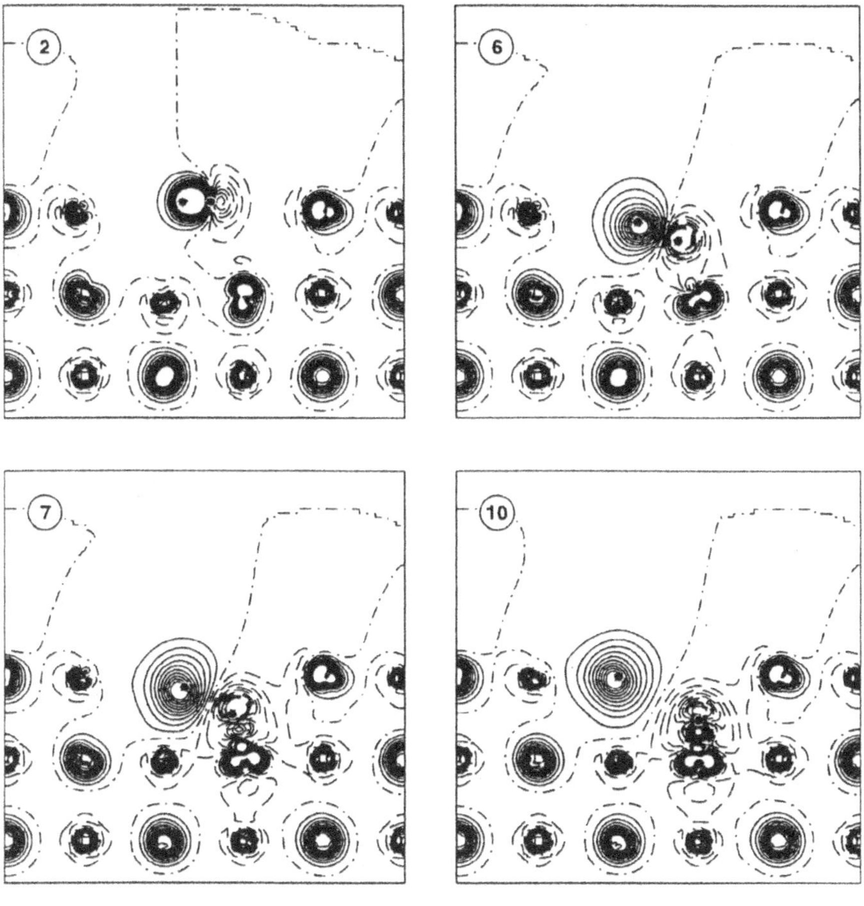

Figure 5. Difference density maps (total minus superposition of isolated species) for different configurations along the reaction path leading to the Tβ configuration. The numbers identifying each plot are as in figure 4. Consecutive iso-difference-density curves differ by 0.005 $|e|/\text{Bohr}^3$. The black dots mark the position of the two hydrogen nuclei.

angle of about 30 degrees with the surface, its positive moiety is close to the position it will assume in the hydroxyl group, and the distance between the two nuclei is 3.0 Bohr (1.40 in the molecule, 4.72 in the equilibrium configuration). The electrostatic field in this configuration is very high. The fact that the energy of the transition state is only slightly higher than that of the dissociated molecule appears justified. The case is different for the reaction at the tub. Consider first the formation of the Tα species. The approach of the molecule to O(α) follows a similar pattern as before but the electrostatic field there is much lower, and the hydride ion must go a long way before reaching a region of sufficiently high and positive potential. The Tβ case is still different and will be described in more detail with reference to figure 4. The molecule approaches the surface horizontally and penetrates into the tub in correspondence of the missing oxygen (steps 0-2), reaching a relative minimum: this position can be reached easily because of the small molecular size and of the relatively large spacing left open by the removal of the oxide ion. Next, the molecule goes deeper into the tub, bends toward O(β), becomes longer and progressively more polarized by reaching the transition state configuration (steps 3-6). This process continues and the molecule dissociates by forming a hydroxyl group at O(β) and a hydride ion (steps 7-8). Finally, the latter comes back to the missing oxygen position (steps 9-10). The transition state configuration (6) corresponds to an elongated molecule (H–H distance 2.20 Bohr) near the Mg ion at the bottom of the tub, almost parallel to the tub axis. The electrostatic field there is enormous, and the molecule experiences a very high polarization which makes the subsequent cleaving of the H–H bond much easier. Figure 5 gives further insight into the mechanism just described by providing difference density maps (total density minus superposition of isolated H, O^{2} , Mg^{2+}, densities) of selected configurations along the reaction path. It appears that the transition state already corresponds to almost complete ionization. The partial de-polarization of ions at the border of the tub at the end of the dissociation process is also apparent.

4. Conclusions

Among the dissociated configurations here considered only Tβ can be generated at low temperature, while the others are excluded for two different reasons: at the isolated anion vacancy the activation energy is relatively small but the process is endothermic; at the tub an enormous barrier prevents the Tα configuration to be appreciably populated. It is clear from the present analysis that an accurate description of the electrostatic potential in the proximity of the defects is mandatory for obtaining reliable information about their activity in this kind of processes. Embedded cluster techniques like those here employed seem particularly appropriate in this respect.

Acknowledgments

The present work is part of a project coordinated by A. Zecchina and cofinanced by the Italian MURST (Cofin98, Area 03). Financial support from the Istituto Nazionale di Fisica della Materia (INFM) through a PAIS project is gratefully acknowledged.

References

1. Coluccia, S., Boccuzzi, F., Ghiotti, G. and Morterra, C. (1982) Infrared Study of Hydrogen Adsorption on MgO, CaO and SrO, *J. Chem. Soc., Farad. Trans. I* **78**, 2111-2119.
2. Ito, T., Watanabe, T., Tashiro, T., Toi, K. (1989) Reaction of preadsorbed Methane with Oxygen over Magnesium Oxide at low Temperatures, *J. Chem. Soc., Farad. Trans. I* **85**, 2381-2395.
3. Hutchings, G.J., Woodhouse, J.R. and Scurrell, M.S. (1989) Partial oxidation of Methane over Oxide Catalysts, *J. Chem. Soc., Farad. Trans. I* **85**, 2507-2523.
4. D'Ercole, A., Giamello, E. and Pisani, C. (1999) Embedde-Cluster Study of Hydrogen Interaction with an Oxygen Vacancy at the Magnesium Oxide Surface, *J. Phys. Chem.* **103**, 3872-3876.
5. D'Ercole, A. and Pisani. C. (1999) Ab-initio Study of Hydrogen Dissociation at a Surface Divacancy on the (001) MgO Surface, *J. Chem. Phys.* **111**, 9743-9753.
6. Paganini, M.C., Chiesa, M., Giamello, E., Coluccia, S., Martra, G., Murphy, D.M. and Pacchioni, G. (1999) Colour Centres at the Surface of Alkali-Earth Oxides. A new Hypothesis on the Location of Surface Electron Traps, *Surf. Sci.* **421**, 246-262.
7. Pisani, C. (1993) Embedded-Cluster Techniques for the Quantum-Mechanical Study of Surface Reactivity, *J. Molec. Catal.* **82**, 229-252.
8. Pisani, C., Birkenheuer, U., Corà, F., Nada, R. and Casassa, S. (1996) EMBED96 User's manual, *Università di Torino*, Torino.
9. Ojamäe, L. and Pisani, C. (1998) Theoretical Characterization of Divacancies at the Surface and in Bulk MgO, *J. Chem. Phys.* **109**, 10984-10995.
10. Dovesi, R., Saunders, V.R., Roetti, C., Causà, M. and Harrison, N.M. (1998) CRYSTAL98 User's manual, *Università di Torino*, Torino.

A DFT STUDY OF CO ADSORPTION ON Ni^{II} IONS 3-FOLD COORDINATED TO SILICA

D. COSTA, M. KERMAREC, M. CHE, G. MARTRA[1]
Laboratoire de Réactivité de Surface, CNRS, UMR 7609, Université P. et M. Curie, 75252 Paris Cedex 05, France

AND

Y. GIRARD AND P. CHAQUIN
Laboratoire de Chimie Théorique, Université P. et M. Curie, 75252 Paris Cedex 05, France

[1]*Present address: Dipartimento di Chimica IFM, Università di Torino, Via P. Giuria 7, 10125 Torino, Italy*

Abstract. A cluster approach was used to model the amorphous silica as a tridentate ligand of the Ni^{II}_{3c} site, using the density functionnal theory. The geometry of mono- and dicarbonyl Ni^{II} complexes, identified by IR spectroscopy, was studied. Cluster models of Ni bound ions were investigated versus the net charge of the complex and the coordination of Ni^{II} ions. Simple ligands (OH^-, H_2O) were first used to model SiO^- and SiOH or Si-O-Si fragments. A cluster composed of the (SiO^-, SiOH, Si-O-Si) groups corresponding to a 1+ charged Ni^{II}_{3c} complex reproduced well the NiO distances found by EXAFS and gave satisfying calculated CO frequencies for the mono- and the dicarbonyl complexes. With silica clusters of larger size (cycles), the CO bond length was very close to that of free CO in neutral complexes whereas it decreases in the 1+ charged complexes.

1. Introduction

Quantum chemical calculations are being increasingly applied to predict the interaction of an adsorbate with specific catalytic sites [1]. The density functional theory (DFT) has proven its efficiency in predicting geometries, binding energies and IR frequencies for transition metal carbonyl complexes [2-9]. Cluster models were already used to model by ab initio calculations the adsorption of CO on Lewis sites on MgO [10] and to investigate the interaction of Ni and Pd atoms on the same support [11]. The cluster approach was successfully used on Cu-exchanged zeolites [12]. The model uses H_2O ligands to represent the zeolite

J. Maruani et al. (eds.), New Trends in Quantum Systems in Chemistry and Physics, Volume 2, 257–268.

framework and provides insight into the interaction of zeolite bound Cu ions. In addition, theoretical calculations of vibrational frequencies were performed on mono and di-carbonyl (nitrosyl) Cu complexes of various charge. Indeed, CO and NO are currently used to probe the degree of unsaturation of the coordination shell and consequently the reactivity of supported transition metal ions in their oxidized and reduced states.

The present investigation reports theoretical calculations using the DFT approach of Ni binding and of CO interaction with Ni^{II} ions in a cluster model of silica. The interest of producing these silica-supported isolated Ni^{II}_{3c} ions is that they are selectively photoreducible by hydrogen into isolated Ni^{I}_{3c} ions [13, 14], suggested to be active for olefin oligomerization [15]. Several techniques contributed to characterize Ni^{II} species in Ni/SiO_2 materials prepared by different methods. A recent EXAFS study showed that, depending on the deposition mode of Ni^{II} ions, several Ni^{II} species may coexist on the surface of the silica support [16]. Exchange in ethanediamine aqueous solutions was selected as being the best route to produce selectively, after an activating treatment at 973K, isolated 3-fold coordinated Ni^{II} ions (Ni^{II}_{3c}). Diffuse Reflectance VUV Studies (DRS) confirmed the 3-fold coordination of Ni^{II} [17]. From the EXAFS results it was possible to draw the geometry of the starting isolated Ni^{II}_{3c} precursor, which corresponds to a distorted site with a distribution of long (d (Ni-O) = 1.95 - 2.04 Å) and short (d (Ni-O) = 1.74-1.76 Å) Ni-O distances [16]. The reactivity of the Ni^{II}_{3c} species was tested using the CO probe molecule in a recent IR investigation [18]. The adsorption of CO was performed at 77K in order to avoid any reduction of the cation. Mono- and dicarbonyl Ni^{II} complexes were evidenced. No tricarbonyl Ni^{II} complexes were observed.

To obtain more insight into the geometry of those carbonyl Ni^{II} complexes identified by IR, CO binding energies in the carbonyl complexes, and the interaction of the Ni^{II} ions with the support, a modelling of selected small molecular clusters simulating the amorphous silica surface and of the carbonyl complexes is done using quantum DFT calculations. We have chosen the cluster approach rather than the periodic method because silica is amorphous and is composed of several types of cycles. In addition, the activating thermal treatment undergone in the presence of Ni may modify the silica surface in an unknown way. This is the reason why we have first chosen fragments of silica, then cycles of silica in the cluster approach.

2. Methodology for modelling calculations

Calculations were performed with Gaussian 94 [19]. The geometrical parameters of the carbonyl complexes were optimized independently at the DFT level using the Becke-Three Parameter Hybrid Method with the Lee-Yang-Paar Correlation Functional (B3LYP) [20] and checked by vibrational analysis. In some cases (mentioned in the text) geometricel constraints were imposed. The basis set used was a double zeta plus polarization basis set (DZVP2) [21]. The results were not corrected for basis set superposition error (BSSE) or zero point energy.

3. Results and Discussion

3.1. METAL-CARBONYL $Ni(CO)_4$, $Fe(CO)_5$, $Cr(CO)_6$ COMPLEXES

As a first test, geometries and vibrational spectra of known complexes such as $Ni(CO)_4$, $Fe(CO)_5$, $Cr(CO)_6$ were optimized and their IR frequencies calculated. Table 1 reports the calculated CO and MC bond lengths and the ν_{CO} of $Ni(CO)_5$, $Cr(CO)_6$ and $Fe(CO)_5$ in comparison with experimental data.

TABLE 1. Calculated CO and MC bond lengths (Å) and ν_{CO} (cm^{-1}) for $Ni(CO)_4$, $Cr(CO)_6$ and $Fe(CO)_5$ in comparison with experimental data

Complex	CO bond length Exp.	CO bond length Calc.	MC bond length Exp.	MC bond length Calc.	$\nu_{CO}^{Exp.}$ [4] (anharmonic)	$\nu_{CO}^{Calc.}$ (harmonic)	$\nu_{CO}^{Exp.}$ / $\nu_{CO}^{Calc.}$
CO	1.128	1.146			2143	2169	0.988
$Ni(CO)_4$	1.13	1.153	1.825	1.831	2067, 2154	2098, 2173	0.985, 0.991
$Cr(CO)_6$	1.14	1.157	1.914	1.922	2018, 2020, 2113	2049, 2070, 2162	0.985, 0.976, 0.977
$Fe(CO)_5$	1.152	1.158	1.807	1.817	2036, 2060	2061, 2087	0.988, 0.987

The experimental MC and CO bond lengths are well reproduced by the calculations. For the CO stretching mode, the ratio of the experimental anharmonic (2143 cm^{-1}) to the harmonic frequency (2170 cm^{-1}) is equal to 0.988. The calculated harmonic □ν_{CO} reproduces well the harmonic CO frequency. The ν_{CO}^{Exp} / ν_{CO}^{Calc} ratios obtained fit this value with an accuracy equal to or better than 1%, showing that the method used for the frequency calculation is accurate enough. Hence, the chosen method appears to be adequate for our study.

For convenience, in the following sections the calculated harmonic CO frequencies are scaled with this factor (x 0.988) in order to be compared with the experimental anharmonic values.

3.2. CARBONYL Ni^{II} COMPLEXES WITH OH^-, H_2O LIGANDS

Simple ligands as OH^- or H_2O can be used to represent the silica surface: the negatively charged ≡SiO⁻ (silanolate) may be represented by the OH^- ligand, whereas H_2O may represent the neutral sites ≡SiOH (silanol) and bridging oxygen (≡Si-O-Si≡). The influence of the total charge of the monocarbonyl complex (0 to 2+) and of the number of ligands other than CO (1 to 3 ligands) on the Ni^{II} ion was studied. The ground state of Ni^{II} is singlet. Results obtained are reported in Table 2.

TABLE 2. CO stretching frequencies calculated for various monocarbonyl Ni^{II} complexes after geometry optimization

Monocarbonyl complex	Calculated ν_{CO} * (cm^{-1})
$[Ni^{II} (H_2O) (CO)]^{2+}$	2327
$[Ni^{II} (H_2O, H_2O) (CO)]^{2+}$	2307
$[Ni^{II} (H_2O, H_2O, H_2O) (CO)]^{2+}$	2284
$[Ni^{II} (OH^-)(CO)]^{1+}$	2263
$[Ni^{II} (OH^-, H_2O) (CO)]^{1+}$	2237
$[Ni^{II} (OH^-, H_2O, H_2O) (CO)]^{1+}$	**2224**
$[Ni^{II} (OH^-, OH^-)(CO)]^{0}$	2142
$[Ni^{II} (OH^-, OH^-, H_2O) (CO)]^{0}$	2132

* experimental frequency = 2201 cm^{-1}

Figure 1 illustrates the evolution of the calculated ν_{CO} versus the charge and coordination number of the Ni^{II} ion in the monocarbonyl complex. It appears that for a fixed coordination number the charge of the complex has a considerable effect on the CO frequency (about one hundred of cm^{-1}) whereas, for a fixed formal charge of the complex, the addition of a neutral ligand such as H_2O has a smaller effect (about ten cm^{-1}). ν_{CO} is shown to decrease linearly with increased coordination numbers.

Figure 1. ν_{CO} as a function of the charge of the monocarbonyl complex and of the Ni coordination number (ligands other than CO).

Whatever the coordination number of the Ni^{II} ion, a neutral complex is an inadequate model since no positive shift of the frequency with respect to the frequency of the free CO molecule (2143 cm^{-1}), as shown by the experimental data, is calculated. On the other hand, the 2+ charged complex should also be excluded because of a too high positive shift. The 1+ positively charged complexes give the more satisfying results for Ni^{II}_{3c} site; the CO value found for the prevalent Ni^{II}_{3c} site (2224 cm^{-1}) fits the experimental one (2201 cm^{-1}) with an accuracy of 1%.

This first study suggests that the tricoordinated Ni^{II} surface site has an overall charge of 1+ and that after CO adsorption the Ni^{II} keeps its 3-fold coordination to the silica surface.

3.3. MODELLING THE Ni^{II}_{3c} SITE AT THE SILICA SURFACE

To go one step further in the modelling of the silica surface we have performed calculations with the silica groups as ligands of the Ni^{II} ion. The modelling of the $[Ni^{II}_{3c}]^{1+}$ site requires one anionic and two neutral ligands. We have chosen a site with one ≡SiO⁻, one ≡SiOH and one ≡Si-O-Si≡ □ligands. The ground state is triplet, as silica groups induce a weak ligand field. After geometry optimization, three Ni-O distances are found at 1.70, 1.83 and 2.02 Å, respectively, in agreement with the 1+ charged model involving three different ligands. The Ni-O distance corresponding to the silanolate ligand is the shortest, in agreement with the strong electrostatic interaction. This cluster model is schematized hereafter as $Si_4O_3H_{13}$. The remaining dangling bonds on each Si atom are saturated with H atoms.

Indeed, the intermediate Ni-O distance at 1.83Å is not observed by EXAFS data which show one short (□1.75 Å) and one long Ni-O (□ 2.00 Å) distances. This is not surprising, owing to the high signal to noise exhibited by our samples and the very weak EXAFS signal. Thus, the very disordered distribution of the Ni-O distances may hinder the intermediate Ni-O distance.

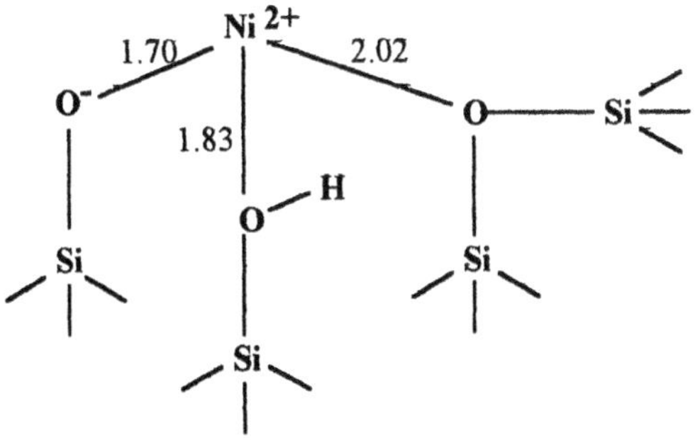

The average calculated Ni-Si distance (3.2 Å) is consistent with the experimental one measured from EXAFS (3.14 Å). The carbonyl complexes were then modelled, keeping Si-Si distances constant during the optimization of the carbonyl complex, in order to take into account the presence of constraints induced by the silica bulk.

3.4. MONOCARBONYL $[(CO)\ Ni^{II}\ (Si_4O_3H_{13})]^{1+}$ COMPLEX

The stable state is singlet, due to the high field induced by CO. The geometry is square-planar like, with a slightly tetrahedral distortion (Figure 2). As with (OH, H_2O) ligands, the Ni^{II} ion remains tricoordinated to the support. The results concerning the geometry of the complex are reported in Table 3 and compared with the $[Ni^{II}_{3c}\ (OH^-)(H_2O)_2(CO)]^{1+}$ complex.

The change of ligands does not modify markedly the CO bond length, which is lower than in the free CO molecule, in agreement with experimental results (a frequency higher than that of the free CO frequency indicates a shorter C-O distance), and the Ni-C-O angle is slightly lower than 180°: the Ni-CO moiety is very slightly bent and points away from the surface.

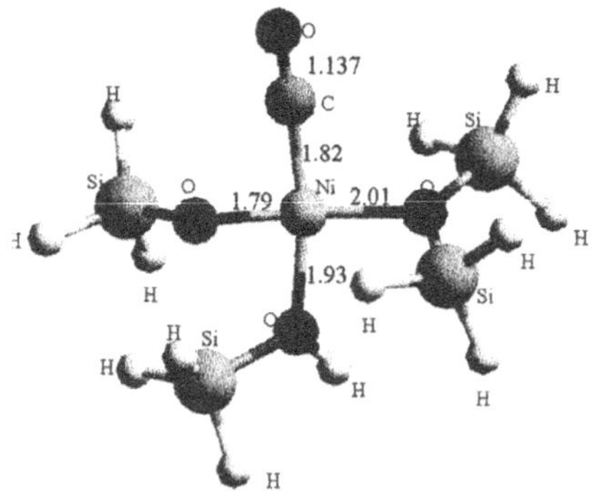

Figure 2. Optimized geometry of $[Ni^{II}\ (Si_4O_3H_{13})\ (CO)]^{1+}$.

Whatever the models used, the formation of the Ni^{II}-(CO) complex is exothermic. The binding energy of the carbonyl complex with the $Si_4O_3H_{13}$ cluster is 9 $kcal.mol^{-1}$. Taking into account the BSSE and zero point energy corrections that are expected to be less than 5 $kcal.mol^{-1}$, the value of the corrected CO binding energy will decrease to about 5-6 $kcal.mol^{-1}$, which is consistent with the observed reversible CO adsorption.

TABLE 3. Optimized geometries of the monocarbonyl Ni^{II}_{3c} complexes

complex	CO bond length (Å)	NiC bond length (Å)	Ni-C-O angle (°)
CO	1.146		
$[Ni^{II}\,(OH^-, H_2O, H_2O)(CO)]^{1+}$	1.136	1.887	179
$[Ni^{II}\,(Si_4O_3H_{13})\,(CO)]^{1+}$	1.137	1.820	177

3.5. DICARBONYL $[(CO)_2(Ni^{II}Si_2O_2H_7)]^{1+}$ COMPLEX

The minimum in the potential energy surface corresponding to a tricoordinated complex $[Ni^{II}\,(OH^-, H_2O, H_2O)\,(CO)_2]^{1+}$ - and $[Ni^{II}\text{-}(Si_4O_3H_{13})\,(CO)_2]^{1+}$ - could not be reached: in both cases, the structure evolved towards a the dicoordinated $Ni^{II}_{2c}\text{-}(CO)_2$ complex, with a square planar symmetry as shown in Figure 3 for the silica cluster.

To model the silica-like cluster bound to the dicarbonyl Ni^{II} complex, two fragments are needed. This cluster can be obtained by eliminating the weaker Si-containing ligand in the above- mentioned $Si_4O_3H_{13}$ cluster. We have chosen to exclude the siloxane group as it is located at the largest Ni-O distance. This new cluster ($Si_2O_2H_7$) representing the silica surface is shown in Figure 3.

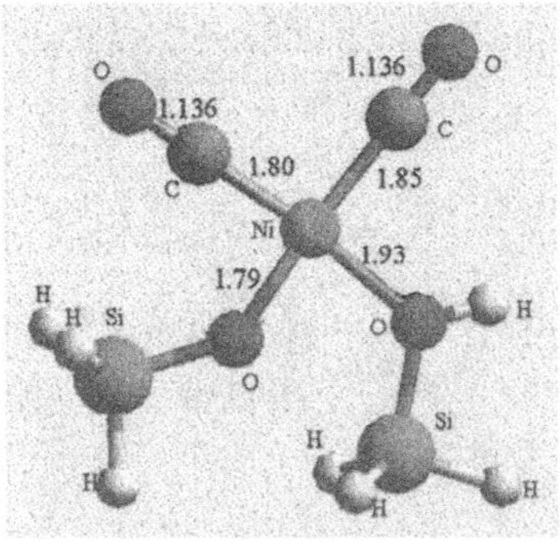

Figure 3. Optimized geometry of $[Ni^{II}\,(Si_2O_2H_7)\,(CO)_2]^{1+}$; the Si atoms are kept fixed.

Table 4 reports on the results obtained for the optimized geometry of the dicarbonyl complex formed on the Ni^{II} ion dicoordinated to one OH^- (SiO^-) and one H_2O (SiOH) ligands. The value of the CO binding energy is -34 kcal.mol^{-1} for $Si_2O_2H_7$ the cluster. These values are higher than those observed for the monocarbonyl complex. This suggests that dicarbonyl species form in competition with monocarbonyl, as observed experimentally [18].

TABLE 4. Optimized parameters for dicarbonyl Ni^{II}_{2c} complexes

Complex	CO bond length (Å)	NiC bond length (Å)	CO-CO angle (°)
free CO	1.146		
$[Ni^{II}(OH^-, H_2O)(CO)_2]^{1+}$	1.135 1.136	1.797 1.866	97°
$[Ni^{II}(Si_2O_2H_7)(CO)_2]^{1+}$	1.136 1.136	1.806 1.853	95°

There is no significant difference between the two CO bond lengths in the dicarbonyl complex but the two Ni-C distances are not equal. This may be attributed to the difference in charges carried by the two oxygen atoms of the support.

3.6. TRICARBONYL Ni^{II} COMPLEX

IR results show that the tricarbonyl complex is not formed. To check this result modelling was performed. It was not possible to reach a minimum in the potential energy surface for the tricarbonyl Ni^{II} complex with a nickel dicoordinated to the cluster: the structure evolves toward a dicarbonyl complex. Even if a ligand displacement of the silica surface is assumed by adding a third CO molecule as shown by the reaction:

$$CO + [Ni^{II}(SiOH,SiO^-)(CO)_2]^{1+} \rightarrow [Ni^{II}(SiO^-)(CO)_3]^{1+} + SiOH,$$

the formation of the tricarbonyl Ni^{II} complex is not favored: this reaction is endothermic (26 kcal.mol^{-1}).

3.7. FREQUENCY CALCULATIONS

Table 5 reports on the CO frequency calculations. It may be noticed that the calculated CO frequencies are higher than the experimental values as usually found using the harmonic approximation. For the monocarbonyl NiIIcomplex, the CO frequency calculated with the Si_4O_3 model is 2215 cm^{-1} which differs by 14 cm^{-1}

(0.6%) from the experimental value. This result is satisfying enough suggesting that our cluster model of silica-supported Ni^{II}_{3c} ions is reasonable. The neutral model Ni-$Si_4O_3H_{12}$ was also used. The calculated ν_{CO} was not in agreement with experimental results (calculated frequency 1952 cm^{-1} for an experimental frequency of 2201 cm^{-1}), confirming that the complex formed may not be neutral.

For the dicarbonyl, one frequency is higher than the CO frequency in the monocarbonyl complex whereas the other one is lower, as observed experimentally. The difference with the experimental value is about 1.3%. However, the calculated monocarbonyl frequency is close to the frequency of the antisymmetric dicarbonyl vibration and this in contrast to the experimental results. Work is under way to find a dicarbonyl structure which give calculated CO frequencies close to the experimental values.

TABLE 5. Calculated and measured CO stretching frequencies for carbonyl Ni^{II} complexes

Complex	Calc. ν_{CO} (cm^{-1})	Exp. ν_{CO} (cm^{-1})	Calc. $\delta\nu$ (cm^{-1})	Exp. $\delta\nu$ (cm^{-1})
Ni^{II}_{3c} monocarbonyl				
$[Ni^{II}(OH^-, (H_2O)_2)(CO)]^{1+}$	2224	2201	82	58
$[Ni^{II}(Si_4O_3H_{13})(CO)]^{1+}$	**2215**	**2201**	**73**	**58**
Ni^{II}_{2c} dicarbonyl				
$[Ni^{II}(OH^-, H_2O)(CO)_2]^{1+}$	2263 (+26) 2244 (+7)	2207 (+6) 2180 (-21)	121 102	64 37
$[Ni^{II}(Si_2O_2H_7)(CO)_2]^{1+}$	**2233 (+18)*** **2210 (-5)***	**2207 (+6)*** **2180 (-21)***	**91** **68**	64 37

* The numbers ( ) indicate the shifts of the measured and calculated frequencies of the dicarbonyl complex with respect to the values of the corresponding monocarbonyl complex

The calculated stretching frequencies, which are not measurable experimentally owing to the strong absorption of the support, are 440 cm^{-1} for the monocarbonyl and 408, 456 cm^{-1} for the dicarbonyl complexes, respectively.

3.8. EFFECT OF THE SIZE OF THE CLUSTER

A geometry optimization of the monocarbonyl Ni^{II} complex was performed with a model larger than $Si_4O_3H_{13}$, where the terminal H were replaced by OH groups ($Si_4O_3H(OH)_{12}$). The CO bond length was 1.138 Å, which is only 0.001 Å higher than the value found when using $Si_4O_3H_{13}$. This confirms that a small cluster approach may be accurate enough to reproduce the CO bond length. Similar results were found by other authors [12] for the replacement of H by OH terminal atoms.

The silica groups used in this study belong to surface silica cycles. In a previous work [22], one of our group tested the ability of several cycles representing the silica surface to reproduce the structure of the Ni^{II}_{3c} site. One of them, the $Si_5O_8H_8$ cluster corresponding to a ten-membered ring with two adjacent SiO^- (silanolate groups) and one SiOH corresponds to a neutral Ni^{II} -($Si_5O_8H_8$) complex and reproduces well the geometry of the Ni^{II}_{3c} site.

However, after geometry optimization of the neutral [Ni^{II} ($Si_5O_8H_8$) (CO)] complex, the CO bond length, 1.143 Å, did not reproduce the correct CO bond length (related to ν_{CO}) whereas the 1+ charged complex [Ni^{II} ($Si_5O_8H_9$) (CO)]$^{1+}$ led to a CO bond length of 1.136 Å very close to that found with the $Si_4O_3H_{13}$ cluster (1.137 Å) (Figure 4).

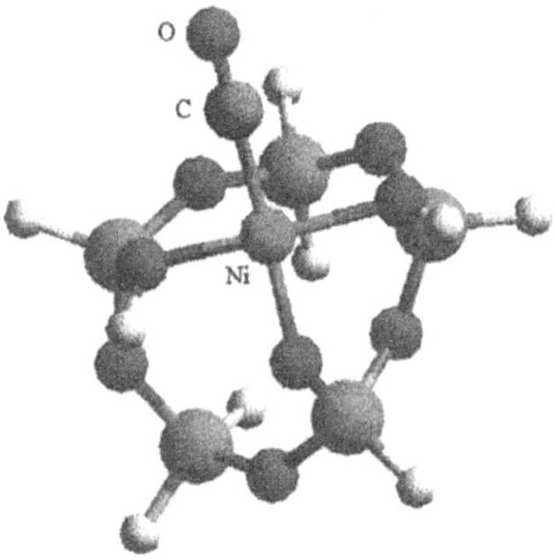

Figure 4. Optimized geometry of [Ni^{II} -($Si_5O_8H_9$)]$^{1+}$-(CO); the Si atoms are kept fixed.

Thus, the 1+ charge of the Ni^{II}_{3c} site is confirmed. This involves the presence of one SiO^- group. This hypothesis is supported by considering the energetic costs of the deprotonation of two vicinal SiOH. Whatever the cluster taken into account (fragment or larger cycle), DFT calculations show that the deprotonation of the first ligand is slightly endothermic (around 20-30 kcal.mol^{-1}) while the deprotonation of the second silanol is strongly endothermic (around 200-230 kcal.mol^{-1}). This positive charge is likely to be compensated by silanolate groups in the vicinity of the complex. The presence of some SiO^- groups at the surface of silica is not surprising since the deposition of nickel occurs in basic medium favoring the deprotonation of silanol groups [23]. Moreover, the high outgassing treatment undergone by the samples maintains a high (but uncomplete) dehydroxylation degree of silica and prevents protonation of the SiO^- groups.

4. Conclusion

A cluster approach was used to model the amorphous silica as a tridentate ligand of the Ni^{II}_{3c} center. A cluster composed of the (SiO^-, SiOH, Si-O-Si) groups corresponding to a 1+ charged Ni^{II}_{3c} complex reproduced well the NiO distances found by EXAFS and gave satisfying calculated CO frequencies for the mono- and dicarbonylcomplexes. This charge of the surface site may have a role on its reactivity, by for example favoring the adsorption of anions. No tricarbonyl is formed and this is also in agreement with experimental data.

The monocarbonyl forms without any breaking of the bond of nickel with the silica support. On the other hand, the formation of the dicarbonyl implies the breaking of one Ni-O bond, showing the versatility of the silica surface as a tridentate ligand.

Acknowledgements

We warmly thank Dr C. Lepetit and Dr C. Giessner for judicious advise on the modelling study and Dr J. F. Lambert and Professor S. Coluccia for helpful discussions.

References

1. Van Santen, R. A.; Neurock, M. *Catal. Rev. Sci. Eng.* **37** (1995) 557. Watwe, R. M.; Spievak, B. E.; Cortright, R. D.; Dumesic, J. A. *Catal Lett.* **51** (1998) 139.
2. Fan, L.; Ziegler, T. *J. Phys. Chem.* **96** (1992) 6937.
3. Fournier, R. *J.Chem. Phys.* **98** (1993) 8041.
4. Delley, B.; Wrinn, M.; Luthi, H. P. *J. Chem. Phys.* **100** (1994) 5785.
5. Berces, A.; Ziegler, T. *J. Phys. Chem.* **99** (1995) 11417.
6. Adamo, C.; Lelj, F. *J.Chem. Phys.* **103** (1995) 10605.
7. Barone, C.; Adamo, J. *J. Phys. Chem.* **100** (1996) 2094.
8. Joly, H. A.; Manceron, L. *Chem. Phys.* **226** (1998) 61.
9. Blomberg, M. R. A.; Siegbahn, P. E.; Svensson, M. *J.Chem. Phys.* **104** (1996) 9546.
10. Pisani, C.; Dovesi, R.; Nada, R.; Tamiro, S., *Surf. Sci.* **216** (1989) 489. Pacchioni, G; Cogliandro, G.; Bagus, P. S., *Surf. Sci.* **255** (1991) 344. Neyman, K. M.; Ruzankin, S. Ph.; Roesch, N., *Chem. Phys. Letters* **246** (1995) 546. Pelmenschikov, A.G; Morosi, G.; Gamba, A.; Coluccia, S. *J. Phys. Chem.* **99** (1995) 15018. Pelmenschikov, A.G; Morosi, G.; Gamba, A.; Coluccia, S.; Martra, G.; Paukshtis, E. A. *J. Phys. Chem.* **100** (1996) 5011. Nygren, M. A.; Petterson L. G. M. *J. Chem. Phys.* **105** (1996) 9339. Pelmenschikov, A.G.; Morosi, G.; Gamba, A.; Coluccia, S. *J. Phys. Chem.* **102** (1998) 2226.
11. Lopez, N.; Illas, F. *J. Phys. Chem. B* **1998**, *102*, 1430. Lopez, N.; Illas, F.; Pacchioni, G. *J. Am. Chem. Soc.* (1999), in press.
12. Ramprasad, R.; Schneider, W. F.; Hass, K. C.; Adams, J. B. *J. Phys. Chem. B* **101** (1997) 1940. Schneider, W. F.; Hass, K. C.; Ramprasad, R.; Adams, J. B. *J. Phys. Chem. B* **100** (1996) 6032.

13. Kazansky, V. B.; Elev, I. V.; Shelimov, B. N.; *J. Mol. Catal.* **21** (1983) 265.
14. Bonneviot, L.; Olivier, D.; Che, M.; *J. Mol. Catal.* **21** (1983) 415.
15. Lepetit, C.; Kermarec, M.; Olivier, D.; J. *Mol. Catal.* **51** (1989) 73.
16. Carriat, J. Y.; Che, M.; Kermarec, M.; Verdaguer, M.; Michalowicz, A. *J. Am. Chem. Soc.* **120** (1998) 2059.
17. Carriat, J. Y. Thesis Université Pierre et Marie Curie Paris (1994).
18. Martra, Costa, Che, Manceron, and Kermarec, to be published.
19. Gaussian 94, revision B.1: Frisch, M. B.; Trucks, G. W.; Schlegel, H. B.; Gill, P. M. W.; Johnson, B. G.; Robb, M. A.; Cheeseman, J. R.; Keith, T.; Petersson, G. A.; Montgomery, J. A.; Raghavachari, K.; Al-Laham, M.; Azakrzewsky, V. G.; Ortiz, J. V.; Foresman, J. B.; Cioslowski, J.; Stefanov, B. B.; Nanayakhara, A.; Challacombe, M.; Peng, C. Y.; Ayala, P. Y.; Cheng, W.; Wong, M. W.; Andres, J. L.; Reploge, E. S.; Gomberts, R.; Martin, R. L.; Fox, D. J.; Binkley, J. S.; Defrees, J. S.; Baker, J.; Stewart, J.; Head-Gordon, M.; Gonzalez, C.; Pople, J. A. Gaussian, Inc., Pittsburgh, PA (1995).
20. Becke, A., D. *J. Chem. Phys.* **98** (1993) 5648. Lee, C.; Yang, W.; Parr, R. G. *Phys. Rev B* **37** (1988) 785.
21. Gotbout, N.; Salahub, D. R.; Andselm, J.; Wimmer, E. *Can. J. Chem.* **70** (1992) 560.
22. Garrot, J. M.; Lepetit, C.; Che, M.; Chaquin, P. *J. Phys. Chem.*, submitted.
23. Lambert, J. F.; Che, M. *Stud. Surf. Sci. Catal.* **109** (1997) 91.

A THEORETICAL STUDY OF STRUCTURE AND REACTIVITY OF TITANIUM CHLORIDES

C. MARTINSKY AND C. MINOT
Laboratoire de Chimie Théorique, UMR 7616 CNRS,
Boîte 137, Tour 23-22, Université P. et M. Curie,
4, place Jussieu, 75252 Paris Cedex 05, France

Abstract. The study of various TiL_n species (L=Cl, H, CH_3) is a prerequisite for understanding the mechanism of the Ziegler-Natta reaction. The oxidation state of the titanium chloride active sites in the heterogeneous catalyst remains uncertain. The aim of this study is to provide better comprehension of the bonding in $TiCl_n$ compounds and to compare their reactivity toward simple reactants in order to help investigating adsorption and catalysis. From DFT-GGA calculations we show that the increase in coordination of titanium shortens the TiCl bond, increasing their covalent component. The TiCl bond remains, however, very ionic and very strong. On the contrary, the TiH and TiC bonds are rather weak; this allows the titanium complexes to loose such ligands and thus form hydrocarbons. $TiCl_3$ structures have unusual geometries that do not obey the Valence-Shell Electron-Pair Rules. Such deviations can be understood in terms of molecular orbital analysis. We also provide a topological description of the Electron Localisation Function (ELF) that explains the arrangement of the H atoms in terms of the ligand field around the titanium atom. Lewis bases and Cl radicals react at the Ti centre, $TiCl_3$ being the most reactive structure. Lewis acids and H and CH_3 radicals react with the ligand, initiating their abstraction; then, the products remain in high spin states.

1. Introduction

Ziegler-Natta catalysts are used for the polymerisation of α-olefins such as ethylene and propylene. The process is fast and stereoselective and has been the subject of a great deal of work (see [1] for a recent review). The catalyst is composed of $MgCl_2$, $TiCl_4$, AlR_3 and Lewis bases. Titanium complexes, $TiCl_2$ or $TiCl_3$, or $TiCl_4$ supported on disordered $MgCl_2$ are the active part of the Ziegler-Natta catalyst for polymerisation of olefins. Most of the studies privilege complexes and homogeneous catalysis. The Cossee mechanism [2] involves one active titanium centre to which the olefin binds to form a complex. The structure of the complex has been studied and agosticity has been evoked as inducing the stereoselectivity [3, 4]. The industrial process

J. Maruani et al. (eds.), New Trends in Quantum Systems in Chemistry and Physics, Volume 2, 269–288.

however implies heterogeneous catalysis [5-8]. The main advantage of heterogeneous catalysis is the large turnover that increases the catalyst efficiency. Recently, in Somorjai's group, $TiCl_4$ was deposited on a support from the gas phase and reduced to generate the catalyst, $TiCl_2$ or $TiCl_3$ [9-14]. The oxidation state of the titanium at the active site is uncertain. The catalyst contains several ingredients which all seem to contribute to the activity. Besides, there is an equilibrium between the different gaseous titanium chlorides. $TiCl_4$, a liquid in normal conditions, is a gas above 136°C. In ambient conditions, $TiCl_3$ is a crystal that undergoes disproportionation reaction upon heating with the formation of $TiCl_2$, a crystal with fairly low vapor pressure, and $TiCl_4$.

$$2TiCl_{3\,c} \rightarrow TiCl_{4\,l} + TiCl_{2\,c} \qquad \Delta H=+36.7 \text{ kcal/mol}$$

This reaction is endothermic. There is a wide range of temperatures where the partial pressure of $TiCl_4$ allows the disproportionation [9].

$$2TiCl_{3\,g} \rightarrow TiCl_{4\,g} + TiCl_{2\,g} \qquad \Delta F=-19. \text{ kcal/mol (300K)}$$

Other halides undergo similar behaviour with the following values [15].

$$2TiF_{3\,c} \rightarrow TiF_{4\,c} + TiF_{2\,c} \qquad \Delta H=-62.0 \text{ kcal/mol}$$
$$2TiBr_{3\,c} \rightarrow TiBr_{4\,c} + TiBr_{2\,c} \qquad \Delta H=+14.0 \text{ kcal/mol}$$

The knowledge of the structure of titanium in its different oxidation states (II to IV) should bring some information on its coordination to H, Cl and CH_3. Here, we present optimised structures of titanium hydrides and chlorides using density functional theory with the generalised gradient approximation (DFT-GGA). Next, for the different oxidation states we study the addition of acids, radicals and bases on the chlorides.

2. Computational methods

Calculations were performed using GAUSS94 [16]. We used pseudopotentials [17] so as to limit the size of the calculations and to be able to keep the same basis sets for further calculations on periodic systems. For Ti, the basis set is that used for TiO_2 [18, 19]. In the cases of low oxidation states (Ti and TiL) and of poorly electronegative ligands (L=H or CH_3), we used 3 gaussians for the 4sp orbitals to improve the description of the s electrons (table 1). This is not necessary when L=Cl since the electronic density on the titanium centre is much less important; in fact, a single gaussian for the 4sp orbitals (ζ = 0.484) leads to a lower energy. For the chlorine atoms, we used the PS-31G basis set [20] with an additional d function (ζ = 0.25).

We have started by Hartree-Fock (HF) calculations and next used the HF density matrices as guesses to run DFT-GGA calculations, using the Perdew and Wang (PW) functional [21]. The dissociation energies of H_2 and Cl_2 calculated by the DFT method, 102.4 kcal/mol and 80.2 kcal/mol are close to experimental values, 104.2 kcal/mol and 58.0 kcal/mol [15], whereas HF / MP4 values remain far from them. For the crystal bulk, we used the CRYSTAL program [22] using an Hartree-Fock Hamiltonian and a post Hartree-Fock estimation of the correlation using the Perdew-Wang functional; the basis sets are the same as in the molecular calculations and optimisations have been performed using the steepest descent method [23].

To test the basis set effects, we have performed a series of calculations with the 6-311+G(2df) basis. The differences in the geometries are very small (for TiH_3^+, the TiH bonds are 1.63 Å instead of 1.61 Å; the H-Ti-H angle is 58° instead of 60°). The difference is much larger when the HF is compared with the DFT for the same basis set (the H-Ti-H angle becomes 88°4). For $TiCl_4$, the Ti-Cl distance calculated with our basis set, 2.18 Å, is closer to the experimental value, 2.17 Å [24], than that obtained with the all-electron calculation, 2.12 Å. May be artificially, our optimised basis set is more realistic than larger ones.

For the Electron Localisation Function (ELF) analyses, we have used the 6-311G** basis set with the PW functional to be able to include the core basins in the representation.

TABLE 1. Basis set for Ti

exponent	**s coefficient**	**p coefficient**
3D *(ZSCALE=1)*		
25.1436	0.023466	
6.7976	0.134297	
2.1625	0.350609	
0.7499	0.480195	
3D'		
0.2182344	1.	
4SP (*ZSCALE=0.7982796*)		
0.7122640246	-0.3088441215	-0.12154686
0.2628702203	0.0196064117	0.571522760
0.1160862609	1.131034442	0.549894947

3. Structure and bonding in the TiH_n and $TiCl_n$ series

Results of the calculations for the TiL_n (L=H, Cl, CH_3) neutral structures are displayed in Tables 2-4, which provide dissociation energies, bond lengths, Mulliken charges and overlap populations for these compounds.

For Ti-H_n the charge on Ti increases with n; however, the Ti-H bond becomes more and more covalent. The increase of Q(Ti) = -nQ(H) comes from the increase in n while Q(H) decreases. As a result, the titanium atom in TiH_4 is much more acidic and much less basic than in TiH_2.

TABLE 2. Dissociation energy defined as D.E. = $E_{TiH_n} - E_{TiH_{n-1}} - E_H$ of a Ti-H bond (in kcal/mol)

	D.E.	Ti-H (Å)	Q(Ti), Q(H)	OP(TiH)
TiH (quartet)	-72.3	1.937	0.273, -0.273	0.247
TiH_2 (triplet)	-57.1	1.904	0.471, -0.236	0.266
TiH_3 (doublet)	-48.8	1.858	0.540, -0.180	0.271
TiH_4 (singlet)	-47.1	1.751	0.332, -0.083	0.288

All the unsaturated structures that we have investigated are in high spin states. The Ti atom is a triplet s^2d^2. TiH, $^4\Phi$ $[\sigma^2\sigma^1\pi^1\delta^1]$ and TiCl, $^4\Phi$ $[\sigma^2\sigma^2\pi^2\pi^2\sigma^1\pi^1\delta^1]$ are quartets. The ground state of TiH_2, $^3B_1$, is bent with a H-Ti-H angle of 120.9° in agreement with the MRCI/SA-CASSCF calculations [25] while $TiCl_2$ is linear ($^3\Sigma g$). TiHCl and $TiCH_3Cl$ are intermediate with angles of 136.8° and 126.6° respectively. TiH_3 and $TiCl_3$ are $^2A'_1$ doublets with a D_{3h} symmetry. TiL_4 is a singlet with a tetrahedral geometry.

TABLE 3. Dissociation energy defined as D.E. = $E_{TiMe_n} - E_{TiMe_{n-1}} - E_{Me}$ of a Ti-C bond (in kcal/mol)

	D.E.	Ti-C (Å)	Q(Ti), Q(CH_3)	OP(TiC)
Ti(CH_3) (quartet)	-71.4	2.315	0.396, -0.396	0.193
$Ti(CH_3)_2$ (triplet)	-55.2	2.265	0.728, -0.364	0.199
$Ti(CH_3)_3$ (doublet)	-56.9	2.159	0.814, -0.271	0.218
$Ti(CH_3)_4$ (singlet)	-65.7	2.084	0.769, -0.256	0.214

The Ti-Cl bond always appears to be much stronger than the Ti-H bond. Titanium chlorides are stable relative to the loss of a Cl ligand while titanium hydrides should thermodynamically decompose. Indeed (table 4) TiCl bond strengths, ~ 100/125 kcal/mol, are much larger than half of the dissociation energy, $(1/2)E_D$(Cl-Cl) = 40.9 kcal/mol. On the contrary, TiH and TiC bond strengths (tables 2-3), ~47/72 kcal/mol for H and ~55/72 kcal/mol for C, are close to half of the dissociation energies, $(1/2)E_D$(H-H) = 60.4 kcal/mol and $(1/2)E_D$(C-C) = 61.4 kcal/mol. TiH_3 and TiH_4 are metastable relative to dissociation into H_2 (14.8 and 24.8 kcal/mol respectively, assuming no interaction) and TiH_{n-2} in a higher spin state.

Since Ti is electropositive, the TiL bonds are very ionic and their strength should be related to the ease of the electron transfer to the ligand. The large electron affinity of Cl (83.4 kcal/mol) relative to that of H (17.4 kcal/mol) [15] or of C (29.24 kcal/mol) explains the formation of a stronger bond.

TABLE 4. Dissociation energy defined as $E_{TiCl_n} - E_{TiCl_{n-1}} - E_{Cl}$ of a Ti-Cl bond (in kcal / mol). * For bulk crystals the energy represents the average energy for a Ti-Cl bond. The experimental value for the Ti-Cl distance in $TiCl_4$ (vapour phase) is 2.18 Å [24]

	D.E.	Ti-Cl (Å)	Q(Ti), Q(Cl)	OP(TiCl)
TiCl (quartet)	-126.8	2.38	-	-
$TiCl_2$ (triplet)	-112.3	2.29	1.418,-0.709	0.101
$TiCl_3$ (doublet)	-100.2	2.21	1.585,-0.528	0.121
$TiCl_4$ (singlet)	-103.7	2.17	1.523,-0.381	0.170
$TiCl_2$ (crystal bulk)	89.9*	2.683	1.517,-0.758	0.054
$TiCl_3$ (crystal bulk)	-110.4*	2.464	1.392,-0.464	0.112

Ti-L bond distances decrease with the coordination. At first, this trend could seem unexpected since the natural ideas of saturation and bond order conservation suggest that the bonding capacity of an atom decreases with the number of ligands as for C-H bonds in the series from divalent to tetravalent carbon atoms (hybridation sp, sp^2 and sp^3). A first explanation is the decrease of the ionic radius of the titanium atom. However, the bonds are not completely ionic. An increase of the covalent character with the coordination also explains the shortening of Ti-L. As the oxidation state of the titanium increases from +1 to +4, the charge on the titanium atom increases, the difference in electronegativity between Ti and L decreases (atomic levels become closer in energy); it follows that the Ti-L bonds become more covalent. Relative to other neutral TiL_n species, the Ti-L bond has the smallest covalent character in TiL and the largest one in TiL_4; this is confirmed by the charges on the ligand (large for TiL, weak for TiL_4) and by Mulliken overlap populations (small for TiL, large for TiL_4; this trend persists when fixed distances are assumed). The OPs are smaller for L=Cl than for L=H or CH_3; however they are far from being negligible in $TiCl_4$ and the shortening of the bonds are nearly equal for L=Cl (8.8%) and for L=H (9.6%). The same trend is observed for the comparison of the solids ($TiCl_3$ vs. $TiCl_2$). The same is also true for the comparison of TiO, Ti_2O_3 and TiO_2 [26].

Crystals appear to be very stable compared to molecules. However, the average Ti-Cl bond energies for $TiCl_2$ and $TiCl_3$ crystals are comparable to those of $TiCl_n$ molecules. The crystals are very ionic compared with molecules; the bond lengths are longer and the charges are larger. The lattice parameters obtained by the DFT-GGA optimisations (3.652 and 5.897 Å for $TiCl_2$, 6.120, 17.50 Å, z = 0.079 for $TiCl_3$) are close (identical for $TiCl_3$) to the experimental values (3.561, 5.875 Å for $TiCl_2$, 6.12, 17.5 Å, z = 0.079 for $TiCl_3$) [43,44].

4 . Discussion on the non-VSEPR geometries of TiH_3

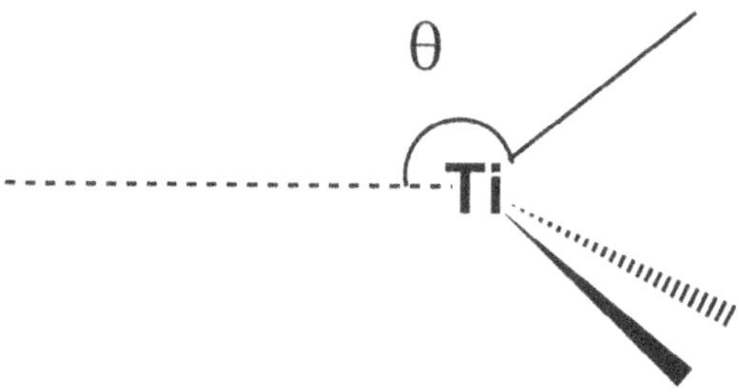

Figure 1. The pyramidalisation angle, θ.

According to the Valence Shell Electron Pair Rules (VSEPR) rules [27, 28] TiH_3^+ (d^0) should be planar (D_{3h}) and TiH_3^- (d^2) pyramidal (at least for the singlet state). Results from the calculations are opposite. The cation with three valence electron pairs (we consider the singlet state in this section) is pyramidal as similar systems [3, 29, 30]. In the HF calculations, the pyramidalisation angle, θ (see figure 1), is 120.8°, leading to H-Ti-H = 96°, so that the structure appears as an ML_3 fragment of an octahedron. With the DFT calculation, pyramidalisation is even more pronounced (θ = 146.8°). $TiCl_3^+$ is less pyramidal (θ = 113°) due to the repulsion between the chlorine atoms. With the full electron basis set, it is nearly planar. TiH_3^- with 8 valence electrons has a planar structure.

4.1. THE MO ANALYSIS

CH_3^+ is known to be planar (sp^2 hybridisation) in order to maximize the σ bonding between the 2s, $2p_x$ and $2p_y$ orbitals of the carbon and the 1s(H) (see figure 2). This opens a gap with the $2p_z$ (non-bonding LUMO). Similarly, for TiH_3^+ the ML_3 environment, with three ligands in the directions x, y and z, intensifies the σ-bonding interaction between 4s, $3d_{z^2}$ and $3d_{x^2-y^2}$ orbitals of the titanium and 1s(H) and opens a gap with the non-bonding d orbitals (LUMOs).

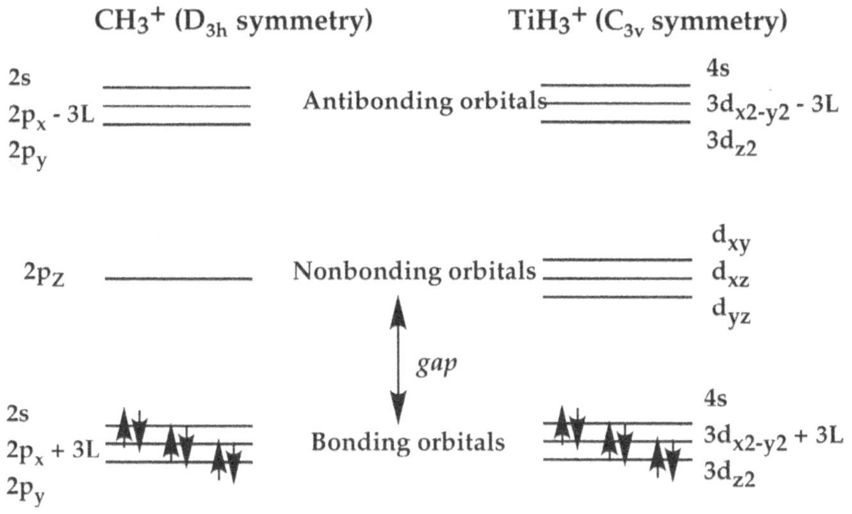

Figure 2. MO splitting for CH_3^+ (planar structure, z is perpendicular to the plane) and TiH_3^+ (partial octahedron, x, y and z are the directions of the ligands).

In the C_{3v} geometry, TiH_3^+ possesses 3 MOs with A and E symmetry. According to a Walsh diagram shown in figure 3 ($D_{3h} \rightarrow C_{3v}$), an optimum is obtained for the H-Ti-H = 90° (θ = 125.3°) which is close to the optimum for the E set. The A orbital, less sensitive to θ, starts by being slightly destabilised by the pyramidalisation, since the H atoms move from an in-phase overlap with the torus of the $3d_{z}2$ orbital approaching the nodal cone. The average valence orbital level shows a minimum slightly inferior to θ=125.3°, the value associated with the ML_3 fragment of an octahedron. In the HF calculation, this geometry represents the optimum for the total energy. In the DFT calculation, the pyramidalisation is more pronounced (θ = 146.8°) in a region where the ligand-ligand (H-H) interactions play the dominant role; A goes down and E goes up. Correlation stabilizes the H-H approach and increases the pyramidalisation. On the contrary, in $TiCl_3^+$, the repulsion between the chlorine atoms prevents them from getting closer and the pyramidalisation is reduced.

The ligand-ligand interaction is revealed when the asymmetry between the three ligands is allowed. Then, two H ligands strongly interact to form a σ-bond; they interact with the titanium that is mostly bound to the third one. As for TiH^+, the ground state of the $(H_2)TiH^+$ complex is a triplet (section 5.1). The C_{3v} singlet structure is thus very high in energy, 53.8 kcal/mol above that of the triplet $(H_2)TiH^+$. The interest of the C_{3v} structures is mostly topological. They allow to analyse the coordination in ML_3 structures where the M-L bond is stronger than Ti-H. Note however that the cationic species have the largest tendency to decompose.

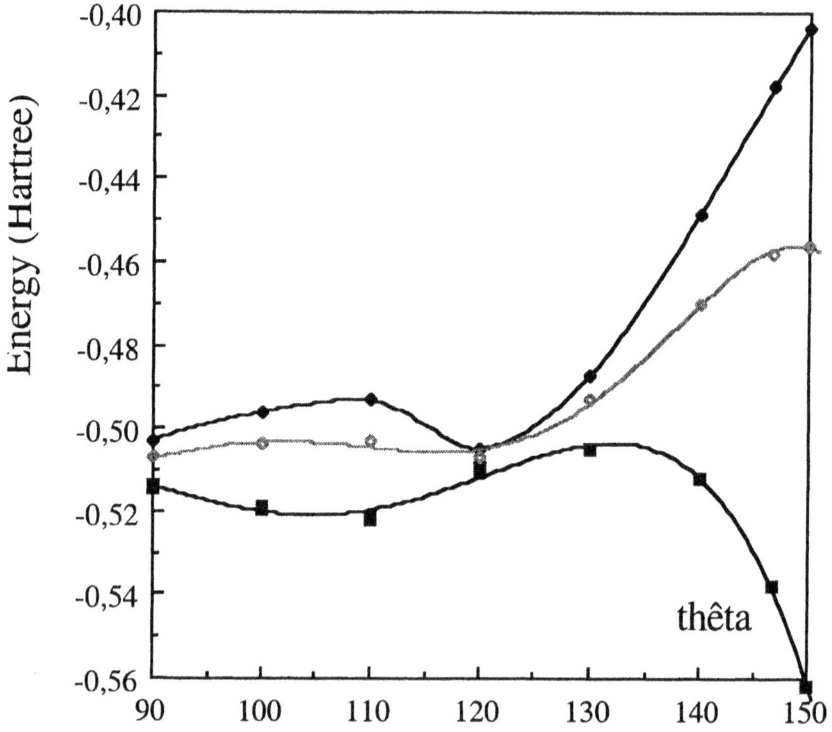

Figure 3. Walsh diagram for TiH_3^+ obtained by the DFT calculations. The upper curve is the energy level of the E orbitals and the lower curve is that of the A orbital. The average level also represented shows a minimum slightly below θ=125.3°, the value associated to the MH_3 fragment of an octahedron. In the HF calculation, this also represents the optimum for the total energy. In the DFT calculation, the optimum for the total energy is at 146.8° when the 3 H get closer to each other. It is associated with a dominant increase of the potential energy and to a ligand-ligand interaction.

Neutral species are planar (doublet state with D_{3h} symmetry). In addition to 3 stabilizing interactions ($4s-3d_{z^2}$, $3d_{xy}$ and $3d_{x^2-y^2}$ orbitals; z is now the C_3 axis) a 4th, $4s+3d_{z^2}$, populated by a single electron must now have some bonding character. The $4s-3d_{z^2}$ atomic orbital is a flat hybrid that participates in the most bonding and antibonding A1 molecular orbitals. The $4s+3d_{z^2}$ atomic orbital is an elongated hybrid that is weakly bonding and is occupied by the single electron. The planar geometry is the result of this new interaction. The geometry has changed to accommodate the extra electron in a bonding orbital that belongs to the nonbonding set in the pyramidalised set. The trend for the ligand-ligand interaction is less pronounced than for the cationic species. The doublet state is metastable and should decompose to H_2 and a quadruplet TiH; this reaction represents 14.8 kcal / mol only and is not easily obtained from the D_{3h} symmetry.

The negative species, $TiCl_3^-$, is not fundamentally different. The energy level of the elongated hybrid is not sufficiently different from the E" level so that it could take two electrons. The best situation is therefore a triplet with one electron in the E" set. This leads to a Jahn-Teller situation, and the structure distorts to C_{2v} symmetry (an A2 state with a Y geometry and a B1 state with a T geometry, see figure 4). This is also the case for the corresponding hydride, TiH_3^-, even though the E" set does not overlap with the orbitals of the hydrogen atoms.

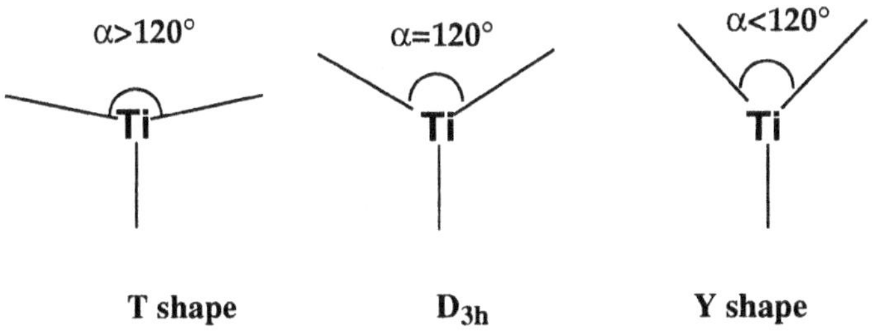

Figure 4. For TiH_3^-, the $^3B_1$ state has a T geometry (α=123°4 and one TiH bond is smaller than the other two, 2.020 Å vs. 2.011 Å); the $^3A_2$ state has a Y geometry (α=114°7 and one TiH bond is longer than the other two, 2.059 Å vs. 2.059 Å).

4.2. THE ELF ANALYSIS

Bader and Gillespie [31] have explained exceptions from the VSEPR rules in terms of core distortion using an analysis of the laplacian of the electronic density [32]. VSEPR first assumes that the electron density at the core is spherical. TiH_3^+ is pyramidalised and looks like an incomplete octahedron with only three ligands. The valence electron density extends in the outer shell of the core, imposing to the core basin a octahedral arrangement that is responsible for the pyramidalisation.

We propose an analysis of the cation structure using the topological analysis of ELF (Electron Localisation Function) in different geometries. The original feature of this topological method allows to visualize the octahedral ligand field around the titanium atom. This method, introduced by Becke and Edgecombe [33] defines a partition of the molecular space by measuring how efficient the Pauli repulsion is at a given point in space [34-37]. The Pauli repulsion is related to the electron delocalisation in term of kinetic energy excess. The formal expression of ELF is :

$$\eta(r) = 1 / [1 + (\frac{D_\sigma(r)}{D_0(r)})^2]$$

where $D_0(r) = 2.871\ \rho(r)^{5/3}$ is the kinetic energy density of a homogeneous electron gas of density $\rho(r)$ and $D_\sigma(r)$ is the excess kinetic energy density due to the Pauli repulsion :

$$D_\sigma(r) = T_S[\rho(r)] - \frac{|\nabla\rho(r)^2|}{8\rho(r)}$$

where $T_S[\rho(r)]$ is the positive definite kinetic energy density of the actual fermionic system and $\frac{|\nabla\rho(r)^2|}{8\rho(r)}$, the von Weizsäcker functional [38], is the positive definite kinetic energy of particle with density $\rho(r)$ in the absence of Pauli repulsion. The constant 2.871 is the Fermi constant in atomic units.

The molecular space is divided into core and valence basins, the latter being classified according to their connectivity to the core basins as monosynaptic basins, associated with electron pairs, or dissynaptic basins, associated with covalent bonds. Moreover, an integration of the electronic population over each basin gives the number of electrons in each of them and the fluctuation between basins is related to the electronic delocalisation [39].

We present the results for the planar (D_{3h}) and the pyramidalised (C_{3v}) geometry. Mostly because of the extension of the d orbitals, the core basins of Ti are not spherical. In the pyramidalised geometry (figure 5b), the TiH basins face the critical points of the titanium core basins showing some covalency; in the planar geometry (figure 5a), this is the opposite; the electrons are weakly localised along the Ti-H directions, corresponding to a TiH bond that remains very ionic. The integration of the electronic population over the basin gives a strong positive charge on Ti (+1.8). The covalent contribution is larger for the pyramidalised shape, with a smaller positive charge (+1.5) on Ti. The fluctuation study shows that the electron population of the Ti-H basins is larger in the pyramidalised structure (14.3 %) than in the planar one (7.5%). For the intermediate geometry, on ML_3 fragment of an octahedron, the fluctuation value, 9.5%, is close to that for the planar geometry; up to this degree of pyramidalisation, the system remains mostly ionic and the electron redistribution allowing the strong pyramidalisation in the DFT calculation mainly occurs beyond this limit.

$TiCl_3^+$ (singlet) has been found to be nearly planar; in the ELF representation for the D_{3h} geometry there are no Ti-Cl basins and the lone pair basins of Cl do not face the critical points of the Ti core. Thus, the ionicity of the Ti-Cl bond can be clearly seen (figure 6).

$TiCl_2CH_3^+$(singlet) is calculated to be pyramidal [3], even if the optimised parameters are sensitive to the calculation method [29].

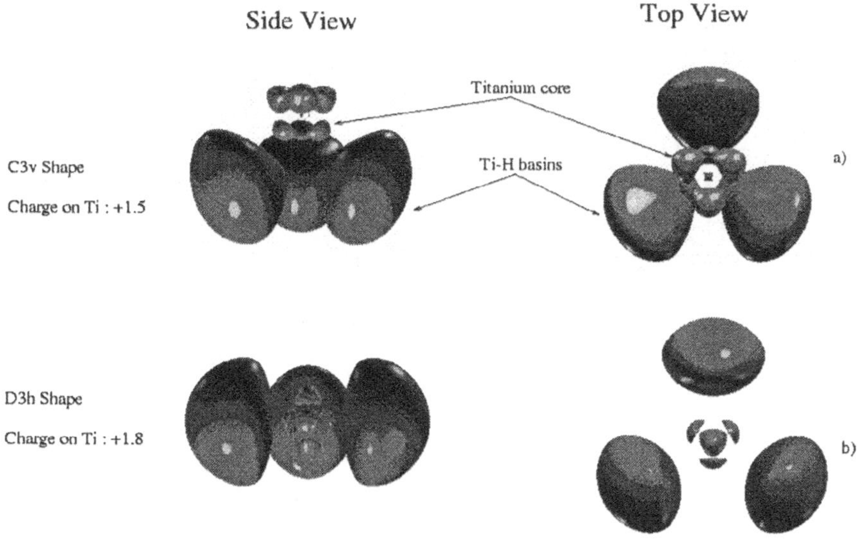

Figure 5. Tridimensional visualisation of the ELF function for TiH_3^+ for the D_{3h} and C_{3v} symmetry. The ML_3 topology of the C_{3v} species is clearly seen in the side view. The orientation of the core basins vs. the Ti-H dissynaptic basins appears in the top view.

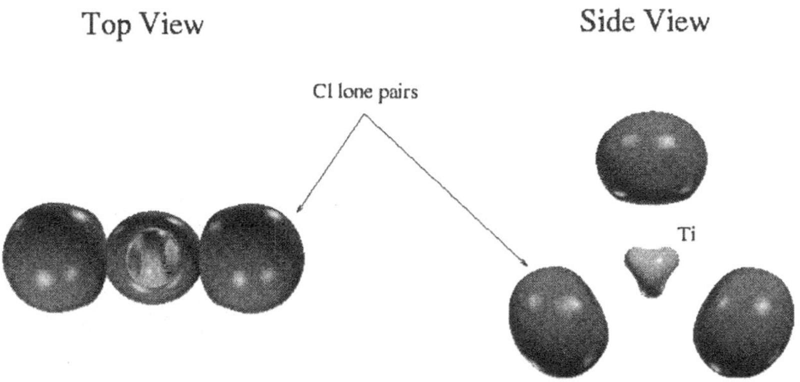

Figure 6. ELF representation of $TiCl_3^+$. The ionic character of Ti-Cl bond can be clearly seen. The basins of the titanium atom are not facing those of the ligands like in TiH_3^+ with D_{3h} symmetry.

5. Reactivity

In this section, we probe the reactivity of titanium chloride vs. ion-radicals and molecules that represent Lewis acids or bases.

TABLE 5. Heat of reaction of basic, radicalar and acidic species on $TiCl_n$ (in kcal/mol, defined positive when exothermic)

	CH_3^-	Cl^-	NH_3	H	CH_3	Cl	CH_3^+	H^+
$TiCl_2$	122.8	76.5	44.3	36.6	54.	100.2	104.5	166.3
$TiCl_3$	146.7	90.7	56.5	44.0	65.	103.7	96.2	128.9
$TiCl_4$	132.4	65.7	23.2	43.6	5.9	1.9	86.2	99.3

5.1. BASIC PROPERTY OF THE TITANIUM CHLORIDES

We have considered the addition of two positive ions, CH_3^+ and H^+ to test the basic properties of $TiCl_n$ compounds.

The reactivity of $TiCl_2$, the chloride that has the smallest coordination and which possesses two valence electrons, is potentially large. $TiCl_2$ is isolobal to a methylene species and, thus, could be thought of as an amphoteric compound with both a possibility to give or receive electrons in binding a new ligand to the titanium atom. From the spin state, Ti(II) appears to be poorly basic. The donation of an electron pair requires the promotion to a singlet state that costs 174 kcal/mol for $TiCl_2$. Indeed, relative to acidic species, the titanium centre is not the active site. Acids preferentially bind to the chlorine ligand and the most stable add-product remains in a triplet state. The proton prefers to bridge two Cl ligands (figure 7a, triplet state) than to form a Ti-H bond (figure 7b, singlet state). The acidic species, bridging two chlorine ligands, allows the formation of a sequence of atoms with opposite charges, favorable to the growth of an ionic structure. Indeed, in the $TiCl_2$ crystal structure (or the dimer structure), a $TiCl_2$ unit binds to another one through bridging Cls and does not form a Ti-Ti "double bond". CH_3^+ binds to a single chlorine ligand.

The singlet states obtained by addition of CH_3^+ or H^+ to $TiCl_2$ and electron donation from Ti(II) to the cation are secondary minima, 5.1 kcal/mol and 39.7 kcal/mol above the triplet state, respectively. Structures are pyramidal; the eclipsed form, which is not the best, shows the agostic effect [3, 4]. In the ground state (triplet), the methyl is bound to a chlorine atom ($d_{C\text{-}Cl}$ = 1.86 Å) that is distant from the titanium atom (d_{TiCl} = 2.46 Å).

For $TiCl_3$ and $TiCl_4$, the chlorine atom is again the atom involved in the bonding. H^+ and CH_3^+ bind to a single chlorine ligand to form an HCl molecule (H-Cl = 1.35Å) in interaction with $TiCl_2^+$ or $TiCl_3^+$. The basicity of $TiCl_2$ is larger than that of the other titanium chlorides (right hand side of table 5) since the charge on the chlorine atom is the largest (table 2).

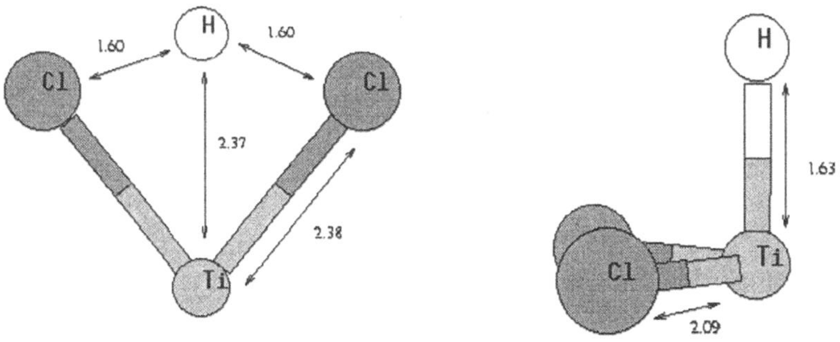

(a) (b)

Figure 7. Structures of the protonated $TiCl_2$. The triplet state (a) is more sta-ble than the singlet state (b) by 39.7 kcal/mol.

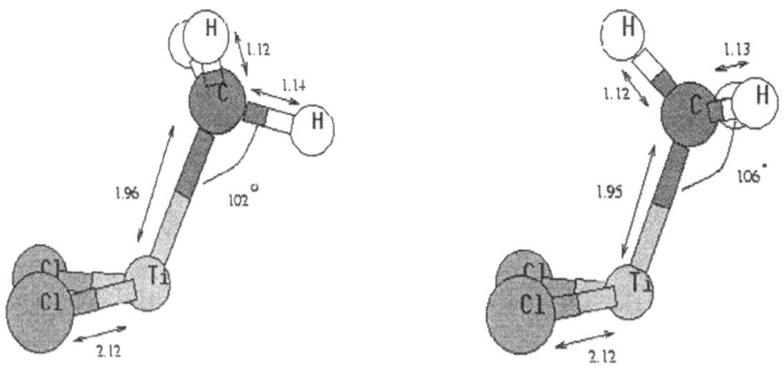

Figure 8. $CH_3 \cdots TiH_2^+$ complexes in singlet states (the conformation that exhibits agosticity, at the left hand side, is 1.3 kcal/mol above the staggered conformation). The ground state is a triplet: $CH_3Cl....TiCl$.

The addition of H^+ to TiHCl leads to TiH_2Cl^+. The optimisation of the triplet state allows to compare the basicity of the H and Cl ligands in TiHCl.

In spite of the larger negative charge on Cl, H^+ binds to H and forms a linear (H_2)-Ti-Cl^+ complex ($d_{Ti(H_2)}$ = 2.23Å). This is coherent with large H-H and Ti-Cl bond strengths compared with H-Cl and Ti-H ones. The binding energy, 205.3 kcal/mol, is larger than that of $TiCl_2$, 166.3 kcal/mol (table 5). We mentioned in section 4.1 that $(H_2)TiH^+$ in a triplet state was also much more stable than the singlet TiH_3^+ with C_{3v} symmetry; the proton binds to the H ligand to remove it. On a positively charged titanium centre, the alkyl ligands do not remain bound to the titanium; they migrate to bind to the ligands.

The optimisation of $TiHCH_3Cl^+$ (triplet state) leads to a methane (with an orientation along a local C_2 axis) bound to Ti-Cl. Alkyl groups alike the hydrogen bind together while Cl remains attached to the titanium atom. The binding energies, 228.0 kcal/mol for CH_3^+/TiHCl and 276.9 kcal/mol for $H^+/TiCH_3Cl$ are superior to that for H^+/TiCHCl. Alkyl ligands are more effective than Cl ligands to fix incoming acidic species.

In conclusion, the $TiCl_n$ compounds are reactive toward acidic species through their ligands. The compounds where Ti is in the lowest oxidation state is the most reactive (n=2 better than n=4). Products are in high spin states. The alkyl ligands are more reactive than the Cl ligands.

5.2. RADICALAR PROPERTY OF THE TITANIUM CHLORIDES

We have considered the addition of three radicals, Cl·, H· and CH_3·, as models for bases. In the Ziegler-Natta reaction, Lewis bases involved are esters of aromatic acids and hindered amines. Bases are supposed either to interact with the support ($MgCl_2$) or to coordinate with Ti [40, 41].

Since $TiCl_2$ and $TiCl_3$ are high spin compounds, one would expect that one of the unpaired electrons from the titanium should couple with those from the radical to build a σ bond. Note that such behaviour was not found in the previous section; the reason is that, when Ti bears a positive charge, the uncoupled electrons are more tightly hold by the nucleus and are less reactive. Both $TiCl_2$ and $TiCl_3$ should be more reactive than $TiCl_4$ that is a closed shell system.

On $TiCl_2$ and $TiCl_3$, all the radicals react at the Ti centre. This is true for Cl· since titanium chlorides are stable relative to the decomposition (section 3). The addition on the saturated compound is weaker than the others. The ligand abstraction is always endothermic :

$$Cl + TiCl_2 \rightarrow Cl_2 + TiCl \text{ (quartet)} \qquad +32.1 \text{ kcal/mol}$$
$$Cl + TiCl_3 \rightarrow Cl_2 + TiCl_2 \text{ (triplet)} \qquad +20.1 \text{ kcal/mol}$$
$$Cl + TiCl_4 \rightarrow Cl_2 + TiCl_3 \text{ (doublet)} \qquad +23.5 \text{ kcal/mol}$$

On $TiCl_2$ and $TiCl_3$, H· and CH_3· also bind to the titanium centre, in spite of the exothermicity of ligand abstraction. The formation of HCl or CH_3Cl is weakly exothermic :

$H + TiCl_2 \rightarrow HCl + TiCl$ (quartet)	-1.1 kcal/mol
$H + TiCl_3 \rightarrow HCl + TiCl_2$ (triplet)	-13.1 kcal/mol
$H + TiCl_4 \rightarrow HCl + TiCl_3$ (doublet)	-9.7 kcal/mol

Similarly :

$CH_3 + TiCl_2 \rightarrow CH_3Cl + TiCl$ (quartet)	-2.8 kcal/mol
$CH_3 + TiCl_3 \rightarrow CH_3Cl + TiCl_2$ (triplet)	-9.3 kcal/mol
$CH_3 + TiCl_4 \rightarrow CH_3Cl + TiCl_3$ (doublet)	-5.9 kcal/mol

The coupling of the unpaired electrons represents a greater energy gain. For instance, HCl...TiCl (quartet) is less stable than $TiCl_2H$ (doublet) by 20.7 kcal/mol. The latter has a C_{2v} symmetry; the Ti-H distance, 1.71 Å is short compared with that in the TiH_3 structure (table 2). The $TiCl_3H$ structure has a C_{3v} symmetry with a short Ti-H distance, 1.66 Å.

On $TiCl_4$, the H· and CH_3· radicals bind to the chlorine atom leading more easily to a decomposition as found by Herzler [42]. H is attached to two chlorine ligands leading to a HCl_2TiCl_2 bridging structure with a C_{2v} symmetry. The reactivity of H· seems remarkably large, $TiCl_4$ being poorly reactive with the neutral species. It behaves as a saturated compound and an increase of the titanium coordination requires charged species. The enthalpy of reaction with the CH_3· radical is weak. The methyl binds to a single chlorine (C-Cl=1.89 Å, Cl-Ti= 2.41 Å).

The addition of H· to TiHCl, much more exothermic than that to $TiCl_2$, involves the reactivity of the H ligand. Since the formation of H_2 is strongly exothermic, the H radical binds to the H ligand rather than to the Cl ligand as protons also do (see previous section) or to the titanium centre :

$H\cdot + TiHCl \rightarrow H_2 + TiCl$ (quartet)	-58.8 kcal/mol
$H\cdot + TiHCl \rightarrow HCl + TiH$ (quartet)	-6.6 kcal/mol

This is coherent with large H-H and Ti-Cl bond strengths compared with H-Cl and Ti-H ones. The interaction energy of H_2...TiCl is very weak; it has been found athermic in spite of large structural modifications (HH = 0.86 Å, TiH = 2.00Å).

5.3. ACIDIC PROPERTY OF THE TITANIUM CHLORIDES

We have considered the addition of two negative ions, CH_3^- and Cl^-, and a neutral Lewis base, NH_3, to test the acidic properties of $TiCl_n$ compounds. These bases bind to the titanium atom that is a Lewis acidic site. For $TiCl_3$ or $TiCl_2$, the unpaired electrons are inactive relative to the addition; the add-products are doublets and triplets respectively. Ti atoms in $TiCl_4$ are in the highest oxidation state, Ti(IV); they bear the largest positive charges and should be, from these considerations, the most reactive toward basic species. However, they also are the most coordinated complexes and this favors the reactivity on $TiCl_3$ or $TiCl_2$ species. Considering successive additions of Cl^- to Ti^{4+}, the last addition ($Cl^- + TiCl_4 \rightarrow TiCl_5^-$) goes beyond the charge compensation. In this respect, the value of 65.7 kcal/mol already seems large compared with the values from table 4.

$TiCl_3$ is slightly more reactive than $TiCl_2$ an $TiCl_4$. Experimentally, $TiCl_4$ is not catalytically active without reduction and, since it is necessary to add $MgCl_2$ to the reaction, $TiCl_4$ is not by itself a sufficient Lewis acid.

5.4. $TiCl_2$ DIMERISATION

The interest in the dimers is double. Dimerisation is a first model for the surface of the solids, increasing the coordination of the atoms without changing the oxidation states. Dimers placed on support (epitactically on the lateral surfaces of $MgCl_2$ crystals or on the corners or edges of crystallites) have also supposed to be the reactive sites for the Ziegler-Natta reaction [40, 41].

Dimerisation could be understood in the light of the previous sections. It involves reciprocal acid-base interactions between the two monomers: donation from the Cl ligand from one monomer (basic centre) to Ti of the other (acidic centre). It results that titanium atoms are connected through bridging chlorine atoms. The unpaired electrons do not participate to the bonding; the Ti-Ti overlap population has negative values. This leads to alternation of titanium cations and chloride anions as in the solid. The TiCl bond distances in the bridge are longer than those involving the terminal Cl's.

The best optimised dimer structures have three (μ_3) bridging ligands with a Cs symmetry. The μ_3 structure is slightly asymmetrical. The structure of $(TiCl_3)_2$ and $(TiCl_4)_2$ are similar with a more pronounced asymmetry. In the μ_3 $(TiCl_4)_2$ structure, the bridging ligand from the symmetry plane appears to be much more bound to the least coordinated titanium (Ti-Cl=2.20 Å) than to the other one (Ti-Cl=2.85 Å) and, therefore, this structure could as well be qualified as a μ_2 structure. Nevertheless, the elongated bond contributes to complete the octahedral arrangement around the most bound titanium centre.

5.5. ADDITION OF C_2H_4

Optimisation of the ethene-$TiCl_n$ complexes leads to π-bonding. The binding energy of ethene is 33, 30 and 17 kcal/mol for $TiCl_2$, $TiCl_3$ and $TiCl_4$ respectively. Ethene behaves as an electron-rich compound binding to the titanium centres as the basic compounds do. There is indeed a small electron transfer from ethene to the $TiCl_n$ complex.

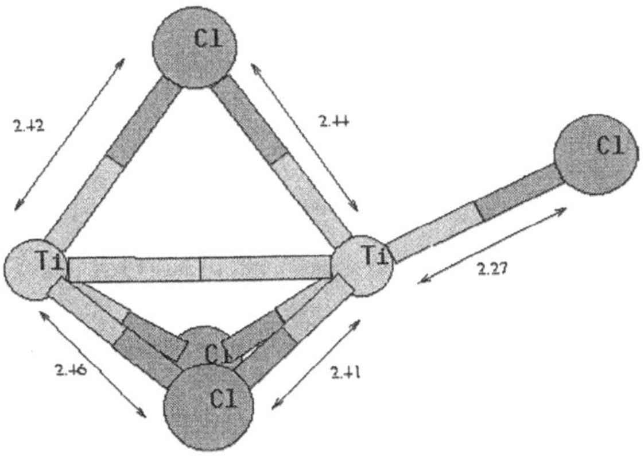

Figure 9 a. Structure of $(TiCl_2)_2$.

The coupling is larger for $(TiCl_2)_2$ or $(TiCl_3)_2$ than for $(TiCl_4)_2$.

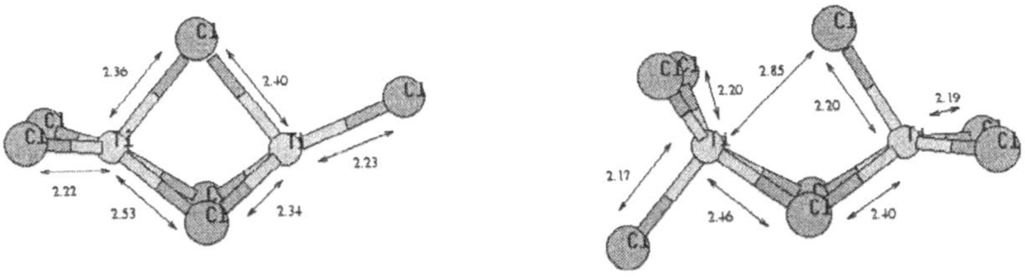

Figure 9, b and c. Structures of $(TiCl_3)_2$ and $(TiCl_4)_2$.

In the $TiCl_2$-C_2H_4 complex (with a C_{2v} symmetry, z is the C_2 axis), the chlorine (yz plane) and two carbon atoms (xz plane) are in different planes. The Cl atoms interact with the $d_{z^2-x^2}$ orbital and πC_2H_4 orbital interacts with the $d_{z^2-y^2}$ orbital so that both of them avoid to overlap with the out-of-phase component of the torus of d_{z^2} orbital in their plane. Such orthogonality cannot be obtained when there are more Cl ligands, in $TiCl_3$-C_2H_4 or $TiCl_4$-C_2H_4. The binding is thus stronger with the $TiCl_2$.

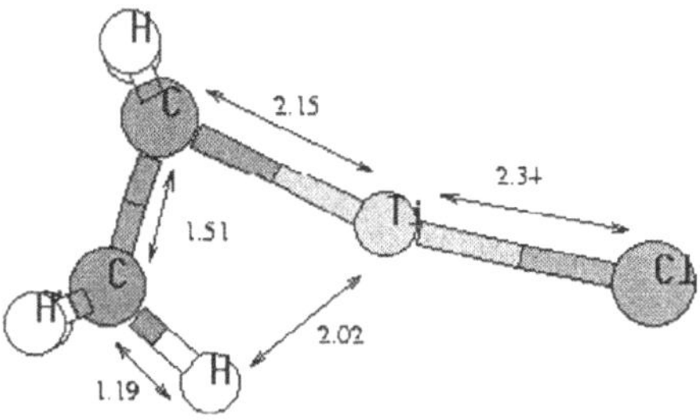

Figure 10. Structure of ClTi-CH_2-CH_3 obtained from by optimisation of TiHCl-C_2H_4 with a π-bonding ethene.

The optimisation of the TiHCl-C_2H_4 complex starting from a π-bonding ethene leads to a final structure, ClTi-CH_2-CH_3, corresponding to an intermediate in the Cossee mechanism. This structure (figure 10) is consistent with the addition of a zwiterrionic form of ethene according to our analysis (sections 5.1 and 5.3); the C^- binds to the titanium centre and the C^+ binds to the H ligand leading to its abstraction : the Ti-H bond is cleaved and the transferred hydrogen becomes the agostic hydrogen of the newly formed methyl group. The Ti-H distance is 2.11 Å and the C-H distance is 1.17 Å (to be compared with 1.12 Å for the two other H atoms). The formation energy of the complex represents 54.8 kcal/mol; again it is verified that the H ligand is more reactive than the Cl ligand.

6. Conclusions

$TiCl_n$ compounds show unfamiliar behaviour. The main specific features are

- The shortening of the bond distances with increasing coordination.
- Non-VSEPR geometry.
- The titanium atom in all oxidation states appears to be an acidic centre.
- Ligands are potentially basic sites. The participation of the ligands to the reactivity toward acidic or radicalar species leads to their abstraction.
- The unpaired electrons of Ti(II) and Ti(III) couple with radicals to build σ bonds.

These features are reasonable. The shortening of bond distances is associated with increased valency. The non-VSEPR geometry is understandable in terms of MO and ELF analysis. The implication of the ligand is associated with abstraction reactions that are thermodynamically favorable and experimentally observed.

It is generally assumed that the Ti atom is the main active site for the Ziegler-Natta reaction. Addition on Ti requires a basic reactant. Otherwise, the ligand is more active. H and R ligands react with cations more easily than Cl since it is less bound to Ti. The R ligands easily react to be detached from the titanium centre. The addition of ethene on TiHCl uses both possibilities: the activity of the Ti site and that of the H ligand. $TiCl_2$ and $TiCl_3$ are more reactive than $TiCl_4$. $TiHCl_2$ and $TiCH_3Cl$ are more reactive than $TiCl_2$.

The addition at the Ti centre should imply a Lewis base as reactant. Radicalar species do not seem as effective as initially thought. The π-electrons of an olefin have both a diradicalar and a zwitterionic component; the development of a negative charge on one side of the olefin is necessary to bind to the Ti centre. We intend to investigate the olefin adsorption and explore the reactivity on solid surfaces.

Acknowledgements

This work has been accomplished in the framework of GDR "Dynamique Moléculaire Quantique Appliquée à la Catalyse"; financial support of the French Ministry of Education, Technology and Research is gratefully acknowledged. We are grateful to Pr. B. Silvi for his help in the ELF analysis. We thank CNRS-IDRIS for computing facilities.

References

1. Bernardi, F., Bottini, A. and Miscione, G. P. (1998) *Organometallics*, **17**, 16-24.
2. Cossee, P. (1966) *Recl. Trav. Chim. Pays-Bas*, **85**, 1151.
3. Axe, F. U. and Coffin, J. M. (1994) *J. Phys. Chem.*, **98**, 2567-2570.
4. Yao, W., Eisenstein, O. and Crabtree, R. (1997) *Inorganica Chimica Acta*, **254**, 105-111.
5. DiDrusco, R. (1984) *Hydrocarb. Proc.*, 113.
6. Giannini, U. (1981) *Makromol. Chem. Suppli.*, **5**, 216.
7. Galli, P., Barbé, P. C. and Noristi, L. (1984) *Angew. Makromol. Chem.*, **120**, 73.
8. Rinaldi, R. (1990) *Chim. Industr. (Italy)*, **234**, 3.
9. Magni, E. and Somorjai, G. A. (1995) *Applied Surf. Sci.*, **89**, 187-195.
10. Magni, E. and Somorjai, G. A. (1995) *Catal. Lett.*, **35**, 205-214.
11. Magni, E. and Somorjai, G. A. (1995) *Surf. Sci.*, **341**, L1078-L1084.
12. Magni, E. and Somorjai, G. A. (1996) *Surf. Sci.*, **345**, 1-16.
13. Magni, E. and Somorjai, G. A. (1996) *J. Phys. Chem.*, **100**, 14786-14793.
14. Magni, E. and Somorjai, G. A. (1996) *Phys. Rev. Lett.*, **77**, 4717-4722.
15. *CRC Handbook of Chemistry and Physics, 58th ed.* (R. C. Weast), CRC Press, Cleveland (1978).
16. Frisch, M. J., Trucks, G. W., Schlegel, H. B., Gill, P. M. W., Johnson, B. G., Robb, M. A., Cheeseman, J. R., Keith, T. A., Petterson, G. A., Montgomery, J. A., Raghavachari, K., Al-Laham, M. A., Zarzewski, V. G., Ortiz, J. V., Foresman, J. B., Cioslowski, J., Stefanov, B. R., Nanayakkara, A., Challacombe, M., Peng, C. Y., Ayala, P. Y., Chen, W., Wong, M. W., Andres, J. L., Replogle, E. S., Gomperts, R., Martin, R. L., Fox, D. J., Binkley, J. S., Defrees, D. J., Baker, J., Stewart, J. P., Head-Gordon, M., Gonzales, C. and Pople, J. A. (1995) *GAUSSIAN* 94 - B.1, Gaussian Inc., Pittsburgh PA, USA.

17. Durand, P. and Barthelat, J. C. (1975) A theoretical Method to Determine Atomic Pseudopotentials for Electronic Structure Calculations of Molecules and Solids *Theor. Chim. Acta*, **38**, 283-302.
18. Fahmi, A. and Minot, C. (1993) A theoretical investigation of the water adsorption on titanium didoxide surfaces *Surf. Science*, **304**, 343-359.
19. Fahmi, A., Minot, C., Silvi, B. and Causa, M. (1993) Theoretical analysis of the structures of titanium dioxide crystals *Phys. Rev. B*, **47**, 11717-11724.
20. Bouteiller, Y., Mijoule, C., Nizam, M., Barthelat, J. C., Daudey, J. P., Pelissier, M. and Silvi, B. (1988) Extended gaussian-type valence basis sets for calculations involving non-empirical core pseudopotentials *Mol. Phys.*, **65**, 295.
21. Perdew, J. P. and Wang, Y. (1992) *Phys. Rev. B*, **45**, 13244.
22. Dovesi, R., Saunders, V. R. and Roetti, C. (1996) *Crystal 95 User Documentation*, University of Torino and SERC Daresbury Laboratory.
23. Dargelos, A. *private communication*.
24. Lister, M. W. and Sutton, L. E. (1941) *Trans. Faraday Soc.*, **37**, 393-406.
25. Kudo, T. and Gordon, M. S. (1995) Molecular and electronic structure of TiH2 *J. Chem. Phys.*, **102**, 6806-6811.
26. Sousa, C. and Illas, F. (1994) Ionic-Covalent transition in titanium oxides *Phys. Rev. B*, **50**, 13974-13980.
27. Gillespie, R. and Hargittai, I. *The VSEPR Model of Molecular Geometry;* (A. a. Bacon), Boston, MA (1991).
28. Gillespie, R. (1992) *J. Chem. Soc. Rev.*, 59.
29. Jolly, C. A. and Marynick, D. S. (1989) Ground-state geometries and inversion barriers for simple complexes of early transition metals *Inorg. Chem.*, 2893.
30. Jolly, C. A. and Marynick, D. S. (1989) The direct insertion Mechanism in Z-N polymerisation *J. Amer. Chem. Soc.*, **111**, 7968-7974.
31. Gillespie, R., Bytheway, I., Tang, T. H. and Bader, R. F. W. (1996) *Inorg. Chem.*, **35**, 3954-3963.
32. Bader, R. F. W. *Atoms in molecules : A Quantum Theory*. Oxford Univ. Press, Oxford (1990).
33. Becke, A. D. and Edgecombe, K. E. (1990) *J. Chem. Phys.*, **92**, 5397.
34. Savin, A., Jepsen, O., Flad, J., Andersen, O. K., Preuss, H. and vonSchnering, H. G. (1992) *Angew. Chem. Int. Ed. Engl.*, **31**, 187.
35. Savin, A., Becke, A. D., Flad, J., Nesper, R., Preuss, H. and von Schnering, H. G. (1991) *Angew. Chem. Int. Ed. Engl.*, **30**, 409.
36. Silvi, B. and Savin, A. (1994) *Nature*, **371**, 683.
37. Silvi, B., Savin, A. and Colonna, F. (1996) *Can. J. Chem.*, **74**, 1088.
38. von Weizsäcker, C. F. (1935) *Z. Phys.*, **96**, 431.
39. Noury, S., Colonna, F., Savin, A. and Silvi, B. (1998) *J. Mol. Struct.*, **450**, 59.
40. Giannini, U., Giunchi, G., Albizzati, E. and Barbè, P. C. *Recent advances in Mechanistic and Synthetic Aspects of polymerisation Series C : vol 215;* (M. Fontanille, A. Guyot and eds.), Reidel (1987).
41. Corradini, P., Barone, V., Fusco, R. and Guerra, G. (1983) A possible model of catalytic sites for the stereospecific polymerisation of alpha-olefins *Gazzetta Chimica Italiana*, **113**, 601-607.
42. Herzler, J. and Roth, P. (1997) Shock tube Study of the reaction of H atoms with TiCl4 *J. Phys. Chem. A*, **101**, 9341-9343.
43. Wyckoff R. W. G. (1982) Crystal Structures **1**, 269, Interscience, Wiley, N.Y.
44. Natta G., Corradini P., Allegra G. (1961) *J. Polymer Sci.* **51**, 399-410.

PHENOMENOLOGICAL DESCRIPTION OF D-WAVE CONDENSATES IN HIGH-T_c SUPERCONDUCTING CUPRATES

E.J. BRÄNDAS[a], L.J. DUNNE[b,c] AND J.N. MURRELL[c]

[a]*Department of Quantum Chemistry*
University of Uppsala, Box 518, Uppsala, S-75120, Sweden
[b]*School of Applied Science*
South Bank University, London SE1 0AA, UK
[c]*School of Chemistry, Physics and Environmental Science*
University of Sussex, Falmer, Brighton, BN1 9QJ, UK

Abstract. The synthesis of many types of high T_c superconducting cuprates occurred over a decade ago but the cause of the superconducting condensation and electronic structure of such materials is still a matter of significant deliberation. Although there is no consensus as to the origin of pairing, there is, on the other hand, a wide agreement about the main behaviour, which a theoretical description should provide. In this paper a theory is presented which accounts, in a straightforward way, for much of the essential behaviour of high T_c superconducting cuprates.

1. Introduction

By the millenium it will be almost 15 years since the finding of high T_c superconducting cuprates but the nature of the superconducting condensate and electronic structure of such materials is still a matter of significant deliberation. In this paper a theory is proposed which accounts, in a simplified way, for many of the essential features of high T_c superconducting cuprates. Although there is no agreement as to the origin of pairing there is, on the other hand, a wide agreement concerning the main behaviour which a theoretical description should provide. The fundamental structural characteristic of $n-$ or $p-$type cuprates are the cuprate layers shown in Figure 1 where the square arrangement of lattice points is centred on a Cu atom shown in black. Every cuprate layer is a four-sided arrangement of $N/2$ lattice points with local C_{4v} point group symmetry about the Cu atoms and with every lattice point with integer index l [1-13]. The inclination for pairs of electrons to evade the region of short range repulsion and to reside in the longer range attractive region of an effective potential can enable the system to adopt a low energy coherent electronic state with superconductive behaviour. Thus our theory presented here is based on the correlation of electrons interacting with a short range screened Coulomb repulsion and perhaps a longer range attraction.

J. Maruani et al. (eds.), New Trends in Quantum Systems in Chemistry and Physics, Volume 2, 289–304.

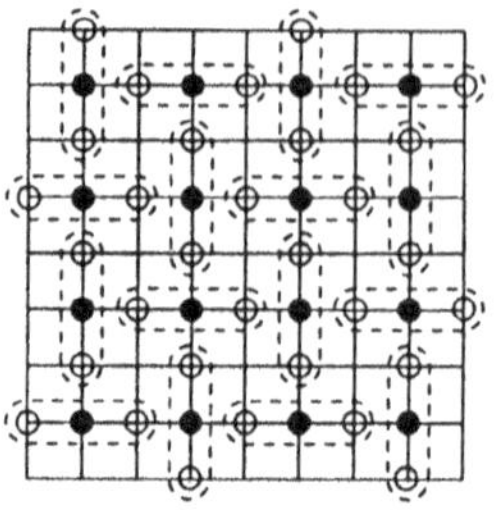

Figure 1. Copper/oxygen layer of Superconducting cuprates. Black and white circles represent copper and oxygen atoms respectively. A pair of E-basis Wannier functions, which are a linear combination of Cu and O atomic orbitals, transforming as (x, y) are localised on every lattice point at the centre of the unit cell.

Our theory is based on an important mathematical argument given by us previously [14-16] but which is simple enough that it bears repetition here. In the usual Bardeen-Cooper-Schrieffer (BCS) theory of superconductivity with attractive off-diagonal matrix elements of the phonon induced electron-electron interaction, it is common to find the stability of the superconducting state explained by a so-called n-fold stabilisation phenomenon. So that, if there are n quasi-degenerate states at energy ε above the Fermi level and these are coupled by an attractive matrix element -**v** then the Hamiltonian matrix has $(n-1)$ eigenvalues at energy ε and one which can be identified as the superconducting state which can cross below the Fermi level at energy $\varepsilon - nV$. However a similar superconducting stabilisation can also occur with repulsive off-diagonal matrix elements. A simple demonstration of this is obtained by reflecting on the real symmetrical eigen-problem shown below.

$$\left(\begin{array}{ccc|ccc} \varepsilon & & \mathbf{0} & & & \\ & \ddots & & & \mathbf{v} & \\ \mathbf{0} & & \varepsilon & & & \\ \hline & & & \varepsilon & & \mathbf{0} \\ & \mathbf{v}^T & & & \ddots & \\ & & & \mathbf{0} & & \varepsilon \end{array}\right) \left(\begin{array}{c} \cdot \\ 1 \\ \cdot \\ \hline \cdot \\ \pm 1 \\ \cdot \end{array}\right) = (\varepsilon \pm \kappa V) \left(\begin{array}{c} \cdot \\ 1 \\ \cdot \\ \hline \cdot \\ \pm 1 \\ \cdot \end{array}\right) \qquad (1)$$

The sub-blocks are of order κ. The lowest eigenvalue at energy $\varepsilon - \kappa V$ splits off from the other eigenvalues at ε and, if this eigenvalue is negative it can cross over and become the ground state. The matrix shown in eq(1) appears to be directly applicable to doped cuprate layers. It can be shown that in both cases above, if the basis functions are pairwise occupied as in a BCS wavefunction, then both lowest states have a macroscopic number of electron pairs all in the same pair state by noting that, following Yang [17-18] or Gor'kov [19] or Coleman [20], the second order reduced electronic density matrices both have one macroscopically large eigenvalue representing a superconducting condensate of pairs. This will be demonstrated in detail below.

2. Flat band electron energy dispersion in superconducting cuprates

The d-wave symmetry [1-13] of the wavefunction describing the condensed electrons in superconducting cuprates can be subjected to a group theoretical analysis [5] which shows that such a symmetry cannot originate in a single band description of the cuprates. Our choice, rationalised in [5], is for a pair of localised E-basis functions in the C_{4v} group [23] associated with each unit cell which are symmetry adapted combinations of Cu e-orbitals and in-plane oxygen p_z orbitals. Apical oxygen p_x and p_y orbitals [8] too have such symmetry and one can combine all these orbitals judiciously to give a pair of E-basis functions on every unit cell.

We assume, following the arguments of [5] that there are a pair of localised Wannier type orbitals centred on each cell l of the form $\{\phi_{l,x}(\mathbf{r}), \phi_{l,y}(\mathbf{r})\}$ transforming together as the E degenerate irreducible representation in the C_{4v} point group. For mathematical expedience every pair of l, x orbitals is assigned a signature $(-1)^{l+1}$ and $(-1)^{l}$ for every pair of l, y orbitals. Hole doping is presumed to occur from a full valence band and electron doping into a vacant conduction band appropriate to the set of Wannier functions.

If the above supposition concerning the pair of E-basis functions on each lattice point is legitimate we should be able to make some statements concerning the electronic energy bands associated with these. Following Alexandrov [8] in the tight binding approximation we attain an X-band and Y-band each with dispersion relations given by

$$E_x(\mathbf{k}) = -2\beta_\sigma \cos(k_x) - 2\beta_\pi \cos(k_y) \tag{2a}$$

$$E_y(\mathbf{k}) = -2\beta_\sigma \cos(k_y) - 2\beta_\pi \cos(k_x) \tag{2b}$$

where β_σ and β_π are empirically selected transfer integrals in the σ and π configurations for nearest neighbour Wannier functions. We will adopt a plausible ratio of $\beta_\sigma/\beta_\pi = 4$ on the basis of crystalline and overlap anisotropies. The sign of this ratio differs from Alexandrov [8] but is required in our theory for a reasonable description of the photoemission data. The electronic energy surface in $\mathbf{k}$-space of uppermost energy levels is constructed by taking the higher of the $E_x(\mathbf{k})$ and $E_y(\mathbf{k})$ for a given $\mathbf{k}=(k_x, k_y)$ where the lattice constant is set to 1. The resulting uppermost energy level dispersion relation for $\beta_\sigma = 0.2$eV is displayed in Figure 2.

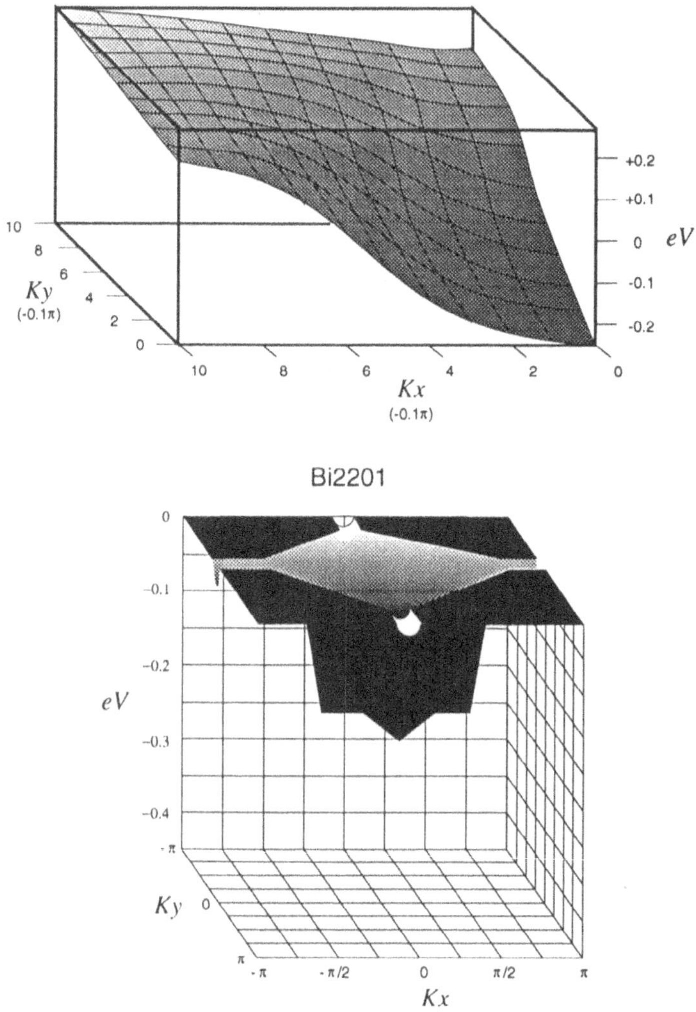

Figure 2. Dispersion relation for uppermost energy levels for the $E_x(\mathbf{k})$, $E_y(\mathbf{k})$ bands discussed in the text for a given $\mathbf{k} = (k_x, k_y)$. The energy surface has fourfold symmetry about the origin. The result is compared with the photoemission data of Shen *et al* [7] also shown in the lower part of the figure.

The important features of the Angular resolved photoemission spectra discussed by Shen *et al* [7] are reproduced. The flatness of the bands which are a widely discussed feature of cuprates [7,6,8] should be noted.

3. Off-diagonal long-range order in cuprate layer electrons

We adopt an extended Hubbard Hamiltonian postulated for the active space of the form

$$\mathcal{H}=\sum_{l,m}\langle l|\mathrm{h}|m\rangle a_l^+ a_m + \frac{1}{2}\sum_{k,l,m,n}\langle kl|\mathrm{v}|mn\rangle a_k^+ a_l^+ a_m a_n \tag{3}$$

where h and v are pertinent one and two body parts of the effective Hamiltonian.

Because of short range repulsion this has positive off-diagonal matrix elements for electron pair transfers between Wannier orbitals. We will presume that the effective pair interaction has the form $\mathrm{v}(i,j) \approx \exp(-\gamma r_{ij})e^2/Dr_{ij}$. This is the same appearance as discussed in [15] where γ is the Thomas-Fermi screening constant calculated for the carriers induced by doping and D is the background dielectric constant as discussed by Mott and Davis [24]. Thus, the effective electron pair interaction is assumed to have a short range repulsive hard core. The symbol $\approx$ above is qualified here so as to characterize the envelope of the function and does not rule out a longer range attractive tail due to bipolaronic coupling or Friedel oscillations. The summation in eq(3) is over cell orbitals and spins and where only on-cell and nearest neighbour matrix elements will be taken into account as significant in a localised basis.

If we suppose that there are N orbitals in the cuprate layer accessible to $2M$ electrons the fractional doping is given by $\rho = M/N$ where $1 \geq \rho \geq 0$ so that for hole doping a full set of cells with $\rho = 1$, and for electron doping $\rho = 0$, will be important. Cell orbitals in the cuprate layer can be occupied by a pair of singlet coupled electrons (a dimer ($\uparrow\downarrow$) or singly occupied by single unpaired electron with down ($\downarrow$) or up ($\uparrow$) spin. Every configuration of electrons arranged in the cuprate layer orbitals can be represented by a Slater determinant and an approximate many electron wavefunction obtained in principle from diagonalisation of the matrix representation of the Hamiltonian in the basis of Slater determinants. Although an exact diagonalisation of such a Hamiltonian matrix is prohibitively arduous we are able to construct a mean-field theory which, with the tentative assumptions of our theory, provides a remarkable phenomenological portrayal of the behaviour of high T_c cuprate superconductors.

Every Slater determinant basis function, represented in the Hamiltonian matrix, corresponds to a unique arrangement of electrons on the cuprate layer and these can be grouped into $N!/(N - M_\uparrow - M_\downarrow)!M_\uparrow!M_\downarrow!$ sets corresponding to the possible ways of arranging the spin down ($\downarrow$) and up ($\uparrow$) spin monomers and to sub-classify the grouping of the remaining $(N - M_\uparrow - M_\downarrow)!/(N - M_\uparrow - M_\downarrow - M_D!)M_D!$ dimer ($\uparrow\downarrow$) configurations into 2 classes resolved by the $\pm$ sign of the product of the occupied dimer orbital signatures as described in detail in [15]. Thus grouped, and neglecting the coupling between blocks the Hamiltonian matrix has the block form shown overleaf. The off-diagonal coupling is most important for pair transfers between Slater determinant basis functions of opposite signature. Every block j has $g_j(M_\uparrow, M_\downarrow) = N!/(N - M_\uparrow - M_\downarrow)!M_\uparrow!M_\downarrow!$

quasi-degenerate states with average energy $(\Delta(X) - V)M_D(1 - 2\rho + X)$ for hole doping and $X(\Delta(X) - V(1 - 2\rho + X))$ for electron doping relative to the completely dissociated normal state described by the lowest block on the right hand side of the Hamiltonian matrix. ($\Delta(X)$ is the screened diagonal Coulomb repulsion for bringing electrons together into the same orbital and $X = M_D/N$ and where V is the sum of nearest neighbour pair transfer matrix elements which is repulsive overall.

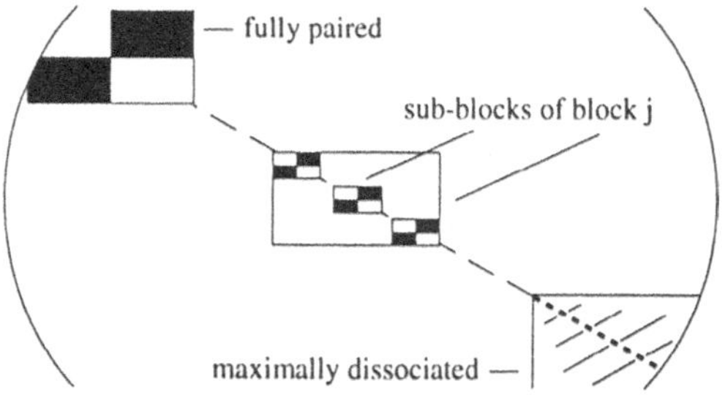

The ground state wavefunction is given by the geminal power

$$(N!/(N-M)!M!)^{-\frac{1}{2}}(1/M!)\left[\sum_l \exp(i\pi l)(a^+_{lx\uparrow}a^+_{lx\downarrow} - a^+_{ly\uparrow}a^+_{ly\downarrow})\right]^M |0\rangle \quad (4)$$

To deduce thermal behaviour of the superconductor excited state energies and wavefunctions are necessary which are obtained as follows. Every block j of the Hamiltonian matrix has $g_j(M_\uparrow, M_\downarrow)$ low-lying states with eigenvectors which are a linear combination of the sub-block degenerate states which may be brought together to give excited state wavefunctions of the form

$$(N!/M_D!(N - M_\uparrow - M_\downarrow - M_D)!M_\uparrow!M_\downarrow!)^{-\frac{1}{2}}\sum_s c_s\psi_s \quad (5)$$

where $|c_s| = 1$ and $s = 1, 2, \ldots, g_j(M_\uparrow, M_\downarrow)$ refers to the sub-block s with eigenvector ψ_s of the block j of our concern. The above wavefunction describes a general state of the superconductor for any degree of pair dissociation up to T_c

It is important to note that every sub-block in the Hamiltonian matrix has the form displayed in eq(1) above. $\Delta(X)$ will depend on the level of dimer dissociation. We will estimate $\Delta(X)$ as in [15] by linearising in the density of condensed electrons derived below to give an average energy $X(1 - 2\rho + X)(\Delta/\rho - V)$ for hole doping and $X(1 - 2\rho + X)(\Delta/(1-\rho) - V)$ for electron doping relative to the completely dissociated normal state described by the lowest block on the right hand side of the Hamiltonian matrix. Δ is the screened diagonal Coulomb repulsion for bringing electrons into the same orbital when its neighbours are fully paired and $X = M_D/N$ where V is the sum of nearest neighbour pair transfer matrix elements. The linearisation procedure adopted here to estimate the doping dependence of $\Delta(X)$ is likely to be reliable at low doping and can implicitly embrace a number of effects not considered here but we do not anticipate this to be a good semblance to reality at high doping.

The second order electronic density matrix can be constructed from the Hamiltonian matrix eigenvector given in eq(5) above to give

$$\begin{pmatrix} M_D/N & -\frac{M_D(N-2M+M_D)}{N(N-1)} & \cdots & \cdots \\ -\frac{M_D(N-2M+M_D)}{N(N-1)} & M_D/N & \cdots & \cdots \\ \vdots & \vdots & \ddots & \vdots \\ \cdots & \cdots & \cdots & M_D/N \end{pmatrix} \times \begin{pmatrix} 1 \\ -1 \\ \vdots \\ \cdots \end{pmatrix}$$

$$= [X + M_D(1-2\rho+X)] \begin{pmatrix} 1 \\ -1 \\ \vdots \\ \cdots \end{pmatrix} \tag{6}$$

This can be seen to have a macroscopically large eigenvalue signifying a condensate of electron pairs. All $g_j(M_\uparrow, M_\downarrow)$ low-lying states have the same density matrix eigenvector and eigenvalue.

Consequently the thermally averaged second order reduced electronic density matrix eigenvector is

$$\begin{aligned} &\psi(\mathbf{r}_1\sigma_1, \mathbf{r}_2\sigma_2) \\ &= \lambda \sum_l (-1)^l [\phi_{lx}(\mathbf{r}_1)\phi_{lx}(\mathbf{r}_2) - \phi_{ly}(\mathbf{r}_2)\phi_{ly}(\mathbf{r}_1)](\uparrow(1)\downarrow(2) - \uparrow(2)\downarrow(1)) \end{aligned} \tag{7}$$

which has $d_{x^2-y^2}$ pair symmetry in the C_{4v} group in conformity with experimental determinations of the condensate symmetry in high T_c superconducting cuprates. The function $\psi(\mathbf{r}_1\sigma_1, \mathbf{r}_2\sigma_2)$ transforms as the B_1 irreducible representation in the C_{4v} group. λ is the macroscopically large reduced second order electronic density matrix eigenvalue corresponding to a population $M_D(1-2\rho+X)$ condensed electron pairs where X is calculated from the Helmholtz free energy F defined in section 4 below.

The distance over which $\psi(\mathbf{r}_1\sigma_1, \mathbf{r}_2\sigma_2)$ stays finite as $|\mathbf{r}_1 - \mathbf{r}_2|$ increases is a measure of the pair size or coherence length and in this theory, in view of the localised basis, this quantity is of the order of a few Angstroms in conformity with experimental observations [5,15].

4. Thermal behaviour of superconducting cuprates up to T_c

Following the arguments of [15] the Helmholtz free energy per orbital, F, is given up to T_c by

$$F = kT[(1 - 2\rho + 2X)\ln(1 - 2\rho + 2X) + 2(\rho - X)\ln(\rho - X)] + [X + \Delta/t(1 - 2\rho + 2X)] \tag{8}$$

where $t = \rho$ when $\rho > 1/2$ and $t = (1 - \rho)$ when $\rho < 1/2$. This is simply derived using the degeneracy factor $g_j(M_\uparrow, M_\downarrow)$ and energy relative to the normal state E_j for each block j given above and the formula $F = -(kT/N)\ln Q$ where $Q = \sum g_j \exp(-E_j/kT)$ is the Canonical partition function and where the logarithm of the sum is evaluated by the method of the maximum term. Therefore minimisation of F with respect to X gives the equilibrium condition as a solution of the non-linear equation

$$2kT\ln[(1 - 2\rho + 2X)/(\rho - X)] + [-V + \Delta/t)(1 - 2\rho + 2X) = 0 \tag{9}$$

The root of this non-linear equation gives the equilibrium value of X and for that reason the density of condensed electrons. The transition temperature can be located by noting that at T_c the equilibrium values of X are constrained by $X = 0$ for electron doping and $X = (2\rho - 1)$ for hole doping corresponding to zero density of condensed electrons.

Insertion of these conditions into eq(9) gives the transition temperature T_c for both doping cases given by

$$kT_c = (V - \Delta/(1 - \rho))(1 - 2\rho)/2\ln[(1 - 2\rho)/\rho] \tag{10a}$$

$$kT_c = (V - \Delta/\rho)(2\rho - 1)/2\ln[(2\rho - 1)/(1 - \rho)] \tag{10b}$$

where eq(10a) refers to electron doping when $\rho < 1/2$ and eq(10b) to hole doping when $\rho > 1/2$. The doping phase diagram implied by eqs(10a,10b) is shown in Figure 3 (which also demonstrates the experimentally well established electron-hole symmetry) and, for plausible values of matrix elements and background dielectric constant given in the caption to Figure 3, our theory depicts superconductivity at high temperatures in the same doping range as for cuprates.

The temperature and doping dependence of the energy gap can also be derived as follows. The energy to remove one electron pair from the condensate at absolute zero is

$$E_g(0) = V - \Delta/(1 - \rho) \tag{11a}$$

$$E_g(0) = V - \Delta/\rho \tag{11b}$$

where eq(11a) refers to electron doping when $\rho < 1/2$ and eq(11b) to hole doping when $\rho > 1/2$. This gives the ratio $E_g(0)/k\mathrm{T}_c$, given by

$$E_g(0)/k\mathrm{T}_c = 2\ln[(1-2\rho)/\rho]/(1-2\rho) \tag{12a}$$

$$E_g(0)/k\mathrm{T}_c = 2\ln[(2\rho-1)/(1-\rho)]/(2\rho-1) \tag{12b}$$

where eq(12a) refers to electron doping when $\rho < 1/2$ and eq(12b) to hole doping when $\rho > 1/2$. These results represent an average of the energy gap over k-space.

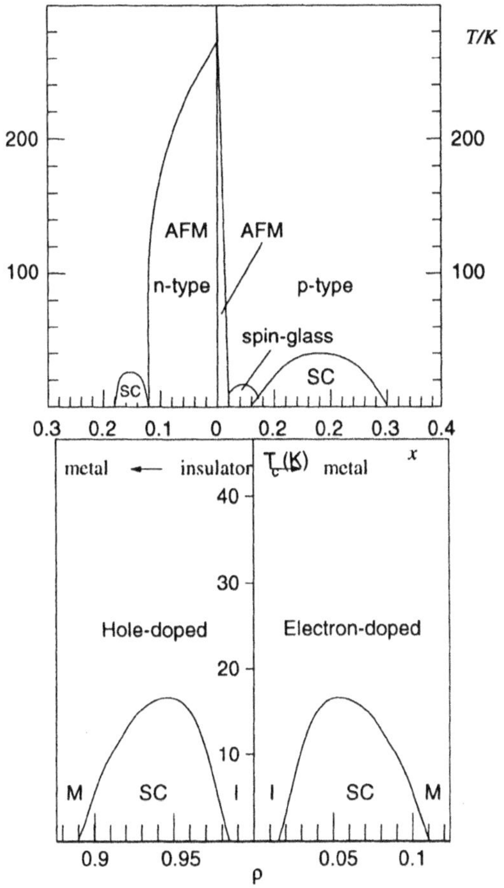

Figure 3. Theoretical phase diagram and doping dependence of the transition temperature predicted by our theory for the parameters $D = 5$, $V = 0.564$eV compared with the well known experimental doping $\mathrm{T_c}$ curve shown above. The superconducting (SC), insulating (I) and metallic phases are characterised by use the Mott-Edwards-Sienko relation [23,24,28] as described in [15].

The predictions of this for hole doping are shown in Figure 4. A comparison can be made with a recent collection of gap data given by Deutscher [4] where the above so-called strong coupling ratio varies from about 20 to 6 and falls with rising doping levels as predicted by our theory.

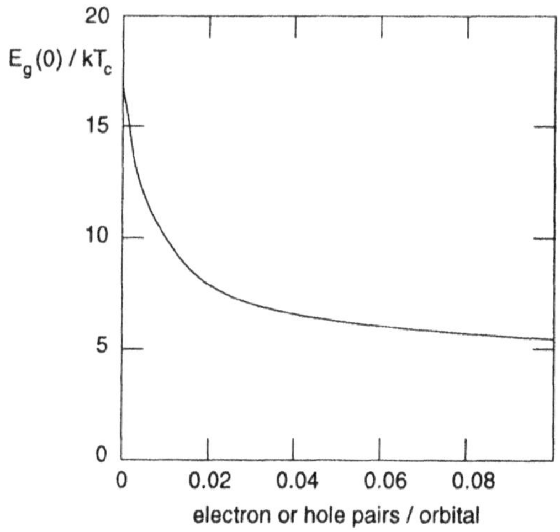

Figure 4. Theoretical gap ratios $E_g(0)/(kT_c)$ as a function of doping for superconducting cuprates.

At finite temperature the energy gap is the difference between the eigenvalues of two neighbouring blocks of the Hamiltonian matrix and given by

$$E_g(T) = (1 - 2\rho + 2X)(V - \Delta/(1 - \rho)) \tag{13a}$$

$$E_g(T) = (1 - 2\rho + 2X)(V - \Delta/\rho) \tag{13b}$$

where eq(13a) refers to electron doping when $\rho < 1/2$ and eq(13b) to hole doping when $\rho > 1/2$.

The ratio $E_g(T)/E_g(0)$ is compared with experimentally determined values in Figure 5 where it can be seen that the theory reproduces the weak temperature dependence of the gap. We expect that an accurate theory with fluctuations included would remove the discontinuity shown in Figure 4 due to short range order in the pairing amplitude in the neighbourhood of T_c and such an effect can account for the trends reviewed by Deutcher in which the gap does not disappear at T_c.

5. Temperature and doping dependence of density of condensed electrons

The prediction that the zero temperature density of condensed electrons should be proportional to $(\rho(1-\rho)$ is tested against experimental observations in Figure

6 where the hole concentration is estimated from the chemical formula of a series of cuprates by valence counting.

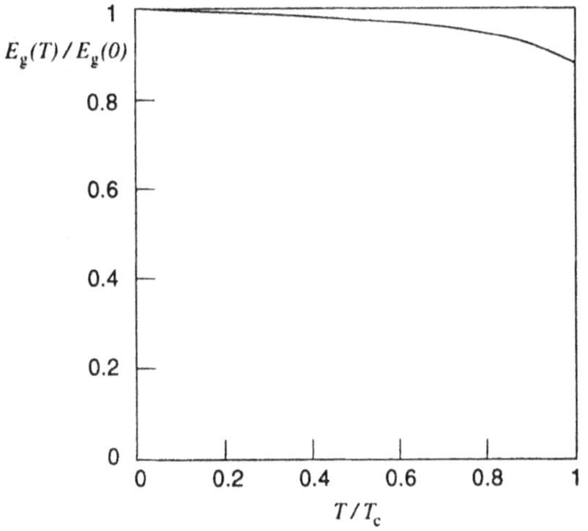

Figure 5. Theoretical temperature dependence of the superconducting energy gap.

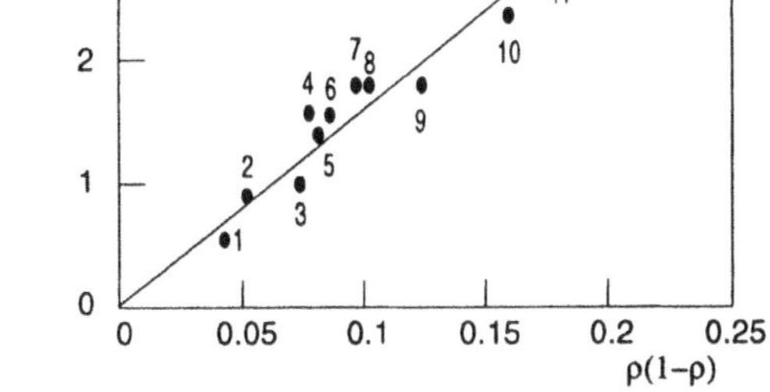

Figure 6. Muon spin relaxation rate, σ, against $2\rho(1-\rho)$ for $La_{2-x}Sr_xCuO_4$ and $YBa_2Cu_3O_y$. The experimental data are from Uemura [27] and $2\rho(1-\rho)$ is estimated from the chemical formula . There are 14 points starting from the left which correspond to the following x or y values. 1. $x = 0.08$, 2. $x = 0.1$, 3. $x = 0.15$, 4. $x = 0.15$, 5. $y = 6.66$ 6. $y = 6.67$, 7. $x = 0.20$, 8. $x = 0.21$, 9. $y = 6.67$, 10. $y = 6.86$, 11. $y = 6.95$, 12. $y = 6.95$, 13. $y = 7.0$, 14. $y = 7.0$, 15. $Y_{0.7}Ca^{2+}_{0.3}Ba_2Cu_3O_7$. The linear fit for this theory differs slightly from one given previously [21].

It can be seen that the prediction is in satisfactory agreement with experimental observations [27]. The temperature dependence of the density of condensed electrons is shown in Figure 7 which has utilised the thermal average of the density of condensed electrons obtained as described above.

The response for low doping closely follows a two-fluid type behaviour for low hole doping but which deviates from this at higher doping levels. Schneider and Keller [26] analysed Muon spin relaxation rate σ data and noted universal trends in the reduced transition temperature T^* for a series of extreme type II superconductors which includes the cuprates and the reduced muon spin relation rate σ^* in whereby

$$T^* = T_c / T_{c,max} \quad \text{and} \quad \sigma^* = \sigma/\sigma_{max} \tag{14}$$

$T_{c,max}$ is the maximum transition temperature in family of materials. σ is known to be proportional to the condensate density. The universal trends found are summarised in Figure 8 and compared with our prediction for the doping sequence given in Figure 3. The density of condensed electrons continues to increase beyond $T_{c,max}$ in the same way as found experimentally. In particular the experimental data and our prediction both have a highest value for σ^* near to 2.

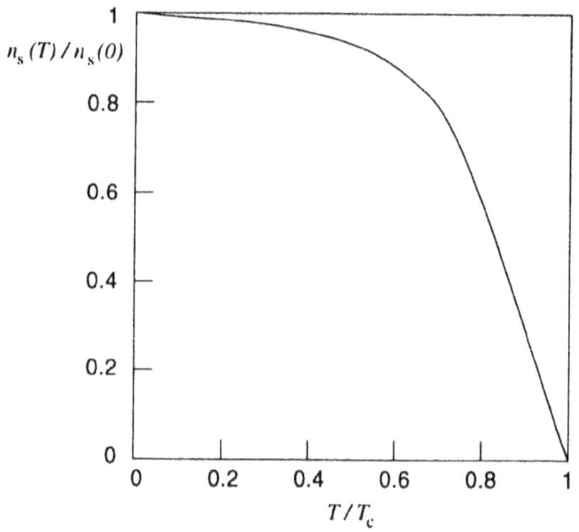

Figure 7. Theoretical temperature dependence of the condensate density.

6. Condensation energy and heat capacity

Heat capacity measurements on superconducting cuprates have been widely undertaken and recently Loram *et al* [25] reported condensation energies for a cuprate superconductor as a function of doping. In our theory the heat capacity per localised orbital (C_v/N_k) is given by

$$C_v/N_k = 2kT \ln[(\rho - X)(1 - 2\rho + 2X)]\frac{d(X)}{d(kT)} \quad (15)$$

which is obtained in the standard way from F given in eq(8). The condensation energy is given above in section 3. The trends shown by the theory are given in Figure 9. The heat capacity in the overdoped regime is predicted to be a few times larger than the underdoped. These unusual features seem to agree with experimental observations [25,29,30].

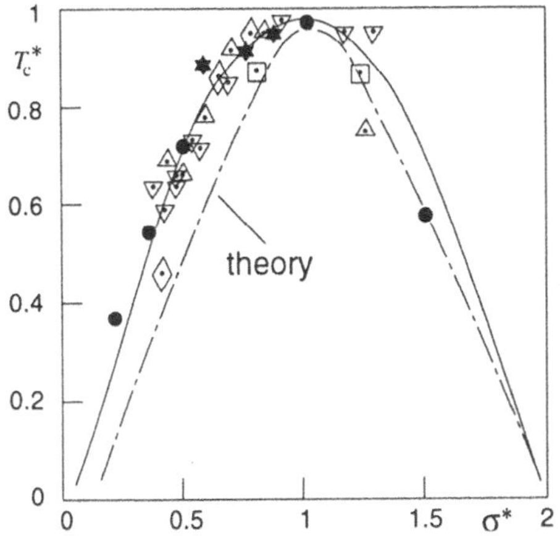

Figure 8. Ratio of Reduced transition temperature to Reduced density of condensed electrons due to Schneider and Keller [26] against the predictions of our theory shown in the lower part of the figure.

7. Temperature dependence of the Knight shift

The Knight shift in NMR measures the local magnetic susceptibility around a nucleus. Following the arguments in [15] the reduced Knight shift $K_s(T)/K_s(T_c)$ for our theory is given by

$$\frac{K_s(T)}{K_s(T_c)} = \frac{[(\rho - X(T))T_c]}{[(\rho - X(T_c))T]} \quad (16)$$

This quantity has been computed and is shown in Figure 10 and seems to provide a good description of the experimental data up to T_c. In order to describe the susceptibility above T_c other singlet excitations not at heart related to the superconductivity can be invoked. Therefore we are in accord with Loram *et al* [26] whose view is that the normal state spin gap is not essentially related to the superconducting pairing. However such excitations in the normal state are yet to be included in our theory but our conjecture is that this will not alter the predicted behaviour by our theory below up to T_c very much. One can conceive a theory where monomer spins are excited into a state above T_c with a higher magnetic moment at a mobility edge thereby relating the spin gap and normal state conductivity along comparable lines to that by Alexandrov [8] and this is extension is underway.

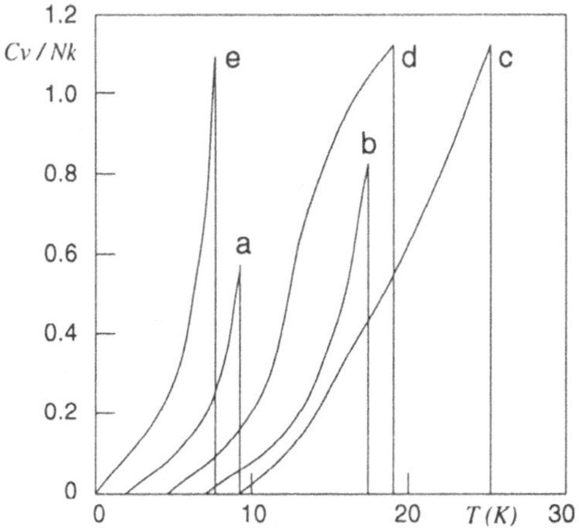

Figure 9. Theoretical heat capacity as a function of doping. (The ρ values are $a = 0.98$, $b = 0.97$, $c = 0.95$, $d = 0.92$, $e = 0.9$).

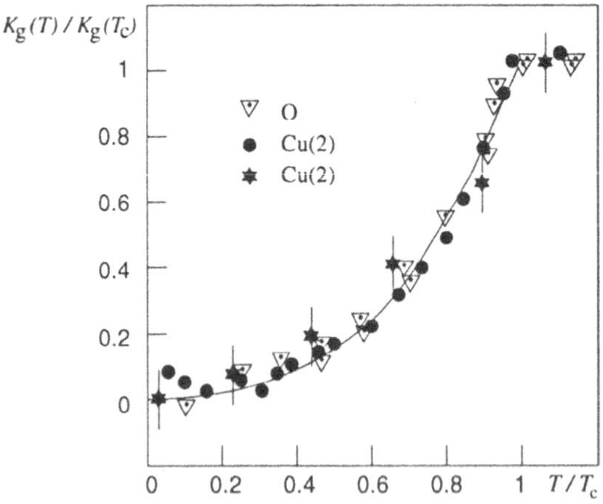

Figure 10. Theoretical Knight shift compared with experimetal data. For full details see [15].

It is pertinent to comment here on the collapse of the Nuclear spin relaxation rate seen below T_c in NMR experiments. When nuclear spins in a material are aligned by an external magnetic field they can relax to equilibrium by causing unpaired electron spin flips. In our theory this process of relaxation should effectively cease well below T_c due to pairing up of electron spins.

8. Summary and Conclusion

This paper describes an extension of the repulsive electron correlation model to embrace the d-wave condensate and other new experimental features. The use of simplifying tentative assumptions renders the major aspects of a complicated problem amenable to analysis. Since the advent of BCS theory we have become used to the idea that attractive off-diagonal matrix elements of a potential can give rise to a so-called 'special eigenvalue problem' [31] and superconductivity. The question must arise then as to the driving force for a superconducting condensation in the model presented here. The key principle must be that electrons seek to avoid each other at short range and take advantage of any long range attractive region of a potential. In our model this effect shows up the sign $(-1)^l$ in the many many electron wavefunction thereby giving a smaller weighting to electronic configurations in which electrons are very close.

All told, the theory makes predictions for weakly doped cuprates for temperatures up to T_c which are in remarkable agreement with experimentation. Our end result is that high temperature superconductivity is primarily an electron correlation effect possibly supplemented by longer range polaronic attraction of the type discussed by Mott and Alexandrov (see [8] for other references). Indeed, it can be argued that this is a theory of unbound bipolarons on a cuprate layer where the Fermion statistics are strictly maintained.

Indeed, it may be possible to develop and hybridise this model with the successful features of other models such as that due to Alexandrov and Mott [8,32] or Jansen and Block [33]. The complexity [34] of cuprate superconductors makes it extremely difficult to describe all features of the cuprates in a single model.

We consider the normal state spin-gap as of indirect significance for high temperature superconductivity but further work is needed to substantiate this standpoint.

Acknowledgments

We have benefited from helpful discussion or correspondence with A. S. Alexandrov, J. R. Cooper, P.P. Edwards, R.L. Johnston, O. Goscinski, N. McAlford, S. Penn.

References

[1] T.M. Rice (1997), *Physica C* **282-287**, xix-xxiii.
[2] B. Batlogg (1997), *Physica C* **282-287**, xxiv-xxx.
[3] D. Pines (1997), *Physica C* **282-287**, 273-278.
[4] G. Deutscher (1999), *Nature* **397**, 410-412.
[5] L.J. Dunne,E.J. Brändas, J. N. Murrell and V. Coropceanu (1998), *Solid State Comm.* **108**, 410-412.
[6] S. G. Ovchinikov (1997), *Uspekhi (Physics)* **40**, 993-1017.
[7] Z.X. Shen, W.E. Spicer, D.M. King, D.S. Dessau and B.O. Wells (1995), *Science* **267**, 343-350.
[8] A.S. Alexandrov (1996), *Phys Rev.* **53 B**, 2863-2869.
[9] D.A. Wollman, D.J. Van Harlingen, W.C. Lee, D.M. Ginsberg and A.J. Leggett (1993), *Phys.Rev.Lett.* **71**, 2134-2137.
[10] M. Sigrist and T.M. Ricer (1995), *Rev.Mod. Physics* **67**, 503-513.
[11] D.J. Van Harlingen (1995), *Rev.Mod. Physics* **67**, 503-513.
& D. J. Van Harlingen (1997) , *Physica* **C 282-287**,128-131.
[12] D. Welscher and J. Ladik (1999), *Int. J. Quant. Chem.* **71**, 285.
[13] C.C. Tsuei and J.R. Kirtley (1997), *Physica* **C 282-287**, 4-12.
[14] L.J. Dunne, J.N. Murrell and E.J. Brändas (1990), *Physica* **C 169**, 501-507
[15] L.J. Dunne (1994), *Physica* **C 223**, 291-312.
[16] L J. Dunne (1997), *Physica* **C 282-287**, 1787-1788.
[17] C.N. Yang (1962), *Rev. Mod. Phys.* **34**, 694-704.
[18] C.N. Yang (1989), *Phys. Rev. Lett.* **63**, 2144-2148.
[19] L.P. Gor'kov (1958), *J. Exp. Teoret. Phys.* **38**, 735-739.
& L.P. Gor'kov (1958) (Translation of ref [19]), *J. Exp. Teoret. Phys.* **7**,505-509.
[20] A.J. Coleman (1967), *Can. J. Phys.* **45**, 1271-1273.
[21] L.J. Dunne and T.P. Spiller (1992), *J.Phys: Condens. Matter* **4**, L563-L566.
& L.J. Dunne and T.P. Spiller (1993) , *J.Phys: Condens. Matter* **5**,145.
[22] L.Jansen and M. Boon (1967) *Theory of Finite Groups. Applications in Physics*, North-Holland Publishing Co., Amsterdam.
[23] P.P. Edwards, T.V. Ramakrishnan and C.N.R. Rao (1995), *J. Phys.Chem.* **99**, 5228-5239.
[24] N. F. Mott and E. A. Davis (1971) *Electronic Processses in Non-Crystalline Materials*, Clarendon Press, Oxford.
[25] J.W. Loram, K.A. Mirza, J.R. Cooper and J.L. Tallon (1997), *Physica* **C 282-287**, 1405-1406.
[26] T. Schneider and H. Keller (1992), *Phys. Rev Lett.* **69**, 3374-3378.
[27] Y.J. Uemura *et al* (1989), *Phys. Rev. Lett.* **62**, 2317-2321.
[28] P.P. Edwards and M.J. Sienko (1987), *Phys.Rev.* **B17**, 2575-2581.
[29] N. Wada, T. Obana, Y. Nakamura and K. Kumaga (1990), *Physica* **B 165-166**, 1341-1342.
[30] R.S. Liu and P.P. Edwards (1993) *Synthesis and Characterisation of High Temperature Superconductors*, Trans. Tech. Publ. Ltd., Switzerland.
[31] C. Kittel (1971) *An Introduction to Solid State Physics*, J. Wiley, New York.
[32] A.S. Alexandrov and N.F. Mott (1994), *Rep. Prog. Phys.* **57**.
[33] L. Jansen and R. Block (1994), *Physica* **212**, 143.
[34] E.J. Brändas (1995) *Dynamics During Spectroscopic Transitions*, Eds E. Lippert and J.D. Macomber, Springer-Verlag, Berlin.

Contents of Volume 1

Part III. Relativistic Formulations and Effects

Part IV. Valence Theory

Part V. Nuclear Motion

Combined Index to Volumes 1 and 2

(Entries are in the form [volume number]:[page number])

Progress in Theoretical Chemistry and Physics

1. S. Durand-Vidal, J.-P. Simonin and P. Turq: *Electrolytes at Interfaces.* 2000
ISBN 0-7923-5922-4
2. A. Hernandez-Laguna, J. Maruani, R. McWeeny and S. Wilson (eds.): *Quantum Systems in Chemistry and Physics.* Volume 1: Basic Problems and Model Systems, Granada, Spain, 1997. 2000 ISBN 0-7923-5969-0; Set 0-7923-5971-2
3. A. Hernandez-Laguna, J. Maruani, R. McWeeny and S. Wilson (eds.): *Quantum Systems in Chemistry and Physics.* Volume 2: Advanced Problems and Complex Systems, Granada, Spain, 1998. 2000 ISBN 0-7923-5970-4; Set 0-7923-5971-2
4. J.S. Avery: *Hyperspherical Harmonics and Generalized Sturmians.* 1999
ISBN 0-7923-6087-7
5. S.D. Schwartz (ed.): *Theoretical Methods in Condensed Phase Chemistry.* 2000
ISBN 0-7923-6687-5
6. J. Maruani, C. Minot, R. McWeeny, Y.G. Smeyers and S. Wilson (eds.): *New Trends in Quantum Systems in Chemistry and Physics.* Volume 1: Basic Problems and Model Systems. 2001 ISBN 0-7923-6708-1; Set: 0-7923-6710-3
7. J. Maruani, C. Minot, R. McWeeny, Y.G. Smeyers and S. Wilson (eds.): *New Trends in Quantum Systems in Chemistry and Physics.* Volume 2: Advanced Problems and Complex Systems. 2001 ISBN 0-7923-6709-X; Set: 0-7923-6710-3

KLUWER ACADEMIC PUBLISHERS – DORDRECHT / LONDON / BOSTON

Batch number: 10761599

Printed by Printforce, the Netherlands